# MATHEMATICAL MODELING
# FOR
# WATER POLLUTION
# CONTROL PROCESSES

# MATHEMATICAL MODELING
# FOR
# WATER POLLUTION
# CONTROL PROCESSES

Edited by

**Thomas M. Keinath**
Associate Professor, Environmental Systems
Engineering, Clemson University, Clemson
South Carolina.

**Martin P. Wanielista**
MPW Associate Professor, Environmental
Systems Engineering Institute, Florida
Technological University, Orlando, Florida.

ANN ARBOR SCIENCE
PUBLISHERS INC
P.O. BOX 1425 • ANN ARBOR, MICH. 48106

# PREFACE

Since the turn of the century, research that has been directed toward the field of environmental engineering has resulted in an extensive array of technology that can be employed to minimize the impact of man on his surrounding environment. Only fairly recently, however, have certain researchers directed their efforts toward mathematical description of a variety of systems, both natural and man-made, that are significant to environmental engineering. Dynamic and steady-state mathematical models which have resulted from the various investigations have been developed to the extent that they can be employed to delineate solutions to real-world questions. Armed with a descriptive and structured model, an environmental engineer can for any system evaluate design and operational alternatives as well as control strategies through a series of simulations, thus placing him in a preferred position in the decision-making process. While models in and of themselves are not a panacea, within certain bounds and limitations, system simulations can provide valuable assistance to the environmental engineer.

Although models which have evolved during the past decades have been extensively employed for description and simulation of laboratory research results, they have only occasionally been employed for the design, operation, and control of man-made systems or for the planning and management of natural systems. Moreover, modeling and simulation procedures have only rarely been used in the classroom as an educational tool. Accordingly, a workshop was formulated by the Association of Environmental Engineering Professors (a) to establish the current status of existing models with emphasis directed toward assumptions, accuracy, limitations, data required for model development, simulation techniques, and applicability; (b) to identify those models that can currently be employed directly for real-world applications; and (c) to define simplified models that can be used by students as an educational tool for mathematically simulating a variety of environmental control systems. Unquestionably it would have been desirable to have included papers regarding all facets of mathematical modeling in environmental engineering. Nonetheless, such an ambitious program could not have been accommodated within the context of the workshop. The scope of the workshop held in Nassau, Bahamas was consequently structured to include only those typical areas regarding process performance models. Authors of national renown who are recognized experts in the selected topical areas were invited to share the fruits of their modeling and simulation efforts with the workshop participants and, in this treatise, with all who are keenly interested in "mathematical modeling of water pollution control processes."

While the first chapter is a modeling and simulation case study of adsorption contactors, it also provides an approach to modeling in general. As such, it serves as the basis for the chapters that follow. Mathematical models for a cross-section of physico-chemical processes including particle aggregation, precipitation, clarification and thickening, oxidation, are then considered sequentially. Steady-state models for both slurry and fixed-film biological reactors are then presented, followed by a monograph on dynamic models and control strategies for biological wastewater treatment systems. The treatise concludes with chapters on modeling concepts for treatment trains and for evaluating the impact of uncertainty and process performance variability.

The cooperation of the authors, the assistance rendered by the Department of Environmental Systems Engineering of Clemson University, and the support of the Association of Environmental Engineering Professors are most genuinely appreciated.

Thomas M. Keinath<br>
August 1975

# CONTENTS

# CHAPTER 1

# MODELING AND SIMULATION OF
# THE PERFORMANCE OF ADSORPTION CONTACTORS

**Thomas M. Keinath**

Department of Environmental Systems Engineering
Clemson University
Clemson, South Carolina

## INTRODUCTION

Use of adsorption contacting systems for industrial and municipal waste-water treatment applications has become more prevalent during recent years. Both the design and operation of adsorption contactors are related to the characteristic breakthrough curves in which the readily measured effluent concentration is plotted as a function of elapsed time. Depending on the ultimate objective of the process, either the breakthrough point or the saturation point represents the design and/or operational target.

Prediction of the breakthrough phenomenon has occupied the attention of numerous investigators during the past three decades. Nevertheless, throughout this period specification of design and operational criteria for large-scale adsorbers have been predicted principally on empirical observations. This is primarily due to the fact that wastewaters have an extremely complex composition which, in turn, substantially complicates modeling and simulation procedures. Any practicable and generally applicable model will undoubtedly have to be of a quasi-empirical nature.

The models developed herein are not intended for direct use in simulating the performance of adsorbers through which wastewaters are passed. They would have to be modified to account for such things as filtration effects before they could be directly employed for this application. Rather, the intent is to provide for the development of a generalized descriptive model

1

which can be employed as an educational tool, in courses on the unit operations and processes of water and wastewater treatment. That is, the dynamic models which have been developed below can be employed by students and faculty alike for simulating the dynamic responses of columnar adsorbers to a variety of input conditions (*e.g.,* adsorber configuration, flow rate, influent concentration, temperature). This will enable one to acquire an understanding of the dynamics of adsorber column performance. Moreover, it provides a rational basis for projecting optimal ranges of design and operational criteria and permits the simulation of control strategies.

## MODELING OF ADSORPTION CONTACTORS

### Material-Balance Relationships

Because adsorption contactors conventionally are of a columnar configuration and, therefore, have concentration profiles in the axial direction, it is necessary to conduct a mass balance over an infinitesimal thickness of bed at a given cross-section. The resultant conservation of mass equation expresses the fact that any loss of solute by the solution passing through that section must equal the gain of solute by the adsorbent contained within that section. For the purpose of this development which is to be employed for educational purposes, it has been assumed that axial dispersion in the bed is negligible compared to bulk flow and that concentration gradients in the radial direction are of minor importance.

Verbally, the materials balance relationship for the packed bed case may be expressed as:

$$\text{Input to Element} = \text{Output from Element} + \text{Adsorption} + \text{Accumulation} \quad (1)$$

For an infinitesimal thickness of bed of unit cross-sectional area the following mathematical formulations can be made for the solution-phase:

| | |
|---|---|
| Input to Element: | $C * U$ |
| Output from Element: | $[C + (\partial c/\partial z) \, dz] \; U$ |
| Adsorption: | $\rho \, (\partial q/\partial t) \, dz \, dA$ |
| Accumulation: | $\epsilon \, (\partial c/\partial t) \, dz \, dA$ |

where

$U$ = solution volumetric flow rate, l/hr
$C$ = solution-phase concentration of solute, mol/l
$q$ = solid-phase concentration of solute, mol/g
$z$ = axial distance, cm
$t$ = time, hr

A = cross-sectional area of column, sq cm
$\epsilon$ = void ration or porosity, dimensionless
$\rho$ = packed-bed density, g/l

Equating in accordance with Equation 1 and assuming unit cross-sectional area, one obtains

$$C * U = \left(C + \frac{\partial c}{\partial z} \, dz\right) U + \rho \left(\frac{\partial q}{\partial t}\right) dz + \epsilon \left(\frac{\partial c}{\partial t}\right) dz$$

which upon simplification becomes

$$U \frac{\partial c}{\partial z} + \epsilon \frac{\partial c}{\partial t} + \rho \frac{\partial q}{\partial t} = 0 \qquad (2)$$

Solution of this partial differential equation (Equation 2) is the essence of the problem of mathematical simulation of the performance of packed-bed adsorption contactors. Such solution may be effected by exact solution of the equations, by the method of finite differences, by the method of characteristics, or by the lumped parameter approach. To obtain an exact solution, it is necessary to make certain simplifying assumptions regarding adsorption equilibria, such as a linear or irreversible isotherm. Although these assumptions are relatively restrictive, they did permit the development of exact solutions by many early investigators including notable developments of Amundson,[1] Kasten, et al.,[2] and Rosen.[3]

Solutions by the methods of finite differences and characteristics have been explicitly detailed by Crank[4] and Acrivos,[5] respectively. Because of certain complexities, however, these are not practically suited for adoption in the classroom. In contrast, the lumped parameter approach can readily be adapted for classroom use through the use of a continuous system modeling program, CSMP/360, produced by IBM.[6] Consequently, all further developments will relate to the lumped parameter approach.

**Lumped Parameter Method**

Use of the lumped parameter approach requires segmentation of the packed-bed of adsorbent into a discrete number of finite elements. Both the liquid- and solid-phase concentrations of solute are assumed to be uniform throughout each element. Further, continuity of mass flows of the solute between adjoining elements must be maintained for both the liquid and solid phases.

*Simple Packed-Bed Case*

For the case in which an adsorption column is charged with an adsorbent and then operated in the unsteady-state until the adsorbent is entirely exhausted, the following formulations can be made for an element of the packed-bed which has been designated as element n in Figure 1.1a:

|  | Solution-Phase | Solid-Phase |
|---|---|---|
| Input to Element | $U * C_{n-1}$ | 0 |
| Output from Element | $U * C_n$ | 0 |
| Adsorption | $\rho * R_A * V$ | $\rho * R_A * V$ |
| Accumulation | $\epsilon (dc_n/dt) V$ | $\rho (dq_n/dt) V$ |

where:

$V$ = volume of the elements, liters
$R_A$ = rate of adsorption, mol/g-hr
$n$ = number of elements

and all other parameters are as defined above.

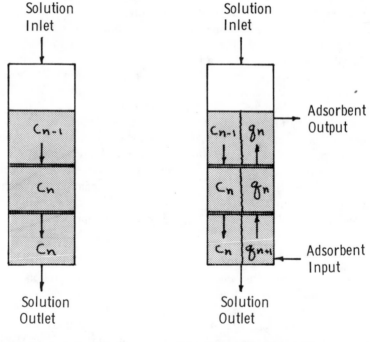

(a)  Simple Packed-Bed Case

(b)  Countercurrent-Flow, Packed-Bed Case

**Figure 1.1.** Elemental segments in a packed-bed, columnar adsorption contactor.

As before, equating in accordance with Equation 1 for the solution-phase:

$$U * C_{n-1} = U * C_n + \rho * R_A * V + \epsilon \, (dc_n/dt) \, V$$

which yields upon rearrangement for the $n^{th}$ element,

$$\frac{dc_n}{dt} = \left(\frac{U}{\epsilon V}\right) \left(C_{n-1} - C_n\right) - \left(\rho/\epsilon\right) \left(R_A\right) \tag{3}$$

Similarly, for the solid-phase the mass balance relationship for the solute becomes

$$0 = 0 - \rho * R_A * V + \rho \, (dq_n/dt) \, V$$

which simplifies to

$$dq_n/dt = R_A \tag{4}$$

The foregoing ordinary differential equations, Equations 3 and 4, can then be solved in a straightforward manner by use of CSMP/360.

## *Countercurrent Flow, Packed-Bed Case*

Packed-bed adsorption contactors may also be operated in a fully counter-current mode, in which the solution is introduced at the top and withdrawn from the bottom while the adsorbent is added at the bottom and removed from the top of the bed. In this manner the solution with the highest concentration of solute is contacted with the most exhausted adsorbent in the bed while the solution which is about to be discharged is contacted with virgin adsorbent for final polishing. The various parameters of the mass balance on the solute in element n as shown in Figure 1.1b are given as follows:

|  | Solution-Phase | Solid-Phase |
|---|---|---|
| Input to Element | $U * C_{n-1}$ | $G * q_{n+1}$ |
| Output from Element | $U * C_n$ | $G * q_n$ |
| Adsorption | $\rho * R_A * V$ | $\rho * R_A * V$ |
| Accumulation | $\epsilon \, (dc_n/dt) \, V$ | $\epsilon \, (dq_n/dt) \, V$ |

where all parameters except G have been defined previously. This term represents the mass rate of flow of adsorbent through the column (g/hr).

It may be observed that for the solution-phase the terms given above are identical to those given for the unsteady-state case. Consequently, the resultant materials balance relationship for the solution-phase of the element is identical to Equation 3. The solid-phase mass balance relationship for the solute, however, may now be stated as

$$G * q_{n+1} = G * q_n - \rho * R_A * V + \rho (dq_n/dt) V$$

which reduces to

$$dq_n/dt = (G/[\rho * V])(q_{n+1} - q_n) + R_A \tag{5}$$

As before, the differential equations, Equations 3 and 5, can be solved to describe the performance of the packed-bed countercurrent adsorber.

It is important to recognize that the set of differential equations (Equations 3 and 5) are extremely stiff. That is, the time constant for the liquid-phase equation is small in contrast to the time constant for the solid-phase equation. Numerical integration of the set of equations requires specification of an integration interval that provides for the stable and accurate integration of the equation that responds most rapidly; *i.e.*, the liquid-phase mass balance.

If, however, the time constants for the two equations are different by several orders of magnitude, one can assume that the equation which responds most rapidly is continuously at steady-state. For this case, the term $dc_n/dt$ in Equation 3 is set to zero and the resulting algebraic equation is solved for $c_n$. Specification of the integration interval is then contingent only on the remaining differential equation—that which responds most slowly (solid-phase mass balance, Equation 5). This serves to materially decrease the computation time required for numerical solution.

**Adsorption Equilibria**

The distribution of solute between liquid- and solid-phases in an adsorbent-solute-solvent system at equilibrium is commonly termed an adsorption isotherm. Adsorption equilibrium data is conventionally presented and correlated by plotting the quantity of solute adsorbed per unit weight of adsorbent, q, as a function of the concentration of solute remaining in solution at equilibrium, C. The several types of distributions between phases that are possible include the irreversible, favorable, linear, and unfavorable cases. These have been shown schematically in Figure 1.2.

Several mathematical formulations that describe adsorption equilibria and that have been widely adopted since their development include the isotherm model originally proposed by Langmuir,

$$q = \frac{Q * b * C}{1 + b * C} \tag{6}$$

that of Freundlich,

$$q = K * C^{1/n'} \tag{7}$$

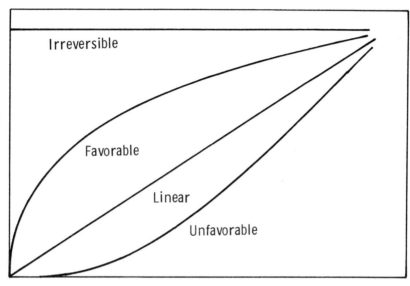

Equilibrium Concentration of Solute in Solution, C

**Figure 1.2.** Basic types of adsorption isotherms

and that of Brunauer, Emmett and Teller,

$$q = \frac{Q * A * C}{(C_S - C) (1 + [A - 1] C/C_S)} \tag{8}$$

where:

  $Q$  = ultimate uptake capacity of adsorbent, mol/g
  $b$  = Langmuir energy term, l/mol
  $K$  = adjustable curve-fitting constant
  $n'$  = adjustable curve-fitting constant
  $A$  = BET energy term, l/mol
  $C_S$ = saturation concentration of solute in solution, mol/l

Of these three, the simple empirical Freundlich expression has probably been the most widely used, primarily because of its simplicity and because it has been successfully employed for correlating data for adsorption of solutes on activated carbon over limited concentration ranges. The Langmuir equation, based on the assumption of monolayer adsorption on a fixed number of equivalent adsorptions sites, has also been useful over limited concentration ranges. In contrast, the BET equation which is based on the assumption of multiple-layer adsorption has not been used as frequently as either the Langmuir or Freundlich models principally because of the difficulty in obtaining suitable values for $C_S'$.

Each of the three models which describes adsorption equilibria, Equations 6, 7 and 8, is, of course, limited for application to single-solute systems. Only the Langmuir model has been extended to account for competitive adsorption equilibria in multiple-solute systems. For the disolute case the extent of adsorption for solute A is given by

$$q_A = \frac{Q_A * b_A * C_A}{1 + b_A * C_A + b_B * C_B} \tag{9}$$

and for solute B the solid-phase equilibrium concentration may be expressed as

$$q_B = \frac{Q_B * b_B * C_B}{1 + b_A * C_A + b_B * C_B} \tag{10}$$

where the constants, $Q_A$, $Q_B$, $b_A$, and $b_B$, are those that are measured in monosolute, pure solution systems. For the general case, where (i) is used to designate the number of solutes in the system, the extent of adsorption for solute (j) is given by

$$q_j = \frac{Q_j * b_j * C_j}{1 + \sum_{i=0}^{i} b_i * C_i} \tag{11}$$

Although this formulation is mathematically simple, it has been experimentally verified only for certain selected disolute competitive systems.[7] Accordingly, it is important to recognize that this relationship should not be employed indiscriminately when simulating the performance of adsorption contactors for competitive, multiple-solute systems. Rather, the relationship must be experimentally validated for the specific multiple-solute/ solvent/adsorbent system before its use can be judged appropriate.

**Kinetics of Interphase Solute Transport**

To relate the dynamically varying solid- and solution-phase concentrations of solute at any time and position in an adsorption contactor, it is necessary to fully understand the kinetics of solute transport. Consequently, the mechanisms of such transport must be considered in detail.

The overall rate of adsorption of solute, from a solvent stream flowing through a bed of porous, granular adsorbent represents the combined effects of (Figure 1.3): (1) diffusion through the boundary layer of fluid surrounding the adsorbent particle (film or external diffusion); (2) diffusion within the pores of the particle (pore diffusion); (3) diffusion along the surface

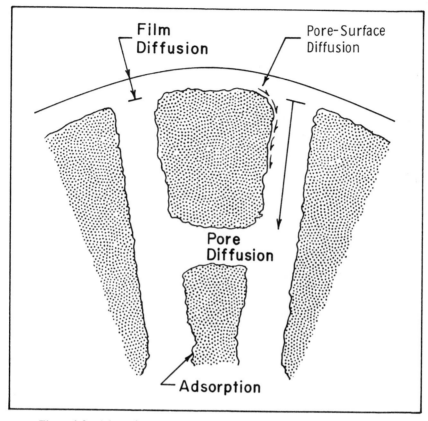

**Figure 1.3.** Adsorption on porous adsorbents—rate limiting mechanisms.

of the pores (pore-surface diffusion); and (4) adsorption on the internal pore surfaces (adsorption). Pore-surface diffusion is distinguished from pore diffusion in that it occurs on the solid side of the phase boundary. That is, the solute may enter a particle of adsorbent directly from the exterior surface by movement in condensed or adsorbed layer along the pore surfaces or it may diffuse through the fluid-phase held in the pores and then be deposited at a stationary location on the pore surface. In pore diffusion, the mass transfer process occurs before the phase change, while in pore-surface diffusion it occurs afterward. Under different operating conditions, the same particle can exhibit either surface- or pore-diffusion behavior, favoring the former if the concentration level in the pore-fluid is low and the latter if the level is high. Of these two transport processes, the more rapid one will control the overall rate of transport, since they act in parallel. Conversely, for either of the two solute transport pathways, film diffusion/pore diffusion/

adsorption or film diffusion/pore-surface diffusion/adsorption, the slowest of the three mass transfer mechanisms will, in each case, limit the overall rate of adsorption because these rate steps occur in a series relationship.

*Adsorption Rate Step*

It has been experimentally determined that surface tension attains equilibrium after a disturbance in approximately 10 msec. This circumstantial evidence indicates that the adsorption process itself is a relatively rapid process and, therefore, is probably not rate controlling. Considerable experimental evidence exists which supports this supposition. Accordingly, adsorption is generally considered as not limiting the overall rate of adsorption. It will, therefore, not be considered further in this chapter.

*Pore-Diffusion*

For a porous adsorbent the rate of diffusion into the liquid in the pores of a spherical particle is given by

$$D_{pore}\left(\frac{\partial^2 C}{\partial r^2} + \frac{2}{r}\frac{\partial C}{\partial r}\right) = \chi\left(\frac{\partial C}{\partial t}\right) + \rho_p\left(\frac{\partial q}{\partial t}\right) \tag{12}$$

where

$D_{pore}$ = pore diffusivity, $cm^2/sec$
$r$ = radial distance within particle, cm
$\chi$ = internal porosity of adsorbent particles, dimensionless
$\rho_p$ = density of adsorbent particle, g/l

The mean concentration of the entire particle of diameter, $D_p$ (cm), is obtained by

$$q = \frac{3}{(D_p/2)^3} \int_0^{D_p/2} q\, r^2\, dr \tag{13}$$

For liquids, the pore diffusivity term ($D_{pore}$) may be approximated by the relationship

$$D_{pore} = \frac{D_\varrho * \chi}{2} \tag{14}$$

where

$D_\varrho$ = diffusivity of solute in solvent, $cm^2/sec$

Although Equations 12, 13 and 14 clearly describe the pore-diffusion rate-limiting case, it is to be noted that solution of these equations in conjunction with either Equations 2 or 3 (materials balance relationships for the columnar adsorption contactor) has not yet been accomplished, either exactly or numerically. This, of course, is due to the fact that an additional independent variable, radial distance within the porous adsorbent particle, is introduced into the mass balance relationship for a packed-bed contactor. Consequently, a linear driving potential is frequently employed as a gross approximation.[8] The rate equation then is

$$R_A = \frac{dq}{dt} = k_p * A_p * (C - C^*) \qquad (15)$$

where

$k_p$ = mass-transfer coefficient (pore), cm/hr
$A_p$ = external interfacial transfer area for adsorbent, $cm^2/l$
$C^*$ = solution-phase concentration of solute considered to be in equilibrium with the outer surface of the adsorbent particle, mol/l

For this approximation, the term $k_p * A_p$ is evaluated by

$$k_p * A_p = \frac{60 * D_{pore}}{D_p^2} (1 - \epsilon) \qquad (16)$$

*Pore-Surface Diffusion*

Diffusion of solute molecules along the surfaces of the pores in a porous adsorbent is described by

$$D_s \left( \frac{\partial^2 q}{\partial r^2} + \frac{2}{r} \frac{\partial q}{\partial r} \right) = \frac{\partial q}{\partial t} \qquad (17)$$

where

$D_s$ = pore-surface diffusivity, $cm^2/sec$

As before, solution of Equation 16 in association with Equation 2 or 3 cannot currently be achieved with the various numerical methods that are available. This equation, therefore, is frequently approximated by the linear-driving force relation of Glueckauf and Coates.[9]

$$R_A = \frac{dq}{dt} = k_s * A_p (q^* - q) \qquad (18)$$

where

$k_s$ = mass-transfer coefficient (pore-surface), cm/hr
$q^*$ = solid-phase concentration of solute considered to be in equilibrium with the instantaneous fluid-phase concentration outside the particle, mol/g

The product $k_s * A_p$ is related to the diffusivity and to the adsorbent particle diameter through the equation

$$k_s * A_p = \frac{60 * D_s}{D_p^2} \qquad (19)$$

Experimentally determined values for $D_s$ are required for solution of Equation 19 such that estimates of the term $K_s * A_p$ can be obtained. Methods of obtaining values for $D_s$ have been detailed explicitly by DiGiano.[10] For aqueous solutions of various substituted phenols and benzenesulfonates, DiGiano[10] determined ratios of $D_s/D_\varrho$ which varied from 0.1 to 0.3.

A somewhat better approximation to the behavior of Equation 17 is given by the quadratic driving potential form postulated by Vermeulen:[11]

$$R_A = \frac{dq}{dt} = k_s * A_p * \psi * \left( \frac{q^{*2} - q^2}{2q - q_o} \right) \qquad (20)$$

where

$q_o$ = initial solid-phase concentration (usually zero), mol/g
$\psi$ = $1/(R + 15 [1 - R]/\pi^2)$
$R$ = separation factor, dimensionless

**Film or External Diffusion**

Transfer of solute molecules from the bulk solution through the hydrodynamic boundary layer surrounding the granular adsorbent particle to its external surface is commonly described by

$$R_A = \frac{dq}{dt} = k_f * A_p (C - C^*) \qquad (21)$$

where

$k_f$ = film-diffusion controlled mass transfer coefficient, cm/hr

Use of Equation 21 does not necessarily imply the existence of a stagnant hydrodynamic boundary layer through which molecular diffusion occurs. Rather, this equation may generally be used for description of mass transfer through the boundary layer. Certainly, this would include the effects of eddy transport in addition to molecular diffusion.

Although various theories including the penetration and free surface models have been advanced to describe transient diffusion through an external boundary layer, the straightforward relationship given by Equation 21 is the only form that is generally applicable for simulation of the dynamics of a columnar adsorber. Because of the difficulty in describing the hydrodynamic conditions in the boundary layer, mass transfer coefficients, $k_f$ are not determined directly but are obtained by correlation of the mass transfer factor, $j_d$, with the Sherwood number, Reynolds number, and Schmidt number as follows:

$$j_d = N_{Sh}/(N_{Re} * N_{Sc}^{1/3}) \qquad (22)$$

where

$j_d = (k_f/U) * (\nu/D_\varrho)^{2/3}$, mass transfer factor, dimensionless

$N_{Sh} = k_f * D_p/D_\varrho$, Sherwood number, dimensionless

$N_{Sc} = \nu/D_\varrho$, Schmidt number, dimensionless

$N_{Re} = \bar{U} * D_p/\nu$, Reynolds number, dimensionless

$\bar{U} = U/A$, velocity of flow, cm/hr

$\nu$ = kinematic viscosity of solution, cm$^2$/sec

Numerous investigators have conducted intensive experimental studies to evaluate the correlation given in Equation 22 for a variety of system types including gas-liquid, liquid-solid, and gas-solid, two-phase systems. Extensive investigations have also focused on single-particle as well as multiparticle systems which are operated either in the packed, expanded, semifluidized, or fully fluidized-bed modes. Several of the more general and widely used correlations of $j_d$ vs. $N_{Re}$ have been listed in Table 1.1. It is to be noted that most correlations have been obtained for a modified Reynolds number $[N_{Re}' = N_{Re}/(1 - \epsilon)]$ which accounts for the porosity within the bed of a multiparticle system.

**Table 1.1. Correlations for Mass Transfer Coefficients ($j_d$)**

| | |
|---|---|
| Chu, *et al.*[12] | (Fluidized & Packed Beds) |
| | $j_d = 5.7 * (N'_{Re})^{-0.78} \quad 30 > N'_{Re} > 1$ |
| | $j_d = 1.77 * (N'_{Re})^{-0.44} \quad 10{,}000 > N'_{Re} > 30$ |
| Pfeffer & Happel[13] | (Low Reynolds Numbers, $0.4 < \epsilon < 1.0$) |
| | $j_d = Be * (N_{Re})^{-0.67}$ |
| where: | $Be = 1.26 [1 - (1 - \epsilon)^{5/3}]^{1/3} [2 - 3(1 - \epsilon)^{1/3} + (1 - \epsilon)^{5/3} - 2(1 - \epsilon)^2]$ |
| Fan, *et al.*[14] | (Semifluidized Beds) |
| | $j'_d = 1.51 * (N'_{Re})^{-0.5}$ |
| where: | $j'_d = N_{Sh} - 2/N_{Re} * N_{Sc}^{1/3}$ |
| Evans & Gerald[15] | $j_d = 2.132 * (N'_{Re})^{-0.512}$     Packed-Bed |
| | $j_d = 1.340 * (N'_{Re})^{-0.468}$     Fluidized-Bed |

Note:     $N_{Re} = \bar{U} * D_p/\nu$

            $N'_{Re} = \bar{U} * D_p/[\nu * (1 - \epsilon)]$

Accordingly, if it has been determined that film diffusion or external transport is rate limiting, then one can obtain a suitable mass transfer coefficient from one of the existing correlations. It is important, however, to pay particular attention to the type of system for which the correlation was obtained and to the procedure employed for calculation of mass transfer coefficients.

## SIMULATION OF THE DYNAMIC PERFORMANCE OF ADSORPTION CONTACTORS

### Program Formulation Using CSMP/360

As was suggested above, solution of the set of differential equations that describe the mass balances for the solution- and solid-phases of an adsorption contactor is facilitated by use of the distributed parameter approach when used in conjunction with CSMP/360. Programs formulated in CSMP/360 generally consist of three segments: initial, dynamic, and terminal. Broadly, the initial segment is intended exclusively for computation of initial condition values. The dynamic segment is normally the most extensive of the model and it includes the complete description of the system dynamics, together with any other computations desired during the run. Conversely, computations to be performed only after completion of the dynamic run are placed into the terminal segment of the model. These will often be simple calculations based on the final value of one or more model variables.

For a simple packed-bed adsorption contactor the differential equations, Equations 3 and 4, describe the system dynamics or the time-variant performance of each segment of the adsorbent bed. Of course, these require input of the appropriate rate relationship for $R_A$ (Equations 15, 18, 20, or 21) and a suitable description of the equilibrium distribution of solute between solution and solid phases (Equations 6, 7 or 8). If it is assumed that mass-transfer is rate-limiting and that the Langmuir equation adequately represents the pertinent adsorption equilibria, then the system dynamics can be formulated as shown below in the dynamic segment of CSMP/360.

The first equation listed for the first element of the packed adsorbent bed is the differential equation that derives from the mass balance on the solution-phase (Equation 3), while the third equation is the differential equation which describes the solid-phase mass balance (Equation 4). The second and fourth equations in the program listing provide for numerical integration of the first and third equations, respectively. Use is made of the fifth equation, Langmuir isotherm, in the first element to calculate the solution-phase concentration of solute at a point immediately adjacent to the adsorbent particle. Blocks of equations for each subsequent element of the packed bed are identical to those for the first element with the exception that the

```
DYNAMIC
***                                                                            ***
***      DIFFERENTIAL EQUATIONS FOR FIRST ELEMENT                              ***
***                                                                            ***
         C1DCT=(U/(EPSI*V))*(CC-C1)-((KF*A*RHO)/EPSI)*(C1-CSTAR1)*1.E-3
         C1=INTGRL(ICC1,C1DOT)
         Q1DOT=KF*A*(C1-CSTAR1)*1.E-3
         Q1=INTGRL(ICQ1,Q1DOT)
         CSTAR1=Q1/(QMAX*B-Q1*B)
***                                                                            ***
***      DIFFERENTIAL EQUATIONS FOR SECOND ELEMENT                             ***
***                                                                            ***
         C2DOT=(U/(EPSI*V))*(C1-C2)-((KF*A*RHO)/EPSI)*(C2-CSTAR2)*1.E-3
         C2=INTGRL(ICC2,C2DOT)
         Q2DOT=KF*A*(C2-CSTAR2)*1.E-3
         Q2=INTGRL(ICQ2,Q2DOT)
         CSTAR2=Q2/(QMAX*B-Q2*B)
***                                                                            ***
***      DIFFERENTIAL EQUATIONS FOR THIRD ELEMENT                             ***
***                                                                            ***
         C3DOT=(U/(EPSI*V))*(C2-C3)-((KF*A*RHO)/EPSI)*(C3-CSTAR3)*1.E-3
         C3=INTGRL(ICC3,C3DOT)
         Q3DOT=KF*A*(C3-CSTAR3)*1.E-3
         Q3=INTGRL(ICQ3,Q3DOT)
         CSTAR3=Q3/(QMAX*B-Q3*B)
***                                                                            ***
```

output of the solution-phase concentration of the first element becomes the input to the second element, the second to the third, and so on. Accordingly, only five equations are required for solution of the system dynamics, although the adsorbent bed must be lumped into a sufficient number of elements to account for the dispersion characteristics of the columnar reactor

Initial conditions for the second and fourth equations of each element of the bed are generally specified in the initial segment of the CSMP/360 program using an INCON label as is shown in the program listing which follows:

```
INITIAL
***                                                                            ***
***      INITIAL CONDITIONS FOR SOLUTION PHASE CONCENTRATION                   ***
***                                                                            ***
         INCON ICC1=0.
         INCON ICC2=0.
         INCON ICC3=0.
         INCON ICC4=0.
         INCON ICC5=0.
         INCON ICC6=0.
         INCON ICC7=0.
         INCON ICC8=0.
         INCON ICC9=0.
         INCON ICC10=0.
***                                                                            ***
***      INITIAL CONDITIONS FOR SOLID PHASE CONCENTRATION                      ***
***                                                                            ***
         INCON ICQ1=0.
         INCON ICQ2=0.
         INCON ICQ3=0.
         INCON ICQ4=0.
         INCON ICQ5=0.
         INCON ICQ6=0.
         INCON ICQ7=0.
         INCON ICQ8=0.
         INCON ICQ9=0.
         INCON ICQ10=0.
```

Likewise, parameters that must be specified for the simulation are introduced in the initial segment by use of the PARAM statement as can be observed in the listings below:

```
***                                                                      ***
***    PARAMETER INPUTS REQUIRED ARE DEFINED AS FOLLOWS                   ***
***                                                                       ***
***    CARBON ---  MASS OF CARBON IN BED (GRAMS)                          ***
***    A --------  EXTERNAL TRANSFER AREA FOR CARBON (SQ CM/GRAM)         ***
***    DP -------  DIAMETER UF CARBON PARTICLES (CM)                      ***
***    RHO ------  PACKED BED DENSITY (GRAMS/LITER)                       ***
***    EPSI -----  PACKED BED POROSITY (DIMENSIONLESS)                    ***
***    QMAX -----  LANGMUIR ULTIMATE UPTAKE CAPACITY (MOLES/GRAM)         ***
***    B --------  LANGMUIR ENERGY TERM (LITERS/MOLE)                     ***
***    CO -------  SOLUTION-PHASE CONCENTRATION OF SOLUTE (MOLES/LITER)***
***    DL -------  DIFFUSIVITY OF SOLUTE (SQ CM/SEC)                      ***
***    AREA -----  CROSS-SECTIONAL AREA OF COLUMN (SQ CM)                 ***
***    U --------  SOLUTION VOLUMETRIC FLOW RATE (LITERS/HOUR)            ***
***    GAMMA ----  KINEMATIC VISCOSITY OF WATER (SQ CM/SEC)               ***

                   ***
                   ***    PARAMETER INPUTS FOLLOW
                   ***
                          PARAM CARBON=250.
                          PARAM A=150.
                          PARAM DP=0.03
                          PARAM RHO=381.
                          PARAM EPSI=0.45
                          PARAM QMAX=8.469E-4
                          PARAM B=3.023E5
                          PARAM CO=1.E-4
                          PARAM DL=6.21E-6
                          PARAM AREA=20.25
                          PARAM U=60.
                          PARAM GAMMA=8.64E-3
```

Moreover, certain parameters that are not time-variant and must be calculated prior to the simulation are evaluated in the initial segment. A listing of those parameters that must be determined by calculation follows:

```
***                                                                      ***
***    CALCULATED INPUT PARAMETERS ARE DEFINED AS FOLLOWS                 ***
***                                                                       ***
***    V --------  TOTAL VOLUME OF PACKED BED (LITERS)                    ***
***    UBAR -----  SUPERFICIAL VELOCITY OF FLOW THROUGH BED (CM/SEC)      ***
***    NREMOD ---  MODIFIED REYNOLDS NUMBER (DIMENSIONLESS)               ***
***    NSC ------  SCHMIDT NUMBER (DIMENSIONLESS)                         ***
***    JD -------  MASS TRANSFER FACTOR (DIMENSIONLESS)                   ***
***    KF -------  MASS TRANSFER COEFFICIENT (CM/HOUR)                    ***
***                                                                       ***
***    INPUT PARAMETER CALCULATIONS FOLLOW                                ***
***                                                                       ***
       V=CARBON/(RHO*10.)
       UBAR=(U*0.277777)/AREA
       NREMOD=(UBAR*DP)/(GAMMA*(1.-EPSI))
       NSC=GAMMA/DL
       JD=5.7*NREMOD**(-0.78)
       KF=(JD*UBAR/NSC**(2./3.))*3600.
***                                                                      ***
```

If, in contrast, certain parameters are time-dependent, they must be included in the dynamic segment of the model, such that at any time during the integration the parameter is evaluated for its current numerical value.

In addition to the foregoing listings in which the dynamic model is fully specified, it is necessary also to provide for several execution and output control statements. Execution control statements are used to specify certain items relating to the actual simulation; *e.g.*, integration method, run time, integration interval, relative error, output times. An example of the execution control statements used in the current simulation studies is shown below.

```
***                                                             ***
***   EXECUTION CONTROL STATEMENTS                             ***
***                                                             ***
METHOD SIMP
TIMER DELT=0.00025,FINTIM=40.,PRDEL=0.5,OUTDEL=0.5
```

The first of two control statement cards specifies the type of integration method to be employed during the simulation. For this example, Simpson's Rule (SIMP) integration method was utilized. Other available integration methods are listed in the CSMP/360 User's Manual.[6] On the Timer statement one specifies the integration interval (DELT), the total duration of the simulation run (FINTIM), and simulation data output intervals (PRDEL & OUTDEL).

For output of the simulation data one has available several options as may be observed in the following listing:

```
***                                                             ***
***   OUTPUT CONTROL STATEMENTS                                ***
***                                                             ***
      PREPAR C1,C2,C3,C4,C5,C6,C7,C8,C9,C10
      PRINT C1,C6,Q1,Q6,C2,C7,Q2,Q7,C3,C8,Q3,Q8,C4,C9,Q4,Q9,...
      C5,C10,Q5,Q10,CO,QT
      PRTPLT C1(0.0,1.E-4),C2(0.0,1.E-4),C3(0.0,1.E-4),C4(0.0,1.E-4)
      PRTPLT C5(0.0,1.E-4),C6(0.0,1.E-4),C7(0.0,1.E-4),C8(0.0,1.E-4)
      PRTPLT C9(0.0,1.E-4),C10(0.0,1.E-4)
      PRTPLT Q1(0.0,1.E-3),Q2(0.0,1.E-3),Q3(0.0,1.E-3),Q4(0.0,1.E-3)
      PRTPLT Q5(0.0,1.E-3),Q6(0.0,1.E-3),Q7(0.0,1.E-3),Q8(0.0,1.E-3)
      PRTPLT Q9(0.0,1.E-3),Q10(0.0,1.E-3)
      PRTPLT QT(0.0,1.E-3)
LABEL SIMULATION OF ADSORPTION BREAKTHROUGH PROFILE
TITLE SIMULATION OF ADSORPTION BREAKTHROUGH PROFILE
***                                                             ***
```

A statement entitled PRINT causes a line-print output of the variables listed behind a PRINT label at an interval of PRDEL. A sample of this output is shown below.

SIMULATION OF ADSORPTION BREAKTHROUGH PROFILE

| TIME | = | 0.0 | CO | = | 1.0000E-04 | C4 | = | 0.0 |
|---|---|---|---|---|---|---|---|---|
| | | | C1 | = | 0.0 | C5 | = | 0.0 |
| | | | C2 | = | 0.0 | C6 | = | 0.0 |
| | | | C3 | = | 0.0 | C7 | = | 0.0 |
| TIME | = | 5.0000E-01 | CO | = | 1.0000E-04 | C4 | = | 8.2700E-07 |
| | | | C1 | = | 3.0174E-05 | C5 | = | 2.4930E-07 |
| | | | C2 | = | 9.0990E-06 | C6 | = | 7.5148E-08 |
| | | | C3 | = | 2.7433E-06 | C7 | = | 2.2653E-08 |
| TIME | = | 1.0000E 00 | CO | = | 1.0000E-04 | C4 | = | 8.5516E-07 |
| | | | C1 | = | 3.0491E-05 | C5 | = | 2.5966E-07 |
| | | | C2 | = | 9.2698E-06 | C6 | = | 7.8830E-08 |
| | | | C3 | = | 2.8160E-06 | C7 | = | 2.3930E-08 |
| TIME | = | 1.5000E 00 | CO | = | 1.0000E-04 | C4 | = | 8.8657E-07 |
| | | | C1 | = | 3.0894E-05 | C5 | = | 2.7105E-07 |
| | | | C2 | = | 9.4722E-06 | C6 | = | 8.2845E-08 |
| | | | C3 | = | 2.8989E-06 | C7 | = | 2.5315E-08 |

Another option for simulation output is designated by the PRTPLT (PRTPLOT) label, which causes print-plotting of the value of a variable as a function of time at interval of OUTDEL. The output caused by this statement is shown below:

SIMULATION OF ADSORPTION BREAKTHROUGH PROFILE                                        PAGE    1

|  |  | MINIMUM 0.0 | C1 VERSUS TIME | MAXIMUM 1.0000E-04 |
|---|---|---|---|---|
| TIME | C1 | I | | I |
| 0.0 | 0.0 | + | | |
| 5.0000E-01 | 3.0174E-05 | ---------------+ | | |
| 1.0000E 00 | 3.0491E-05 | ---------------+ | | |
| 1.5000E 00 | 3.0894E-05 | ---------------+ | | |
| 2.0000E 00 | 3.1422E-05 | ---------------+ | | |
| 2.5000E 00 | 3.2144E-05 | ---------------+ | | |
| 3.0000E 00 | 3.3185E-05 | ----------------+ | | |
| 3.5000E 00 | 3.4804E-05 | ----------------+ | | |
| 4.0000E 00 | 3.7614E-05 | ------------------+ | | |
| 4.5000E 00 | 4.3377E-05 | ----------------------+ | | |
| 5.0000E 00 | 5.8210E-05 | ------------------------------+ | | |
| 5.5000E 00 | 8.8658E-05 | ------------------------------------------------+ | | |
| 6.0000E 00 | 9.9156E-05 | -----------------------------------------------------+ | | |
| 6.5000E 00 | 9.9612E-05 | ------------------------------------------------------+ | | |
| 7.0000E 00 | 9.9612E-05 | ------------------------------------------------------+ | | |
| 7.5000E 00 | 9.9612E-05 | ------------------------------------------------------+ | | |
| 8.0000E 00 | 9.9612E-05 | ------------------------------------------------------+ | | |
| 8.5000E 00 | 9.9612E-05 | ------------------------------------------------------+ | | |
| 9.0000E 00 | 9.9612E-05 | ------------------------------------------------------+ | | |
| 9.5000E 00 | 9.9612E-05 | ------------------------------------------------------+ | | |
| 1.0000E 01 | 9.9612E-05 | ------------------------------------------------------+ | | |
| 1.0500E 01 | 9.9612E-05 | ------------------------------------------------------+ | | |
| 1.1000E 01 | 9.9612E-05 | ------------------------------------------------------+ | | |
| 1.1500E 01 | 9.9612E-05 | ------------------------------------------------------+ | | |
| 1.2000E 01 | 9.9612E-05 | ------------------------------------------------------+ | | |

**Simulations of Packed-Bed Adsorption Contactors**

To illustrate the use of CSMP/360 for solution of the requisite materials-balance, equilibria, and rate relationships for the simulation of adsorption column system dynamics, four specific examples have been developed that clearly demonstrate the modeling techniques involved. These have been selected for application in undergraduate- and graduate-level classes in water and wastewater pollution control processes. Moreover, the models presented can also be employed for simulation of the performance of laboratory-scale adsorption contactors when adsorption isotherm data, which has been obtained experimentally in the laboratory, is used as input information. It is to be noted further that the models developed herein are applicable for description of ion exchange column dynamics.

The examples discussed below do not incorporate all combinations and permutations of column dynamics as a function of system design and operation parameters. Rather, the examples have been selected to illustrate general approaches to the solution of several major classes of problems. When these have been mastered, one can readily modify and adapt the existing models to any specific set of operational conditions or system configuration.

Each example listed below has been developed with the use of several assumptions regarding adsorption kinetics and equilibria. First, it has been assumed that external or film diffusion limits the rate of transfer of solute molecules from the bulk solution to the internal adsorbent surfaces. In this respect, use has been made of the $j_d$ vs. $N'_{Re}$ correlation as presented by Chu, *et al.*[12] to determine suitable mass transfer coefficients. Alternately, another correlation could be employed, provided that the system for which simulations are desired corresponds to the general configuration and operational mode of the system for which the correlation was developed; *i.e.*, packed-bed, semifluidized-beds, fluidized-beds. An additional assumption that has been employed is that the Langmuir model for adsorption equilibria adequately represents the equilibrium distribution between solid- and liquid-phases of the solutes that have been used for the current simulations. For simulations on adsorption dynamics, any model that accurately describes adsorption isotherm data can be used. In fact, by use of the function generator listing of CSMP/360 one could, indeed, input actual adsorption isotherm data points rather than attempt to fit a model to the data.

*Case Number 1:  Simple Packed-Bed (Single-Solute)*

The simplest of the four examples has been formulated for a two-inch I.D. laboratory-scale adsorption contactor which contains 250 g of granular activated carbon and is operated in the packed-bed mode. The solution, which is applied to the contactor at a volumetric flow rate of 60 l/hr contains only dinitro-o-sec-butylphenol at a concentration of 100 $\mu$mol/l. For this case all input parameters are constant (non-time variant). The computer listing which describes the system and specifies the input parameters has been reproduced below.

```
*     THOMAS M KEINATH *** AEEP WORKSHOP (BAHAMAS)                        *
*     MODELING OF ADSORPTION COLUMNS, DISTRIBUTED PARAMETER APPROACH      *
*     CASE NO. 1 --- SINGLE SOLUTE SYSTEM, CONSTANT INPUTS                *
*     PACKED BED COLUMN WHICH CONTAINS 250 GRAMS OF CARBON                *
*     DNOSBP IS SOLUTE, CONCENTRATION=100 MICROMOLES PER LITER            *
*     SOLUTION VOLUMETRIC FLOW RATE=60 LITERS PER HOUR                    *
*     ASSUMPTION --- MASS TRANSFER IS RATE LIMITING                       *
*     BED HAS BEEN DIVIDED INTO TEN ELEMENTS                              *
*                                                                         *


  ***                                                                 ***
  ***   PARAMETER INPUTS REQUIRED ARE DEFINED AS FOLLOWS             ***
  ***                                                                 ***
  ***   CARBON --- MASS OF CARBON IN BED (GRAMS)                     ***
  ***   A -------- EXTERNAL TRANSFER AREA FOR CARBON (SQ CM/GRAM)     ***
  ***   DP ------- DIAMETER OF CARBON PARTICLES (CM)                 ***
  ***   RHO ------ PACKED BED DENSITY (GRAMS/LITER)                  ***
  ***   EPSI ----- PACKED BED POROSITY (DIMENSIONLESS)               ***
  ***   QMAX ----- LANGMUIR ULTIMATE UPTAKE CAPACITY (MOLES/GRAM)    ***
  ***   B -------- LANGMUIR ENERGY TERM (LITERS/MOLE)                ***
  ***   CO ------- SOLUTION-PHASE CONCENTRATION OF SOLUTE (MOLES/LITER)***
  ***   DL ------- DIFFUSIVITY OF SOLUTE (SQ CM/SEC)                 ***
  ***   AREA ----- CROSS-SECTIONAL AREA OF COLUMN (SQ CM)            ***
  ***   U -------- SOLUTION VOLUMETRIC FLOW RATE (LITERS/HOUR)       ***
  ***   GAMMA ---- KINEMATIC VISCOSITY OF WATER (SQ CM/SEC)          ***


              ***
              ***   PARAMETER INPUTS FOLLOW
              ***
              PARAM CARBON=250.
              PARAM A=150.
              PARAM DP=0.03
              PARAM RHO=381.
              PARAM EPSI=0.45
              PARAM QMAX=8.469E-4
              PARAM B=3.023E5
              PARAM CO=1.E-4
              PARAM DL=6.21E-6
              PARAM AREA=20.25
              PARAM U=60.
              PARAM GAMMA=8.64E-3
```

Generally, the model for this case was formulated in precise accord with an earlier section of this paper entitled, *Program Formulation Using CSMP/360*. The computer program is listed in its entirety in Appendix A. Program statements denoted by brackets have been inserted to cause output of data by the Calcomp plotter in addition to the line printer output which is normally obtained as specified by the PRINT and PRTPLT statements. The bracketed statements can be removed from the main program without changing any other program statements. This will serve only to eliminate output of the Calcomp plots.

Dynamic simulations for this case are shown in Figures 1.4, 1.5, and 1.6 as prepared by the Calcomp plotter. Effluent concentrations for each of 10 elements of the segmented bed have been plotted in Figure 1.4 while the respective solid-phase concentration/time profiles for each element have been plotted in Figure 1.5. The total solid-phase concentration for the entire bed plotted in Figure 1.6 shows a typical solid-phase uptake curve.

*Case Number 2: Simple Packed-Bed,*
*Time Variant Flow Input*

This example is identical to the first with the exception that one of the input parameters (solution volumetric flow rate) is time dependent and has been specified to vary sinusoidally. The program description and parameter input specification have been listed below, while a complete program listing has been included in Appendix B. Again, bracketed statements are those that specify Calcomp output.

Because flow rate is, in this case, a time-dependent variable, the program formulation has been modified slightly. That is, all parameter calculations that become time-dependent in response to a dynamic flow input ($N'_{Re}$, $j_d$, $\overline{U}$, $k_f$) must be removed from the initial and placed into the dynamic segment of the CSMP/360 statement structure as is shown following the program description. In doing so, the simulation proceeds using current values for the time-dependent parameters.

It should be noted that a variety of time-dependent signal inputs are available within CSMP/360 including, among others, functions labeled as ramp, step, impulse, pulse, random or arbitrary. The reader is referred to the CSMP/360 User's Manual[6] for a description of these functions. By use of these, virtually any type of time variant input can be attained.

Calcomp output plots of the dynamic simulation for this case are shown in Figures 1.7 through 1.10. Figure 1.7 is a plot of the time varying solution volumetric flow rate, which is observed to vary sinusoidally as was specified by the problem statement. Figures 1.8-1.10 correspond to Figures 1.4-1.6, respectively, for Case 1. These are noted to have the same general shape as those for Case 1, although deviations in the curves due to the sinusoidal flow input are observed on close inspection.

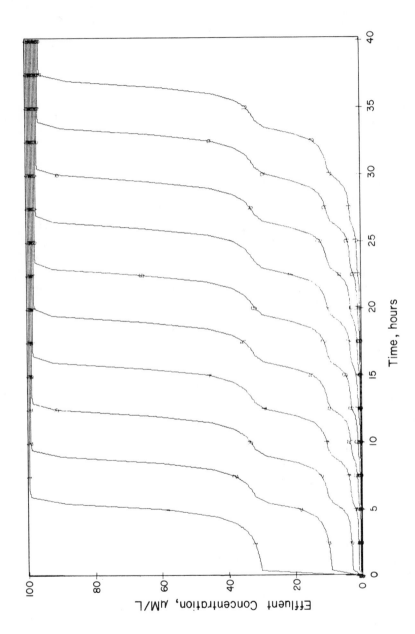

Figure 1.4. Effluent concentration/time profiles for Case 1.

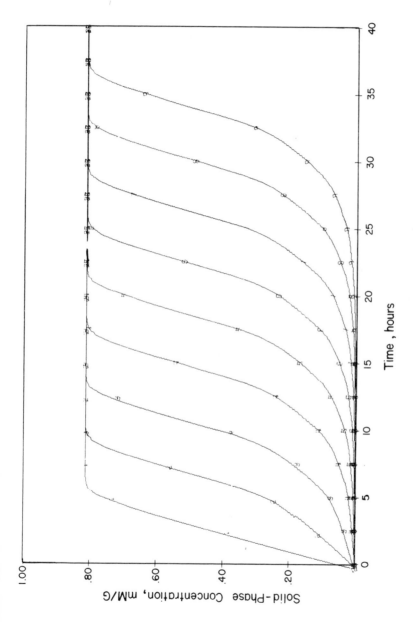

Figure 1.5. Solid-phase concentration/time profiles for Case 1.

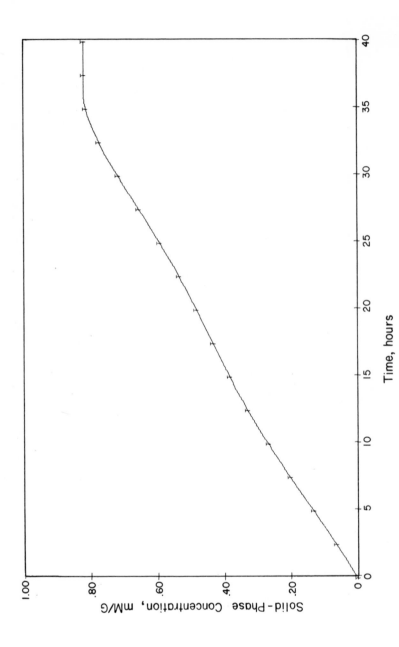

Figure 1.6. Total solid-phase concentration/time profile for Case 1.

```
*     THOMAS M KEINATH *** AEEP WORKSHOP (BAHAMAS)                        *
*     MODELING OF ADSORPTION COLUMNS, DISTRIBUTED PARAMETER APPROACH     *
*     CASE NO. 2 --- SINGLE SOLUTE SYSTEM, SIN WAVE FLOW INPUT           *
*     PACKED BED COLUMN WHICH CONTAINS 250 GRAMS OF CARBON               *
*     DNOSBP IS SOLUTE, CONCENTRATION=100 MICROMOLES PER LITER           *
*     MEAN SOLUTION VOLUMETRIC FLOW RATE=60 LITERS PER HOUR              *
*     AMPLITUDE OF SIN WAVE FLOW INPUT=10 LITERS PER HOUR                *
*     FREQUENCY OF SIN WAVE FLOW INPUT=1 CYCLE PER DAY                   *
*     ASSUMPTION --- MASS TRANSFER IS RATE LIMITING                      *
*     BED HAS BEEN DIVIDED INTO TEN ELEMENTS                             *
*                                                                        *
```

```
***   PARAMETER INPUTS REQUIRED ARE DEFINED AS FOLLOWS                ***
***                                                                   ***
***   CARBON --- MASS OF CARBON IN BED (GRAMS)                        ***
***   A -------- EXTERNAL TRANSFER AREA FOR CARBON (SQ CM/GRAM)       ***
***   DP ------- DIAMETER OF CARBON PARTICLES (CM)                    ***
***   RHO ------ PACKED BED DENSITY (GRAMS/LITER)                     ***
***   EPSI ----- PACKED BED POROSITY (DIMENSIONLESS)                  ***
***   QMAX ----- LANGMUIR ULTIMATE UPTAKE CAPACITY (MOLES/GRAM)       ***
***   B -------- LANGMUIR ENERGY TERM (LITERS/MOLE)                   ***
***   CO ------- SOLUTION-PHASE CONCENTRATION OF SOLUTE (MOLES/LITER)***
***   DL ------- DIFFUSIVITY OF SOLUTE (SQ CM/SEC)                    ***
***   AREA ----- CROSS-SECTIONAL AREA OF COLUMN (SQ CM)               ***
***   UMEAN ---- MEAN SOLUTION VOLUMETRIC FLOW RATE (LITERS/HOUR)     ***
***   UVAR ----- AMPLITUDE OF SIN WAVE FLOW INPUT (LITERS/HOUR)       ***
***   GAMMA ---- KINEMATIC VISCOSITY OF WATER (SQ CM/SEC)             ***
***   XOR ------ CALCOMP PLOTTING ORIGIN FOR X-DIRECTION              ***
***   YOR ------ CALCOMP PLOTTING ORIGIN FOR Y-DIRECTION              ***
***                                                                   ***
***   PARAMETER INPUTS FOLLOW                                         ***
***                                                                   ***
      PARAM CARBON=250.
      PARAM A=150.
      PARAM DP=0.03
      PARAM RHO=381.
      PARAM EPSI=0.45
      PARAM QMAX=8.469E-4
      PARAM B=3.023E5
      PARAM CO=1.E-4
      PARAM DL=6.21E-6
      PARAM AREA=20.25
      PARAM UMEAN=60.
      PARAM UVAR=10.
      PARAM GAMMA=8.64E-3
      PARAM XOR=0.0
      PARAM YOR=0.0
***                                                                   ***
```

```
DYNAMIC
***                                                                      ***
***     CALCULATED INPUT PARAMETERS ARE DEFINED AS FOLLOWS                ***
***                                                                      ***
***     V -------- TOTAL VOLUME OF PACKED BED (LITERS)                    ***
***     UBAR ----- SUPERFICIAL VELOCITY OF FLOW THROUGH BED (CM/SEC)      ***
***     NREMOD --- MODIFIED REYNOLDS NUMBER (DIMENSIONLESS)               ***
***     NSC ------ SCHMIDT NUMBER (DIMENSIONLESS)                         ***
***     JD ------- MASS TRANSFER FACTOR (DIMENSIONLESS)                   ***
***     KF ------- MASS TRANSFER COEFFICIENT (CM/HOUR)                    ***
***     U -------- SOLUTION VOLUMETRIC FLOW RATE (LITERS/HOUR)            ***
***                                                                      ***
***     INPUT PARAMETER CALCULATIONS FOLLOW                               ***
***                                                                      ***
        V=CARBON/(RHO*10.)
        UBAR=(U*0.277777)/AREA
        NREMOD=(UBAR*DP)/(GAMMA*(1.-EPSI))
        NSC=GAMMA/DL
        U=UMEAN+UVAR*(SINE(0.0,0.2618,0.0))
        JD=5.7*NREMOD**(-0.78)
        KF=(JD*UBAR/NSC**(2./3.))*3600.
***                                                                      ***
```

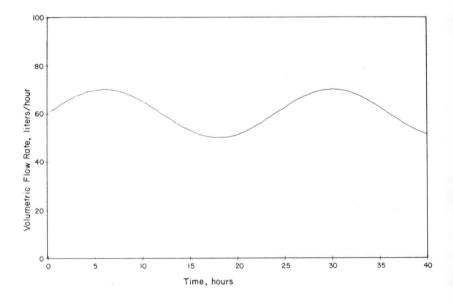

**Figure 1.7.** Variation of solution volumetric flow rate with time.

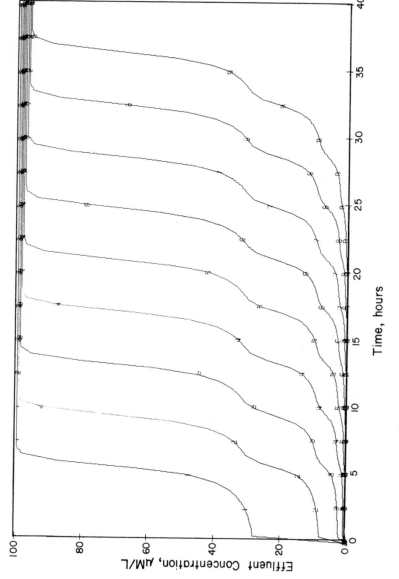

**Figure 1.8.** Effluent concentration/time profiles for Case 2.

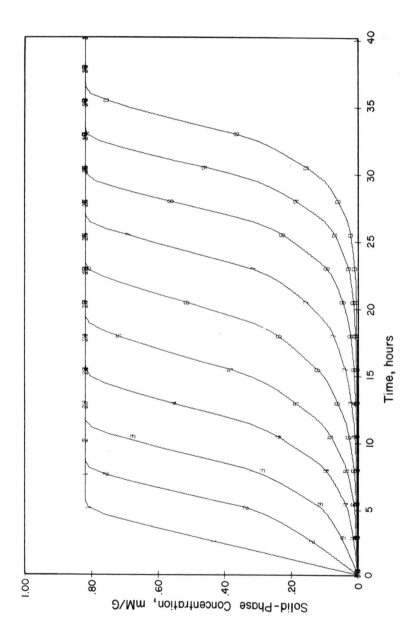

**Figure 1.9.**  Solid-phase concentration/time profiles for Case 2.

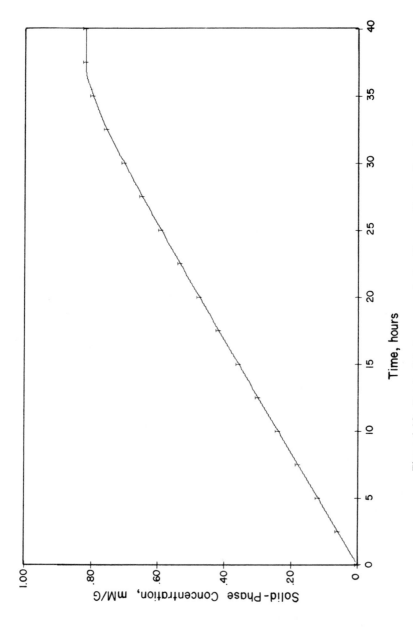

**Figure 1.10.** Total solid-phase concentration/time profiles for Case 2.

*Case Number 3: Single-Solute Countercurrent-Flow System*

In the previous two examples, the column run was initiated by charging the contactor with 250 g of granular activated carbon. Thereafter, no additional adsorbent was added to the bed. For this case, the bed is also charged with 250 g of the granular adsorbent at the outset of the run. In addition, however, active carbon is continuously added to and withdrawn from the bed at a rate of 7 g/hr. All other conditions are as specified for Case 1. A listing of the problem statement and the parameter input specification have been listed below, while a complete program listing has been given in Appendix C.

```
*       THOMAS M KEINATH *** AEEP WORKSHOP (BAHAMAS)                       *
*       MODELING OF ADSORPTION COLUMNS, DISTRIBUTED PARAMETER APPROACH     *
*       CASE NO. 3 --- SINGLE SOLUTE COUNTERCURRENT SYSTEM                 *
*       PACKED BED COLUMN WHICH CONTAINS 250 GRAMS OF CARBON               *
*       DNOSBP IS SOLUTE, CONCENTRATION=100 MICROMOLES PER LITER           *
*       SOLUTION VOLUMETRIC FLOW RATE=60 LITERS PER HOUR                   *
*       ASSUMPTION --- MASS TRANSFER IS RATE LIMITING                      *
*       BED HAS BEEN DIVIDED INTO TEN ELEMENTS                             *
*                                                                         *


***                                                                   ***
***     PARAMETER INPUTS REQUIRED ARE DEFINED AS FOLLOWS              ***
***                                                                   ***
***     CARBON --- MASS OF CARBON IN BED (GRAMS)                      ***
***     CFLO ----- FLOW OF CARBON THROUGH BED (GRAMS/HOUR)            ***
***     A -------- EXTERNAL TRANSFER AREA FOR CARBON (SQ CM/GRAM)     ***
***     DP ------- DIAMETER OF CARBON PARTICLES (CM)                  ***
***     RHO ------ PACKED BED DENSITY (GRAMS/LITER)                   ***
***     EPSI ----- PACKED BED POROSITY (DIMENSIONLESS)                ***
***     QMAX ----- LANGMUIR ULTIMATE UPTAKE CAPACITY (MOLES/GRAM)     ***
***     Q0 ------- INLET SOLID-PHASE CONCENTRATION OF SOLUTE (MOLES/G) ***
***     B -------- LANGMUIR ENERGY TERM (LITERS/MOLE)                 ***
***     CO ------- SOLUTION-PHASE CONCENTRATION OF SOLUTE (MOLES/LITER)***
***     DL ------- DIFFUSIVITY OF SOLUTE (SQ CM/SEC)                  ***
***     AREA ----- CROSS-SECTIONAL AREA OF COLUMN (SQ CM)             ***
***     U -------- SOLUTION VOLUMETRIC FLOW RATE (LITERS/HOUR)        ***
***     GAMMA ---- KINEMATIC VISCOSITY OF WATER (SQ CM/SEC)           ***
***     XOR ------ CALCOMP PLOTTING ORIGIN FOR X-DIRECTION            ***
***     YOR ------ CALCOMP PLOTTING ORIGIN FOR Y-DIRECTION            ***
***                                                                   ***
***     PARAMETER INPUTS FOLLOW                                       ***
***                                                                   ***
        PARAM CARBON=250.
        PARAM CFLO=7.0
        PARAM A=150.
        PARAM DP=0.03
        PARAM RHO=381.
        PARAM EPSI=0.45
        PARAM QMAX=8.469E-4
        PARAM Q0=0.0
        PARAM B=3.023E5
        PARAM CO=1.E-4
        PARAM DL=6.21E-6
        PARAM AREA=20.25
        PARAM U=60.
        PARAM GAMMA=8.64E-3
        PARAM XOR=0.0
        PARAM YOR=0.0
***                                                                   ***
```

```
***                                                                 ***
DYNAMIC
***                                                                 ***
***      DIFFERENTIAL EQUATIONS FOR FIRST ELEMENT                   ***
***                                                                 ***
         C1DOT=(U/(EPSI*V))*(CO-C1)-((KF*A*RHO)/EPSI)*(C1-CSTAR1)*1.E-3
         C1=INTGRL(ICC1,C1DOT)
         Q1DOT=(CFLO/(RHO*V))*(Q2-Q1)+KF*A*(C1-CSTAR1)*1.E-3
         Q1=INTGRL(ICQ1,Q1DOT)
         CSTAR1=Q1/(QMAX*B-Q1*B)
***                                                                 ***
***      DIFFERENTIAL EQUATIONS FOR SECOND ELEMENT                  ***
***                                                                 ***
         C2DOT=(U/(EPSI*V))*(C1-C2)-((KF*A*RHO)/EPSI)*(C2-CSTAR2)*1.E-3
         C2=INTGRL(ICC2,C2DOT)
         Q2DOT=(CFLO/(RHO*V))*(Q3-Q2)+KF*A*(C2-CSTAR2)*1.E-3
         Q2=INTGRL(ICQ2,Q2DOT)
         CSTAR2=Q2/(QMAX*B-Q2*B)
***                                                                 ***
```

For this countercurrent-flow case (Figure 1.1b), Equation 5 must be substituted for Equation 4 for the mass balance on the solid-phase. The program listing for the differential equations for the first two elements of the segmented bed shows this substitution for the solid-phase mass balance relationship (Q1DOT).

This case of countercurrent flow of solid- and solution-phases inherently tends toward steady-state, whereas the previous two cases showed the traditional unsteady-state breakthrough profiles which are commonly observed in laboratory and prototype adsorption installations. Results of this simulation have been Calcomp plotted in Figures 1.11, 1.12 and 1.13. It can be observed that both the solid- and solution-phase concentrations are approaching steady-state values, although steady-state was not attained in the 40-hr simulation run. Further scrutiny of Figure 1.12 shows that the adsorbent which is withdrawn from the bed (Element #1) has attained a steady-state solid-phase concentration in excess of 0.8 mmol/g. Comparison of this value with that obtained for saturation of the bed for Case 1 shows that the adsorbent which is removed from the bed is virtually saturated with DNOSBP. Concomitantly, the concentration of solute in the solution discharged from the tenth element is essentially zero. These observations clearly point out the practical utility of the countercurrent packed-bed contacting mode.

*Case Number 4: Simple Packed-Bed (Disolute System)*

This case has been selected to demonstrate the competitive interactions between two solutes when a disolute solution is passed through a packed-bed adsorption contactor. Functionally, this case is identical to Case 1 with the exception that a second solute, *para*-nitrophenol (PNP), is also

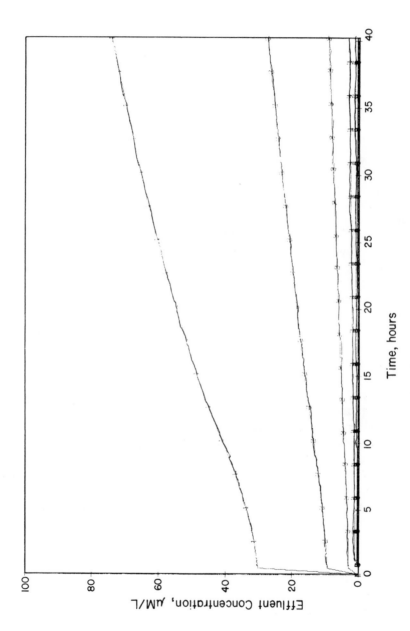

**Figure 1.11.** Effluent concentration/time profiles for Case 3.

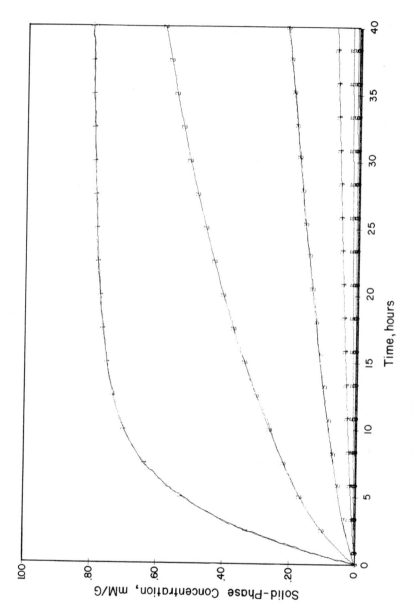

**Figure 1.12.** Solid-phase concentration/time profiles for Case 3.

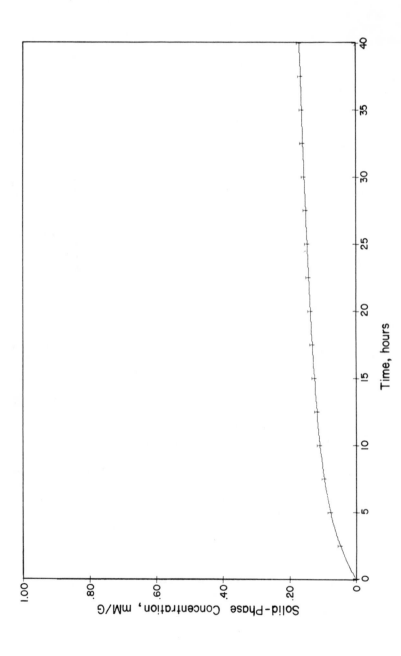

**Figure 1.13.** Total solid-phase concentration/time profiles for Case 3.

added to the column feed at a concentration of 100 μmol. The computer listing that describes the systemic conditions is shown below while a listing of the parameter inputs required for simulation of this special case follows.

```
*    THOMAS M KEINATH *** AEEP WORKSHOP (BAHAMAS)                           *
*    MODELING OF ADSORPTION COLUMNS, DISTRIBUTED PARAMETER APPROACH         *
*    CASE NO. 4 --- DI-SOLUTE SYSTEM                                        *
*    PACKED BED COLUMN WHICH CONTAINS 250 GRAMS OF CARBON                   *
*    SOLUTE A IS DNOSBP (CONCENTRATION=100 MICROMOLES/LITER)                *
*    SOLUTE B IS PNP (CONCENTRATION=100 MICROMOLES/LITER)                   *
*    SOLUTION VOLUMETRIC FLOW RATE=60 LITERS PER HOUR                       *
*    ASSUMPTION --- MASS TRANSFER IS RATE LIMITING                          *
*    BED HAS BEEN DIVIDED INTO NINE ELEMENTS                                *
*                                                                           *
```

```
***                                                                     ***
***    PARAMETER INPUTS REQUIRED ARE DEFINED AS FOLLOWS                  ***
***                                                                     ***
***    CARBON --- MASS OF CARBON IN BED (GRAMS)                          ***
***    A -------- EXTERNAL TRANSFER AREA FOR CARBON (SQ CM/GRAM)         ***
***    DP ------- DIAMETER OF CARBON PARTICLES (CM)                      ***
***    RHO ------ PACKED BED DENSITY (GRAMS/LITER)                       ***
***    EPSI ----- PACKED BED POROSITY (DIMENSIONLESS)                    ***
***    QAMAX ---- LANGMUIR ULTIMATE UPTAKE CAPACITY FOR A (MOLES/GRAM)***
***    QBMAX ---- LANGMUIR ULTIMATE UPTAKE CAPACITY FOR B (MOLES/GRAM)***
***    BA ------- LANGMUIR ENERGY TERM FOR A (LITERS/MOLE)               ***
***    BB ------- LANGMUIR ENERGY TERM FOR B (LITERS/MOLE)               ***
***    CAO ------ SOLUTION-PHASE CONCENTRATION OF A (MOLES/LITER)        ***
***    CBO ------ SOLUTION-PHASE CONCENTRATION OF B (MOLES/LITER)        ***
***    DLA ------ DIFFUSIVITY OF SOLUTE A (SQ CM/SEC)                    ***
***    DLB ------ DIFFUSIVITY OF SOLUTE B (SQ CM/SEC)                    ***
***    AREA ----- CROSS-SECTIONAL AREA OF COLUMN (SQ CM)                 ***
***    U -------- SOLUTION VOLUMETRIC FLOW RATE (LITERS/HOUR)            ***
***    GAMMA ---- KINEMATIC VISCOSITY OF WATER (SQ CM/SEC)               ***


***    PARAMETER INPUTS FOLLOW                                          ***
***                                                                     ***
       PARAM CARBON=250.
       PARAM A=150.
       PARAM DP=0.03
       PARAM RHO=381.
       PARAM EPSI=0.45
       PARAM QAMAX=6.469E-4
       PARAM QBMAX=9.1E-4
       PARAM BA=3.023E5
       PARAM BB=0.27E4
       PARAM CAO=1.E-4
       PARAM CBO=1.E-4
       PARAM DLA=6.21E-6
       PARAM DLB=1.1696E-5
       PARAM AREA=20.25
       PARAM U=60.
       PARAM GAMMA=8.64E-3
       PARAM XOR=0.0
       PARAM YOR=0.0
```

```
***                                                                    ***
***     CALCULATED INPUT PARAMETERS ARE DEFINED AS FOLLOWS              ***
***                                                                    ***
***     V --------- TOTAL VOLUME OF PACKED BED (LITERS)                 ***
***     UBAR ------ SUPERFICIAL VELOCITY OF FLOW THROUGH BED (CM/SEC)   ***
***     NREMOD ---- MODIFIED REYNOLDS NUMBER (DIMENSIONLESS)            ***
***     NSCA ------ SCHMIDT NUMBER FOR A (DIMENSIONLESS)                ***
***     NSCB ------ SCHMIDT NUMBER FOR B (DIMENSIONLESS)                ***
***     JD -------- MASS TRANSFER FACTOR (DIMENSIONLESS)                ***
***     KFA ------- MASS TRANSFER COEFFICIENT FOR A (CM/HOUR)           ***
***     KFB ------- MASS TRANSFER COEFFICIENT FOR B (CM/HOUR)           ***
***                                                                    ***
***     INPUT PARAMETER CALCULATIONS FOLLOW                            ***
***                                                                    ***
        V=CARBON/(RHO*9.)
        UBAR=(U*0.277777)/AREA
        NREMOD=(UBAR*DP)/(GAMMA*(1.-EPSI))
        NSCA=GAMMA/DLA
        NSCB=GAMMA/DLB
        JD=5.7*NREMOD**(-0.78)
        KFA=(JD*UBAR/NSCA**(2./3.))*3600.
        KFB=(JD*UBAR/NSCB**(2./3.))*3600.
***                                                                    ***
```

Structurally, the dynamic model for this case is similar to that for Case 1 except that mass balances for solution- and solid-phases must be written for both solutes that are introduced in the influent to the column. Further, a competitive adsorption equilibrium relationship must be substituted for the single-solute isotherm model. For this simulation the competitive Langmuir model (Equations 9 and 10) has been adopted for description of competitive adsorption equilibria. It is extremely important that the competitive inter-actions with regard to adsorption equilibria must be precisely described such that accurate dynamic simulations can be obtained. The competitive Langmuir model is not broadly applicable in this context. Its suitability must be evaluated for each specific disolute system of interest.

It is vitally important, furthermore, that the extent of adsorption reversibility be properly defined for any competitive adsorption system. Complete reversibility was assumed for these simulations.

The following listing gives the differential equations for the first two elements of the packed bed. As for the previous cases the total program listing has been tabulated in Appendix D.

```
DYNAMIC
***                                                                        ***
***     DIFFERENTIAL EQUATIONS FOR FIRST ELEMENT                           ***
***                                                                        ***
        CA1DOT=(U/(EPSI*V))*(CA0-CA1)-((KFA*A*RHO)/EPSI)*(CA1-CSTA1)*1.E-3
        CA1=INTGRL(ICCA1,CA1DOT)
        CB1DOT=(U/(EPSI*V))*(CB0-CB1)-((KFB*A*RHO)/EPSI)*(CB1-CSTB1)*1.E-3
        CB1=INTGRL(ICCB1,CB1DOT)
        QA1DOT=KFA*A*(CA1-CSTA1)*1.E-3
        QA1=INTGRL(ICQA1,QA1DCT)
        QB1DOT=KFB*A*(CB1-CSTB1)*1.E-3
        QB1=INTGRL(ICQB1,QB1DOT)
        CSTA1=QBMAX*BB*QA1/(QAMAX*QBMAX*BA*BB-QAMAX*QB1*BA*BB-QBMAX*QA1...
        *BA*BB)
        CSTB1=QAMAX*BA*QB1/(QAMAX*QBMAX*BA*BB-QAMAX*QB1*BA*BB-QBMAX*QA1...
        *BA*BB)
***                                                                        ***
***     DIFFERENTIAL EQUATIONS FOR SECOND ELEMENT                          ***
***                                                                        ***
        CA2DOT=(U/(EPSI*V))*(CA1-CA2)-((KFA*A*RHO)/EPSI)*(CA2-CSTA2)*1.E-3
        CA2=INTGRL(ICCA2,CA2DOT)
        CB2DOT=(U/(EPSI*V))*(CB1-CB2)-((KFB*A*RHO)/EPSI)*(CB2-CSTB2)*1.E-3
        CB2=INTGRL(ICCB2,CB2DOT)
        QA2DOT=KFA*A*(CA2-CSTA2)*1.E-3
        QA2=INTGRL(ICQA2,QA2DOT)
        QB2DOT=KFB*A*(CB2-CSTB2)*1.E-3
        QB2=INTGRL(ICQB2,QB2DOT)
        CSTA2=QBMAX*BB*QA2/(QAMAX*QBMAX*BA*BB-QAMAX*QB2*BA*BB-QBMAX*QA2...
        *BA*BB)
        CSTB2=QAMAX*BA*QB2/(QAMAX*QBMAX*BA*BB-QAMAX*QB2*BA*BB-QBMAX*QA2...
        *BA*BB)
***                                                                        ***
```

Simulations for this disolute case have been plotted in Figures 1.14 to 1.17. Figures 1.14 and 1.16 show the solution-phase concentration profiles while Figures 1.15 and 1.17 give the solid-phase profiles for components A and B (DNOSBP and PNP), respectively. These results are particularly interesting in that they clearly show that DNOSBP is preferentially adsorbed near the inlet to the column, while PNP, which has a substantially lower energy of adsorption, passes further into the bed where it is adsorbed essentially as a pure solute. As time proceeds, PNP equilibrates between solid- and solution-phases as a pure solute in the lower regions of the packed-bed. As the adsorption wave of DNOSBP begins to move through the bed, it displaces the PNP until equilibrium is attained in accordance with the competitive Langmuir model. This displacement phenomena is graphically illustrated in Figures 1.16 and 1.17. Certainly, this phenomena is commonly observed in ion exchange contactors, although it has only on several occasions been documented for adsorption contactors.

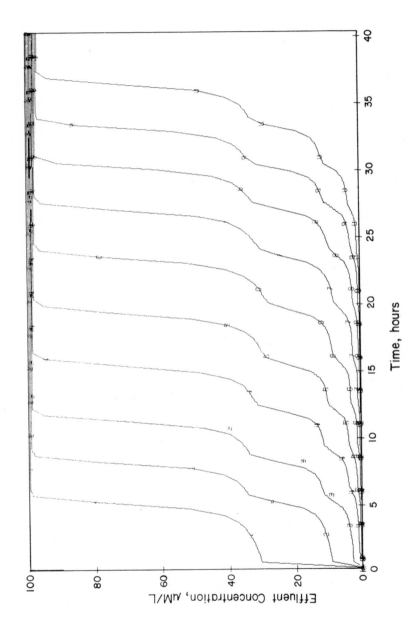

**Figure 1.14.** Effluent concentration/time profiles for DNOSBP; Case 4.

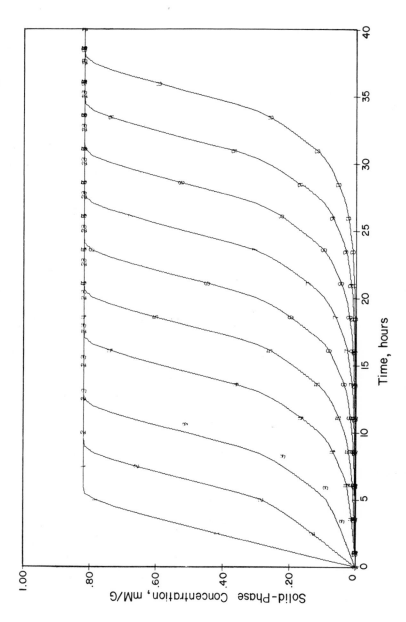

**Figure 1.15.** Solid-phase concentration/time profiles for DNOSBP; Case 4.

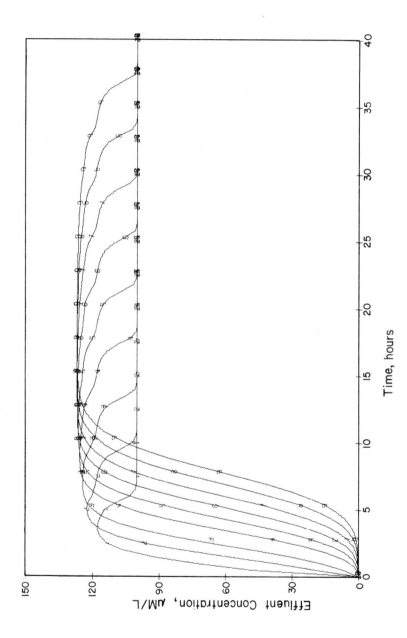

**Figure 1.16.**  Effluent concentration/time profiles for PNP; Case 4.

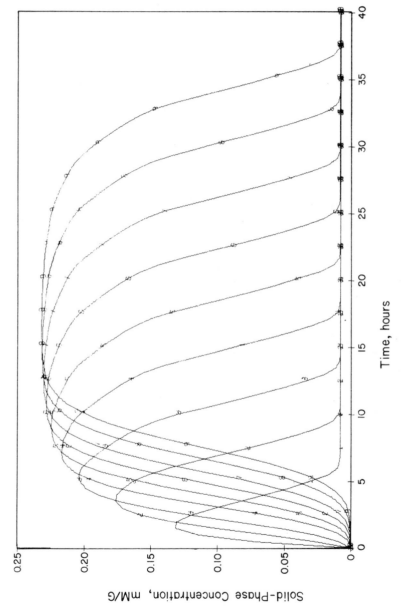

**Figure 1.17.** Solid-phase concentration/time profiles for PNP; Case 4.

## REFERENCES

1. Amundson, N. R. "Solid-Fluid Interactions in Fixed and Moving Beds: Fixed Beds with Large Particles," *Ind. Eng. Chem.* **48**, 35 (1956).
2. Kasten, P. R., L. Lapidus and N. R. Amundson. "Mathematics of Adsorption in Beds: V. Effect of Intraparticle Diffusion in Flow Systems in Fixed Beds," *J. Phys. Chem.* **56**, 683 (1952).
3. Rosen, J. B. "General Numerical Solution for Solid Diffusion into Fixed Beds," *Ind. Eng. Chem.* **46**, 1590 (1954).
4. Crank, J. *Mathematics of Diffusion*. (London: Clarendon Press, 1965).
5. Acrivos, A. "On the Combined Effect of Longitudinal Diffusion and External Mass Transfer Resistance in Fixed Bed Operations," *Chem. Eng. Sci.* **13**, 1 (1960).
6. International Business Machines Corporation. *System/360 Continuous System Modeling Program User's Manual*. Program Number 36A-CX-16X (GH20-0367-4), 5th ed. (1972).
7. Snoeyink, V. L. and J. S. Jain. "Competitive Adsorption on Active Carbon," Presented at the 45th Annual Conference of the Water Pollution Control Federation, Atlanta, Georgia (1972).
8. Perry, J. H., Ed. *Chemical Engineers' Handbook*, 4th ed. (New York: McGraw-Hill Book Co., 1963).
9. Glueckauf, E. and J. I. Coates. "Theory of Chromatography," *J. Chem. Soc.*, 1315 (1947).
10. DiGiano, F. A. "Mathematical Modeling of Sorption Kinetics in Finite and Infinite Bath Systems," Technical Publication T-69-1, University of Michigan (1969).
11. Vermeulen, T. "Theory for Irreversible and Constant-Pattern Solid Diffusion," *Ind. Eng. Chem.* **45**, 1664 (1953).
12. Chu, J. C., J. Kalil and W. A. Wetteroth. "Mass Transfer in a Fluidized Bed," *Chem. Eng. Prog.* **49**, 141 (1953).
13. Pfeffer, R. and J. Happel. "An Analytical Study of Heat and Mass Transfer in Multiparticle Systems at Low Reynolds Numbers," *A. I. Ch. E. J.* **10**, 605 (1964).
14. Fan, L. T., Y. C. Yang and C. Y. Wen. "Mass Transfer in Semi-Fluidized Beds for Solid-Liquid System," *A. I. Ch. E. J.* **3**, 482 (1960).
15. Evans, G. C. and C. F. Gerald. *Chem. Eng. Prog.* **49**, 135 (1953).

## NOMENCLATURE

| Symbol | Description |
| --- | --- |

A        BET energy term, liters/mol
A        cross-sectional area of column, sq cm
b        Langmuir energy term, liters/mol
C        solution-phase concentration of solute, mol/liter
$C_s$    saturation concentration of solute in solution, mol/liter
$D_{pore}$ pore diffusivity, $cm^2/sec$
$j_d$    $(K_f/\bar{U}) * (\nu/D_\varrho)^{2/3}$, mass transfer factor, dimensionless
K        adjustable curve-fitting constant
$n'$     adjustable curve-fitting constant
n        number of elements
$N_{Re}$ $\bar{U} * D_p/\nu$, Reynolds number, dimensionless
$N_{Sc}$ $\nu/D_\varrho$, Schmidt number, dimensionless
$N_{Sh}$ $K_f * D_p/D_\varrho$, Sherwood number, dimensionless
Q        ultimate uptake capacity of adsorbent, mol/g
q        solid-phase concentration of solute, mol/g
$q_o$    initial solid-phase concentration (usually zero), mol/g
R        separation factor, dimensionless
$R_A$    rate of adsorption, mol/g-hr
r        radial distance within particle
t        time, hours
U        solution volumetric flow rate, liters/hr
$\bar{U}$ U/A, velocity of flow, cm/hr
V        volume of the element, liters
z        axial distance, cm

$\epsilon$   void ratio or porosity, dimensionless
$\psi$       $1/(R + 15 (1 - R)\pi^2)$
$\nu$        Kinematic viscosity of solution, $cm^2/sec$
$\rho$       packed-bed density, g/liter
$\rho_p$     density of adsorbent particle, g/liter
$\chi$       internal porosity of adsorbent particles, dimensionless

## APPENDIX A

Program Listing for Case No. 1—
Single-Solute System, Constant Inputs

```
*       THOMAS M KEINATH *** AEEP WORKSHOP (BAHAMAS)                              *
*       MODELING OF ADSORPTION COLUMNS, DISTRIBUTED PARAMETER APPROACH           *
*       CASE NO. 1 --- SINGLE SOLUTE SYSTEM, CONSTANT INPUTS                     *
*       PACKED BED COLUMN WHICH CONTAINS 250 GRAMS OF CARBON                     *
*       DNOSBP IS SOLUTE, CONCENTRATION=100 MICROMOLES PER LITER                 *
*       SOLUTION VOLUMETRIC FLOW RATE=60 LITERS PER HOUR                         *
*       ASSUMPTION --- MASS TRANSFER IS RATE LIMITING                           *
*       BED HAS BEEN DIVIDED INTO TEN ELEMENTS                                  *
*                                                                               *
INITIAL
***                                                                           ***
***     PRELIMINARY DIMENSIONING FOR CALCOMP PLOTTER OUTPUT FOLLOWS           ***
***                                                                           ***
/       DIMENSION C1P(200),C2P(200),C3P(200),C4P(200),C5P(200),C6P(200),
/       1C7P(200),C8P(200),C9P(200),C10P(200),Q1P(200),Q2P(200),Q3P(200),
/       2Q4P(200),Q5P(200),Q6P(200),Q7P(200),Q8P(200),Q9P(200),Q10P(200),
/       3QTP(200),T(200)
/       DIMENSION BUFR(800)
/       CALL PLOTS(BUFR,800)
        STORAGE OUTIME(200)
        FIXED KEY
        KEY=1
        TABLE OUTIME (1-81)=0.0,0.5,1.0,1.5,2.0,2.5,3.0,3.5,4.0,4.5,...
        5.0,5.5,6.0,6.5,7.0,7.5,8.0,8.5,9.0,9.5,10.0,10.5,11.0,11.5,...
        12.0,12.5,13.0,13.5,14.0,14.5,15.0,15.5,16.0,16.5,17.0,17.5,...
        18.0,18.5,19.0,19.5,20.0,20.5,21.0,21.5,22.0,22.5,23.0,23.5,...
        24.0,24.5,25.0,25.5,26.0,26.5,27.0,27.5,28.0,28.5,29.0,29.5,...
        30.0,30.5,31.0,31.5,32.0,32.5,33.0,33.5,34.0,34.5,35.0,35.5,...
        36.0,36.5,37.0,37.5,38.0,38.5,39.0,39.5,40.0
        CLOCK=OUTIME(KEY)
***                                                                           ***
***     INITIAL CONDITIONS FOR SOLUTION PHASE CONCENTRATION                   ***
***                                                                           ***
        INCON ICC1=0.
        INCON ICC2=0.
        INCON ICC3=0.
        INCON ICC4=0.
        INCON ICC5=0.
        INCON ICC6=0.
        INCON ICC7=0.
        INCON ICC8=0.
        INCON ICC9=0.
        INCON ICC10=0.
***                                                                           ***
***     INITIAL CONDITIONS FOR SOLID PHASE CONCENTRATION                      ***
***                                                                           ***
        INCON ICQ1=0.
        INCON ICQ2=0.
        INCON ICQ3=0.
        INCON ICQ4=0.
        INCON ICQ5=0.
        INCON ICQ6=0.
        INCON ICQ7=0.
        INCON ICQ8=0.
        INCON ICQ9=0.
        INCON ICQ10=0.
***                                                                           ***
```

```
***     PARAMETER INPUTS REQUIRED ARE DEFINED AS FOLLOWS               ***
***                                                                    ***
***     CARBON --- MASS OF CARBON IN BED (GRAMS)                       ***
***     A -------- EXTERNAL TRANSFER AREA FOR CARBON (SQ CM/GRAM)      ***
***     DP ------- DIAMETER OF CARBON PARTICLES (CM)                   ***
***     RHO ------ PACKED BED DENSITY (GRAMS/LITER)                    ***
***     EPSI ----- PACKED BED POROSITY (DIMENSIONLESS)                 ***
***     QMAX ----- LANGMUIR ULTIMATE UPTAKE CAPACITY (MOLES/GRAM)      ***
***     B -------- LANGMUIR ENERGY TERM (LITERS/MOLE)                  ***
***     CO ------- SOLUTION-PHASE CONCENTRATION OF SOLUTE (MOLES/LITER)***
***     DL ------- DIFFUSIVITY OF SOLUTE (SQ CM/SEC)                   ***
***     AREA ----- CROSS-SECTIONAL AREA OF COLUMN (SQ CM)              ***
***     U -------- SOLUTION VOLUMETRIC FLOW RATE (LITERS/HOUR)         ***
***     GAMMA ---- KINEMATIC VISCOSITY OF WATER (SQ CM/SEC)            ***
***     XOR ------ CALCOMP PLOTTING ORIGIN FOR X-DIRECTION             ***
***     YOR ------ CALCOMP PLOTTING ORIGIN FOR Y-DIRECTION             ***
***                                                                    ***

***     PARAMETER INPUTS FOLLOW                                        ***
***                                                                    ***
        PARAM CARBON=250.
        PARAM A=150.
        PARAM DP=0.03
        PARAM RHO=381.
        PARAM EPSI=0.45
        PARAM QMAX=8.469E-4
        PARAM B=3.023E5
        PARAM CO=1.E-4
        PARAM DL=6.21E-6
        PARAM AREA=20.25
        PARAM U=60.
        PARAM GAMMA=8.64E-3
        PARAM XOR=0.0
        PARAM YOR=0.0
***                                                                    ***
***     CALCULATED INPUT PARAMETERS ARE DEFINED AS FOLLOWS             ***
***                                                                    ***
***     V -------- TOTAL VOLUME OF PACKED BED (LITERS)                 ***
***     UBAR ----- SUPERFICIAL VELOCITY OF FLOW THROUGH BED (CM/SEC)   ***
***     NREMOD --- MODIFIED REYNOLDS NUMBER (DIMENSIONLESS)            ***
***     NSC ------ SCHMIDT NUMBER (DIMENSIONLESS)                      ***
***     JD ------- MASS TRANSFER FACTOR (DIMENSIONLESS)                ***
***     KF ------- MASS TRANSFER COEFFICIENT (CM/HOUR)                 ***
***                                                                    ***
***     INPUT PARAMETER CALCULATIONS FOLLOW                           ***
***                                                                    ***
        V=CARBON/(RHO*10.)
        UBAR=(U*0.277777)/AREA
        NREMOD=(UBAR*DP)/(GAMMA*(1.-EPSI))
        NSC=GAMMA/DL
        JD=5.7*NREMOD**(-0.78)
        KF=(JD*UBAR/NSC**(2./3.))*3600.
***                                                                    ***
DYNAMIC
***                                                                    ***
***     DIFFERENTIAL EQUATIONS FOR FIRST ELEMENT                       ***
***                                                                    ***
        C1DOT=(U/(EPSI*V))*(CO-C1)-((KF*A*RHO)/EPSI)*(C1-CSTAR1)*1.E-3
        C1=INTGRL(ICC1,C1DOT)
        Q1DOT=KF*A*(C1-CSTAR1)*1.E-3
        Q1=INTGRL(ICQ1,Q1DOT)
        CSTAR1=Q1/(QMAX*B-Q1*B)
***                                                                    ***
***     DIFFERENTIAL EQUATIONS FOR SECOND ELEMENT                      ***
***                                                                    ***
        C2DOT=(U/(EPSI*V))*(C1-C2)-((KF*A*RHO)/EPSI)*(C2-CSTAR2)*1.E-3
        C2=INTGRL(ICC2,C2DOT)
        Q2DOT=KF*A*(C2-CSTAR2)*1.E-3
        Q2=INTGRL(ICQ2,Q2DOT)
        CSTAR2=Q2/(QMAX*B-Q2*B)
***                                                                    ***
***     DIFFERENTIAL EQUATIONS FOR THIRD ELEMENT                       ***
***                                                                    ***
```

```
      C3DOT=(U/(EPSI*V))*(C2-C3)-((KF*A*RHO)/EPSI)*(C3-CSTAR3)*1.E-3
      C3=INTGRL(ICC3,C3DOT)
      Q3DOT=KF*A*(C3-CSTAR3)*1.E-3
      Q3=INTGRL(ICQ3,Q3DOT)
      CSTAR3=Q3/(QMAX*B-Q3*B)
***                                                                       ***
***   DIFFERENTIAL EQUATIONS FOR FOURTH ELEMENT                           ***
***                                                                       ***
      C4DOT=(U/(EPSI*V))*(C3-C4)-((KF*A*RHO)/EPSI)*(C4-CSTAR4)*1.E-3
      C4=INTGRL(ICC4,C4DOT)
      Q4DOT=KF*A*(C4-CSTAR4)*1.E-3
      Q4=INTGRL(ICQ4,Q4DOT)
      CSTAR4=Q4/(QMAX*B-Q4*B)
***                                                                       ***
***   DIFFERENTIAL EQUATIONS FOR FIFTH ELEMENT                            ***
***                                                                       ***
      C5DOT=(U/(EPSI*V))*(C4-C5)-((KF*A*RHO)/EPSI)*(C5-CSTAR5)*1.E-3
      C5=INTGRL(ICC5,C5DOT)
      Q5DOT=KF*A*(C5-CSTAR5)*1.E-3
      Q5=INTGRL(ICQ5,Q5DOT)
      CSTAR5=Q5/(QMAX*B-Q5*B)
***                                                                       ***


***   DIFFERENTIAL EQUATIONS FOR SIXTH ELEMENT                            ***
***                                                                       ***
      C6DOT=(U/(EPSI*V))*(C5-C6)-((KF*A*RHO)/EPSI)*(C6-CSTAR6)*1.E-3
      C6=INTGRL(ICC6,C6DOT)
      Q6DOT=KF*A*(C6-CSTAR6)*1.E-3
      Q6=INTGRL(ICQ6,Q6DOT)
      CSTAR6=Q6/(QMAX*B-Q6*B)
***                                                                       ***
***   DIFFERENTIAL EQUATIONS FOR SEVENTH ELEMENT                          ***
***                                                                       ***
      C7DOT=(U/(EPSI*V))*(C6-C7)-((KF*A*RHO)/EPSI)*(C7-CSTAR7)*1.E-3
      C7=INTGRL(ICC7,C7DOT)
      Q7DOT=KF*A*(C7-CSTAR7)*1.E-3
      Q7=INTGRL(ICQ7,Q7DOT)
      CSTAR7=Q7/(QMAX*B-Q7*B)
***                                                                       ***
***   DIFFERENTIAL EQUATIONS FOR EIGHTH ELEMENT                           ***
***                                                                       ***
      C8DOT=(U/(EPSI*V))*(C7-C8)-((KF*A*RHO)/EPSI)*(C8-CSTAR8)*1.E-3
      C8=INTGRL(ICC8,C8DOT)
      Q8DOT=KF*A*(C8-CSTAR8)*1.E-3
      Q8=INTGRL(ICQ8,Q8DOT)
      CSTAR8=Q8/(QMAX*B-Q8*B)
***                                                                       ***
***   DIFFERENTIAL EQUATIONS FOR NINTH ELEMENT                            ***
***                                                                       ***
      C9DOT=(U/(EPSI*V))*(C8-C9)-((KF*A*RHO)/EPSI)*(C9-CSTAR9)*1.E-3
      C9=INTGRL(ICC9,C9DOT)
      Q9DOT=KF*A*(C9-CSTAR9)*1.E-3
      Q9=INTGRL(ICQ9,Q9DOT)
      CSTAR9=Q9/(QMAX*B-Q9*B)
***                                                                       ***
***   DIFFERENTIAL EQUATIONS FOR TENTH ELEMENT                            ***
***                                                                       ***
      C10DOT=(U/(EPSI*V))*(C9-C10)-((KF*A*RHO)/EPSI)*(C10-CSTA10)*1.E-3
      C10=INTGRL(ICC10,C10DOT)
      Q10DOT=KF*A*(C10-CSTA10)*1.E-3
      Q10=INTGRL(ICQ10,Q10DOT)
      CSTA10=Q10/(QMAX*B-Q10*B)
***                                                                       ***
      QT=(Q1+Q2+Q3+Q4+Q5+Q6+Q7+Q8+Q9+Q10)/10.
***                                                                       ***
***   VARIABLES THAT ARE TO BE PLOTTED BY THE CALCOMP PLOTTER             ***
***   ARE READ INTO ARRAYS IN THE FOLLOWING SEGMENT                       ***
***                                                                       ***
```

```
        NOSORT
        IF(KEEP.EQ.0) GO TO 100
        IF (TIME.LT.CLOCK) GO TO 100
        C1P(KEY)=(C1*1.E6/20.0)+2.0
        C2P(KEY)=(C2*1.E6/20.0)+2.0
        C3P(KEY)=(C3*1.E6/20.0)+2.0
        C4P(KEY)=(C4*1.E6/20.0)+2.0
        C5P(KEY)=(C5*1.E6/20.0)+2.0
        C6P(KEY)=(C6*1.E6/20.0)+2.0
        C7P(KEY)=(C7*1.E6/20.0)+2.0
        C8P(KEY)=(C8*1.E6/20.0)+2.0
        C9P(KEY)=(C9*1.E6/20.0)+2.0
        C10P(KEY)=(C10*1.E6/20.0)+2.0
        Q1P(KEY)=(Q1*1.E3/0.2)+2.0
        Q2P(KEY)=(Q2*1.E3/0.2)+2.0
        Q3P(KEY)=(Q3*1.E3/0.2)+2.0
        Q4P(KEY)=(Q4*1.E3/0.2)+2.0
        Q5P(KEY)=(Q5*1.E3/0.2)+2.0
        Q6P(KEY)=(Q6*1.E3/0.2)+2.0
        Q7P(KEY)=(Q7*1.E3/0.2)+2.0
        Q8P(KEY)=(Q8*1.E3/0.2)+2.0
        Q9P(KEY)=(Q9*1.E3/0.2)+2.0
        Q10P(KEY)=(Q10*1.E3/0.2)+2.0
        QTP(KEY)=(QT*1.E3/0.2)+2.0
        T(KEY)=(OUTIME(KEY)/5.0)+1.0
        KEY=KEY+1
        CLOCK=OUTIME(KEY)
100     CONTINUE
        SORT                                                        ***
***                                                                 ***
***     EXECUTION CONTROL STATEMENTS                                ***
***                                                                 ***
METHOD SIMP
TIMER DELT=0.00025,FINTIM=40.,PRDEL=0.5,OUTDEL=0.5                  ***
***                                                                 ***
***     OUTPUT CONTROL STATEMENTS                                   ***
***                                                                 ***
        PREPAR C1,C2,C3,C4,C5,C6,C7,C8,C9,C10
        PRINT C1,C6,Q1,Q6,C2,C7,Q2,Q7,C3,C8,Q3,Q8,C4,C9,Q4,Q9,...
        C5,C10,Q5,Q10,C0,QT
        PRTPLT C1(0.0,1.E-4),C2(0.0,1.E-4),C3(0.0,1.E-4),C4(0.0,1.E-4)
        PRTPLT C5(0.0,1.E-4),C6(0.0,1.E-4),C7(0.0,1.E-4),C8(0.0,1.E-4)
        PRTPLT C9(0.0,1.E-4),C10(0.0,1.E-4)
        PRTPLT Q1(0.0,1.E-3),Q2(0.0,1.E-3),Q3(0.0,1.E-3),Q4(0.0,1.E-3)
        PRTPLT Q5(0.0,1.E-3),Q6(0.0,1.E-3),Q7(0.0,1.E-3),Q8(0.0,1.E-3)
        PRTPLT Q9(0.0,1.E-3),Q10(0.0,1.E-3)
        PRTPLT QT(0.0,1.E-3)
LABEL SIMULATION OF ADSORPTION BREAKTHROUGH PROFILE
TITLE SIMULATION OF ADSORPTION BREAKTHROUGH PROFILE                 ***
***
TERMINAL                                                            ***
***                                                                 ***
***     ACTUAL PLOTTING BY THE CALCOMP PLOTTER IS ACCOMPLISHED BY   ***
***     THE FOLLOWING STATEMENTS                                    ***
***                                                                 ***
        CALL GRAPH1(XOR,YOR)
        CALL LINE(T,C1P,81,1,113,5)
        CALL LINE(T,C2P,81,1,114,5)
        CALL LINE(T,C3P,81,1,115,5)
        CALL LINE(T,C4P,81,1,116,5)
        CALL LINE(T,C5P,81,1,117,5)
        CALL LINE(T,C6P,81,1,118,5)
        CALL LINE(T,C7P,81,1,119,5)
        CALL LINE(T,C8P,81,1,120,5)
        CALL LINE(T,C9P,81,1,121,5)
        CALL LINE(T,C10P,81,1,112,5)
        CALL PLOT1(15.,0.,-3)
***                                                                 ***
```

```
          CALL GRAPH2(XOR,YOR)
          CALL LINE(T,Q1P,81,1,113,5)
          CALL LINE(T,Q2P,81,1,114,5)
          CALL LINE(T,Q3P,81,1,115,5)
          CALL LINE(T,Q4P,81,1,116,5)
          CALL LINE(T,Q5P,81,1,117,5)
          CALL LINE(T,Q6P,81,1,118,5)
          CALL LINE(T,Q7P,81,1,119,5)
          CALL LINE(T,Q8P,81,1,120,5)
          CALL LINE(T,Q9P,81,1,121,5)
          CALL LINE(T,Q10P,81,1,112,5)
          CALL PLOT1(15.,0.,-3)
***                                                                        ***
          CALL GRAPH2(XOR,YOR)
          CALL LINE(T,QTP,81,1,99,5)
          CALL PLOT1(15.,0.,-3)
***                                                                        ***
      END
      STOP
CCC                                                                        CCC
CCC   SUBROUTINES WHICH GENERATE THE AXES AND LEGENDS FOR EACH             CCC
CCC   OF THE PLOTS ARE DETAILED AS FOLLOWS                                 CCC
CCC                                                                        CCC
CCC   SUBROUTINE FOR CALCOMP OUTPUT OF SOLUTION-PHASE CONCENTRATION        CCC
CCC   VS. TIME FOLLOWS                                                     CCC
CCC                                                                        CCC
          SUBROUTINE GRAPH1(XOR,YOR)
          CALL PLOT1(XOR,YOR,-3)
          CALL AXIS(1.0,2.0,' ',-1,8.0,0.0,0.0,5.0,10.0)
          CALL AXIS(1.0,2.0,' ',+1,5.0,90.0,0.0,20.0,10.0)
          CALL PLOT1(1.0,7.0,+3)
          CALL PLOT1(9.0,7.0,+2)
          CALL PLOT1(9.0,2.0,+2)
          CALL SYMBOL(4.25,1.25,0.20,'TIME, HOURS',0.0,+11)
          CALL SYMBOL(0.5,2.,0.20,'EFFLUENT CONCENTRATION, E-6 M/L',90.,31)
          RETURN
          END
CCC                                                                        CCC
CCC   SUBROUTINE FOR CALCOMP OUTPUT OF SOLID-PHASE CONCENTRATION           CCC
CCC   VS. TIME FOLLOWS                                                     CCC
CCC                                                                        CCC
          SUBROUTINE GRAPH2(XOR,YOR)
          CALL PLOT1(XOR,YOR,-3)
          CALL AXIS(1.0,2.0,' ',-1,8.0,0.0,0.0,5.0,10.0)
          CALL AXIS(1.0,2.0,' ',+1,5.0,90.0,0.0,0.2,10.0)
          CALL PLOT1(1.0,7.0,+3)
          CALL PLOT1(9.0,7.0,+2)
          CALL PLOT1(9.0,2.0,+2)
          CALL SYMBOL(4.25,1.25,0.20,'TIME, HOURS',0.0,+11)
          CALL SYMBOL(0.5,2.,0.2,'SOLID-PHASE CONCENTRATION, MM/G',90.,+31)
          RETURN
          END
      ENDJOB
```

# APPENDIX B

## Program Listing for Case No. 2—
## Single-Solute System, Sin-Curve Flow Input

```
*      THOMAS M KEINATH *** AEEP WORKSHOP (BAHAMAS)                      *
*      MODELING OF ADSORPTION COLUMNS, DISTRIBUTED PARAMETER APPROACH    *
*      CASE NO. 2 --- SINGLE SOLUTE SYSTEM, SIN WAVE FLOW INPUT          *
*      PACKED BED COLUMN WHICH CONTAINS 250 GRAMS OF CARBON             *
*      DNOSBP IS SOLUTE, CONCENTRATION=100 MICROMOLES PER LITER          *
*      MEAN SOLUTION VOLUMETRIC FLOW RATE=60 LITERS PER HOUR             *
*      AMPLITUDE OF SIN WAVE FLOW INPUT=10 LITERS PER HOUR               *
*      FREQUENCY OF SIN WAVE FLOW INPUT=1 CYCLE PER DAY                  *
*      ASSUMPTION --- MASS TRANSFER IS RATE LIMITING                     *
*      BED HAS BEEN DIVIDED INTO TEN ELEMENTS                            *
*                                                                       *
INITIAL
***                                                                   ***
***    PRELIMINARY DIMENSIONING FOR CALCOMP PLOTTER OUTPUT FOLLOWS     ***
***                                                                   ***
/      DIMENSION C1P(200),C2P(200),C3P(200),C4P(200),C5P(200),C6P(200),
/      1C7P(200),C8P(200),C9P(200),C10P(200),Q1P(200),Q2P(200),Q3P(200),
/      2Q4P(200),Q5P(200),Q6P(200),Q7P(200),Q8P(200),Q9P(200),Q10P(200),
/      3QTP(200),T(200),UP(200)
/      DIMENSION BUFR(800)
/      CALL PLOTS (BUFR,800)
       STORAGE OUTIME(200)
       FIXED KEY
       KEY=1
       TABLE OUTIME (1-81)=0.0,0.5,1.0,1.5,2.0,2.5,3.0,3.5,4.0,4.5,...
       5.0,5.5,6.0,6.5,7.0,7.5,8.0,8.5,9.0,9.5,10.0,10.5,11.0,11.5,...
       12.0,12.5,13.0,13.5,14.0,14.5,15.0,15.5,16.0,16.5,17.0,17.5,...
       18.0,18.5,19.0,19.5,20.0,20.5,21.0,21.5,22.0,22.5,23.0,23.5,...
       24.0,24.5,25.0,25.5,26.0,26.5,27.0,27.5,28.0,28.5,29.0,29.5,...
       30.0,30.5,31.0,31.5,32.0,32.5,33.0,33.5,34.0,34.5,35.0,35.5,...
       36.0,36.5,37.0,37.5,38.0,38.5,39.0,39.5,40.0
       CLOCK=OUTIME(KEY)
***                                                                   ***
***    INITIAL CONDITIONS FOR SOLUTION PHASE CONCENTRATION             ***
***                                                                   ***
       INCON ICC1=0.
       INCON ICC2=0.
       INCON ICC3=0.
       INCON ICC4=0.
       INCON ICC5=0.
       INCON ICC6=0.
       INCON ICC7=0.
       INCON ICC8=0.
       INCON ICC9=0.
       INCON ICC10=0.
***                                                                   ***
***    INITIAL CONDITIONS FOR SOLID PHASE CONCENTRATION               ***
***                                                                   ***
       INCON ICQ1=0.
       INCON ICQ2=0.
       INCON ICQ3=0.
       INCON ICQ4=0.
       INCON ICQ5=0.
       INCON ICQ6=0.
       INCON ICQ7=0.
       INCON ICQ8=0.
       INCON ICQ9=0.
       INCON ICQ10=0.
***                                                                   ***
```

```
***    PARAMETER INPUTS REQUIRED ARE DEFINED AS FOLLOWS              ***
***                                                                  ***
***    CARBON --- MASS OF CARBON IN BED (GRAMS)                      ***
***    A -------- EXTERNAL TRANSFER AREA FOR CARBON (SQ CM/GRAM)     ***
***    DP ------- DIAMETER OF CARBON PARTICLES (CM)                  ***
***    RHO ------ PACKED BED DENSITY (GRAMS/LITER)                   ***
***    EPSI ----- PACKED BED POROSITY (DIMENSIONLESS)                ***
***    QMAX ----- LANGMUIR ULTIMATE UPTAKE CAPACITY (MOLES/GRAM)     ***
***    B -------- LANGMUIR ENERGY TERM (LITERS/MOLE)                 ***
***    CO ------- SOLUTION-PHASE CONCENTRATION OF SOLUTE (MOLES/LITER)***
***    DL ------- DIFFUSIVITY OF SOLUTE (SQ CM/SEC)                  ***
***    AREA ----- CROSS-SECTIONAL AREA OF COLUMN (SQ CM)             ***
***    UMEAN ---- MEAN SOLUTION VOLUMETRIC FLOW RATE (LITERS/HOUR)   ***
***    UVAR ----- AMPLITUDE OF SIN WAVE FLOW INPUT (LITERS/HOUR)     ***
***    GAMMA ---- KINEMATIC VISCOSITY OF WATER (SQ CM/SEC)           ***
***    XOR ------ CALCOMP PLOTTING ORIGIN FOR X-DIRECTION            ***
***    YOR ------ CALCOMP PLOTTING ORIGIN FOR Y-DIRECTION            ***
***                                                                  ***
***    PARAMETER INPUTS FOLLOW                                       ***
***                                                                  ***
       PARAM CARBON=250.
       PARAM A=150.
       PARAM DP=0.03
       PARAM RHO=381.
       PARAM EPSI=0.45
       PARAM QMAX=8.469E-4
       PARAM B=3.023E5
       PARAM CO=1.E-4
       PARAM DL=6.21E-6
       PARAM AREA=20.25
       PARAM UMEAN=60.
       PARAM UVAR=10.
       PARAM GAMMA=8.64E-3
       PARAM XOR=0.0
       PARAM YOR=0.0
***                                                                  ***
DYNAMIC
***                                                                  ***
***    CALCULATED INPUT PARAMETERS ARE DEFINED AS FOLLOWS            ***
***                                                                  ***
***    V -------- TOTAL VOLUME OF PACKED BED (LITERS)                ***
***    UBAR ----- SUPERFICIAL VELOCITY OF FLOW THROUGH BED (CM/SEC)  ***
***    NREMOD --- MODIFIED REYNOLDS NUMBER (DIMENSIONLESS)           ***
***    NSC ------ SCHMIDT NUMBER (DIMENSIONLESS)                     ***
***    JD ------- MASS TRANSFER FACTOR (DIMENSIONLESS)               ***
***    KF ------- MASS TRANSFER COEFFICIENT (CM/HOUR)                ***
***    U -------- SOLUTION VOLUMETRIC FLOW RATE (LITERS/HOUR)        ***
***                                                                  ***
***    INPUT PARAMETER CALCULATIONS FOLLOW                           ***
***                                                                  ***
       V=CARBON/(RHO*10.)
       UBAR=(U*0.277777)/AREA
       NREMOD=(UBAR*DP)/(GAMMA*(1.-EPSI))
       NSC=GAMMA/DL
       U=UMEAN+UVAR*(SINE(0.0,0.2618,0.0))
       JD=5.7*NREMOD**(-0.78)
       KF=(JD*UBAR/NSC**(2./3.))*3600.
***                                                                  ***
***    DIFFERENTIAL EQUATIONS FOR FIRST ELEMENT                      ***
***                                                                  ***
       C1DOT=(U/(EPSI*V))*(CO-C1)-((KF*A*RHO)/EPSI)*(C1-CSTAR1)*1.E-3
       C1=INTGRL(ICC1,C1DOT)
       Q1DOT=KF*A*(C1-CSTAR1)*1.E-3
       Q1=INTGRL(ICQ1,Q1DOT)
       CSTAR1=Q1/(QMAX*B-Q1*B)
***                                                                  ***
***    DIFFERENTIAL EQUATIONS FOR SECOND ELEMENT                     ***
***                                                                  ***
       C2DOT=(U/(EPSI*V))*(C1-C2)-((KF*A*RHO)/EPSI)*(C2-CSTAR2)*1.E-3
       C2=INTGRL(ICC2,C2DOT)
       Q2DOT=KF*A*(C2-CSTAR2)*1.E-3
       Q2=INTGRL(ICQ2,Q2DOT)
       CSTAR2=Q2/(QMAX*B-Q2*B)
***                                                                  ***
```

```
***    DIFFERENTIAL EQUATIONS FOR THIRD ELEMENT                          ***
***                                                                       ***
       C3DOT=(U/(EPSI*V))*(C2-C3)-((KF*A*RHO)/EPSI)*(C3-CSTAR3)*1.E-3
       C3=INTGRL(ICC3,C3DOT)
       Q3DOT=KF*A*(C3-CSTAR3)*1.E-3
       Q3=INTGRL(ICQ3,Q3DOT)
       CSTAR3=Q3/(QMAX*B-Q3*B)
***                                                                       ***
***    DIFFERENTIAL EQUATIONS FOR FOURTH ELEMENT                          ***
***                                                                       ***
       C4DOT=(U/(EPSI*V))*(C3-C4)-((KF*A*RHO)/EPSI)*(C4-CSTAR4)*1.E-3
       C4=INTGRL(ICC4,C4DOT)
       Q4DOT=KF*A*(C4-CSTAR4)*1.E-3
       Q4=INTGRL(ICQ4,Q4DOT)
       CSTAR4=Q4/(QMAX*B-Q4*B)
***                                                                       ***
***    DIFFERENTIAL EQUATIONS FOR FIFTH ELEMENT                           ***
***                                                                       ***
       C5DOT=(U/(EPSI*V))*(C4-C5)-((KF*A*RHO)/EPSI)*(C5-CSTAR5)*1.E-3
       C5=INTGRL(ICC5,C5DOT)
       Q5DOT=KF*A*(C5-CSTAR5)*1.E-3
       Q5=INTGRL(ICQ5,Q5DOT)
       CSTAR5=Q5/(QMAX*B-Q5*B)
***                                                                       ***
***    DIFFERENTIAL EQUATIONS FOR SIXTH ELEMENT                           ***
***                                                                       ***
       C6DOT=(U/(EPSI*V))*(C5-C6)-((KF*A*RHO)/EPSI)*(C6-CSTAR6)*1.E-3
       C6=INTGRL(ICC6,C6DOT)
       Q6DOT=KF*A*(C6-CSTAR6)*1.E-3
       Q6=INTGRL(ICQ6,Q6DOT)
       CSTAR6=Q6/(QMAX*B-Q6*B)
***                                                                       ***
***    DIFFERENTIAL EQUATIONS FOR SEVENTH ELEMENT                         ***
***                                                                       ***
       C7DOT=(U/(EPSI*V))*(C6-C7)-((KF*A*RHO)/EPSI)*(C7-CSTAR7)*1.E-3
       C7=INTGRL(ICC7,C7DOT)
       Q7DOT=KF*A*(C7-CSTAR7)*1.E-3
       Q7=INTGRL(ICQ7,Q7DOT)
       CSTAR7=Q7/(QMAX*B-Q7*B)
***                                                                       ***
***    DIFFERENTIAL EQUATIONS FOR EIGHTH ELEMENT                          ***
***                                                                       ***
       C8DOT=(U/(EPSI*V))*(C7-C8)-((KF*A*RHO)/EPSI)*(C8-CSTAR8)*1.E-3
       C8=INTGRL(ICC8,C8DOT)
       Q8DOT=KF*A*(C8-CSTAR8)*1.E-3
       Q8=INTGRL(ICQ8,Q8DOT)
       CSTAR8=Q8/(QMAX*B-Q8*B)
***                                                                       ***
***    DIFFERENTIAL EQUATIONS FOR NINTH ELEMENT                           ***
***                                                                       ***
       C9DOT=(U/(EPSI*V))*(C8-C9)-((KF*A*RHO)/EPSI)*(C9-CSTAR9)*1.E-3
       C9=INTGRL(ICC9,C9DOT)
       Q9DOT=KF*A*(C9-CSTAR9)*1.E-3
       Q9=INTGRL(ICQ9,Q9DOT)
       CSTAR9=Q9/(QMAX*B-Q9*B)
***                                                                       ***
***    DIFFERENTIAL EQUATIONS FOR TENTH ELEMENT                           ***
***                                                                       ***
       C10DOT=(U/(EPSI*V))*(C9-C10)-((KF*A*RHO)/EPSI)*(C10-CSTA10)*1.E-3
       C10=INTGRL(ICC10,C10DOT)
       Q10DOT=KF*A*(C10-CSTA10)*1.E-3
       Q10=INTGRL(ICQ10,Q10DOT)
       CSTA10=Q10/(QMAX*B-Q10*B)
***                                                                       ***
       QT=(Q1+Q2+Q3+Q4+Q5+Q6+Q7+Q8+Q9+Q10)/10.
***                                                                       ***
```

```
***    VARIABLES THAT ARE TO BE PLOTTED BY THE CALCOMP PLOTTER    ***
***    ARE READ INTO ARRAYS IN THE FOLLOWING SEGMENT              ***
***                                                               ***
       NOSORT
       IF(KEEP.EQ.0) GO TO 100
       IF (TIME.LT.CLOCK) GO TO 100
       C1P(KEY)=(C1*1.E6/20.0)+2.0
       C2P(KEY)=(C2*1.E6/20.0)+2.0
       C3P(KEY)=(C3*1.E6/20.0)+2.0
       C4P(KEY)=(C4*1.E6/20.0)+2.0
       C5P(KEY)=(C5*1.E6/20.0)+2.0
       C6P(KEY)=(C6*1.E6/20.0)+2.0
       C7P(KEY)=(C7*1.E6/20.0)+2.0
       C8P(KEY)=(C8*1.E6/20.0)+2.0
       C9P(KEY)=(C9*1.E6/20.0)+2.0
       C10P(KEY)=(C10*1.E6/20.0)+2.0
***                                                               ***
       Q1P(KEY)=(Q1*1.E3/0.2)+2.0
       Q2P(KEY)=(Q2*1.E3/0.2)+2.0
       Q3P(KEY)=(Q3*1.E3/0.2)+2.0
       Q4P(KEY)=(Q4*1.E3/0.2)+2.0
       Q5P(KEY)=(Q5*1.E3/0.2)+2.0
       Q6P(KEY)=(Q6*1.E3/0.2)+2.0
       Q7P(KEY)=(Q7*1.E3/0.2)+2.0
       Q8P(KEY)=(Q8*1.E3/0.2)+2.0
       Q9P(KEY)=(Q9*1.E3/0.2)+2.0
       Q10P(KEY)=(Q10*1.E3/0.2)+2.0
       QTP(KEY)=(QT*1.E3/0.2)+2.0
       UP(KEY)=(U/20.)+2.0
       T(KEY)=(OUTIME(KEY)/5.0)+1.0
       KEY=KEY+1
       CLOCK=OUTIME(KEY)
100    CONTINUE
       SORT

***                                                               ***
***    EXECUTION CONTROL STATEMENTS                               ***
***                                                               ***
METHOD SIMP
TIMER DELT=0.00025,FINTIM=40.,PRDEL=0.5,OUTDEL=0.5
***                                                               ***
***    OUTPUT CONTROL STATEMENTS                                  ***
***                                                               ***
       PREPAR C1,C2,C3,C4,C5,C6,C7,C8,C9,C10
       PRINT C1,C6,Q1,Q6,C2,C7,Q2,Q7,C3,C8,Q3,Q8,C4,C9,Q4,Q9,....
       C5,C10,Q5,Q10,C0,QT
       PRTPLT C1(0.0,1.E-4),C2(0.0,1.E-4),C3(0.0,1.E-4),C4(0.0,1.E-4)
       PRTPLT C5(0.0,1.E-4),C6(0.0,1.E-4),C7(0.0,1.E-4),C8(0.0,1.E-4)
       PRTPLT C9(0.0,1.E-4),C10(0.0,1.E-4)
       PRTPLT Q1(0.0,1.E-3),Q2(0.0,1.E-3),Q3(0.0,1.E-3),Q4(0.0,1.E-3)
       PRTPLT Q5(0.0,1.E-3),Q6(0.0,1.E-3),Q7(0.0,1.E-3),Q8(0.0,1.E-3)
       PRTPLT Q9(0.0,1.E-3),Q10(0.0,1.E-3)
       PRTPLT QT(0.0,1.E-3)
       PRTPLT U(0.0,80.0)
LABEL SIMULATION OF ADSORPTION BREAKTHROUGH PROFILE
TITLE SIMULATION OF ADSORPTION BREAKTHROUGH PROFILE
***                                                               ***
```

```
TERMINAL
***                                                                    ***
***     ACTUAL PLOTTING BY THE CALCOMP PLOTTER IS ACCOMPLISHED BY      ***
***     THE FOLLOWING STATEMENTS                                       ***
***                                                                    ***
        CALL GRAPH1(XOR,YOR)
        CALL LINE(T,C1P,81,1,113,5)
        CALL LINE(T,C2P,81,1,114,5)
        CALL LINE(T,C3P,81,1,115,5)
        CALL LINE(T,C4P,81,1,116,5)
        CALL LINE(T,C5P,81,1,117,5)
        CALL LINE(T,C6P,81,1,118,5)
        CALL LINE(T,C7P,81,1,119,5)
        CALL LINE(T,C8P,81,1,120,5)
        CALL LINE(T,C9P,81,1,121,5)
        CALL LINE(T,C10P,81,1,112,5)
        CALL PLOT1(15.,0.,-3)
***                                                                    ***
        CALL GRAPH2(XOR,YOR)
        CALL LINE(T,Q1P,81,1,113,5)
        CALL LINE(T,Q2P,81,1,114,5)
        CALL LINE(T,Q3P,81,1,115,5)
        CALL LINE(T,Q4P,81,1,116,5)
        CALL LINE(T,Q5P,81,1,117,5)
        CALL LINE(T,Q6P,81,1,118,5)
        CALL LINE(T,Q7P,81,1,119,5)
        CALL LINE(T,Q8P,81,1,120,5)
        CALL LINE(T,Q9P,81,1,121,5)
        CALL LINE(T,Q10P,81,1,112,5)
        CALL PLOT1(15.,0.,-3)
***                                                                    ***
        CALL GRAPH2(XOR,YOR)
        CALL LINE(T,QTP,81,1,99,5)
        CALL PLOT1(15.,0.,-3)
***                                                                    ***
        CALL GRAPH3(XOR,YOR)
        CALL LINE(T,UP,81,1,0,0)
        CALL PLOT1(15.,0.,-3)
***                                                                    ***
END
STOP
CCC
CCC                                                                    CCC
CCC     SUBROUTINES WHICH GENERATE THE AXES AND LEGENDS FOR EACH       CCC
CCC     OF THE PLOTS ARE DETAILED AS FOLLOWS                           CCC
CCC                                                                    CCC
CCC     SUBROUTINE FOR CALCOMP OUTPUT OF SOLUTION-PHASE CONCENTRATION  CCC
CCC     VS. TIME FOLLOWS                                               CCC
CCC                                                                    CCC
        SUBROUTINE GRAPH1(XOR,YOR)                                     CCC
        CALL PLOT1(XOR,YOR,-3)
        CALL AXIS(1.0,2.0,' ',-1,8.0,0.0,0.0,5.0,10.0)
        CALL AXIS(1.0,2.0,' ',+1,5.0,90.0,0.0,20.0,10.0)
        CALL PLOT1(1.0,7.0,+3)
        CALL PLOT1(9.0,7.0,+2)
        CALL PLOT1(9.0,2.0,+2)
        CALL SYMBOL(4.25,1.25,0.20,'TIME, HOURS',0.0,+11)
        CALL SYMBOL(0.5,2.,0.20,'EFFLUENT CONCENTRATION, E-6 M/L',90.,31)
        RETURN
        END
```

```
CCC                                                                    CCC
CCC     SUBROUTINE FOR CALCOMP OUTPUT OF SOLID-PHASE CONCENTRATION     CCC
CCC     VS. TIME FOLLOWS                                               CCC
CCC                                                                    CCC
        SUBROUTINE GRAPH2(XOR,YOR)
        CALL PLOT1(XOR,YOR,-3)
        CALL AXIS(1.0,2.0,' ',-1,8.0,0.0,0.0,5.0,10.0)
        CALL AXIS(1.0,2.0,' ',+1,5.0,90.0,0.0,0.2,10.0)
        CALL PLOT1(1.0,7.0,+3)
        CALL PLOT1(9.0,7.0,+2)
        CALL PLOT1(9.0,2.0,+2)
        CALL SYMBOL(4.25,1.25,0.20,'TIME, HOURS',0.0,+11)
        CALL SYMBOL(0.5,2.,0.2,'SOLID-PHASE CONCENTRATION, MM/G',90.,+31)
        RETURN
        END
CCC                                                                    CCC
CCC     SUBROUTINE FOR CALCOMP OUTPUT OF FLOW RATE VS. TIME FOLLOWS     CCC
CCC                                                                    CCC
        SUBROUTINE GRAPH3(XOR,YOR)
        CALL PLOT1(XOR,YOR,-3)
        CALL AXIS(1.0,2.0,' ',-1,8.0,0.0,0.0,5.0,10.0)
        CALL AXIS(1.0,2.0,' ',+1,5.0,90.0,0.0,20.,10.0)
        CALL PLOT1(1.0,7.0,+3)
        CALL PLOT1(9.0,7.0,+2)
        CALL PLOT1(9.0,2.0,+2)
        CALL SYMBOL(4.25,1.25,0.20,'TIME, HOURS',0.0,+11)
        CALL SYMBOL(.5,2.,.2,'VOLUMETRIC FLOW RATE, LITERS/HOUR',90.,+33)
        RETURN
        END
ENDJOB
 /*
```

## APPENDIX C

Program Listing for Case No. 3—
Single-Solute Countercurrent System

```
*     THOMAS M KEINATH *** AEEP WORKSHOP (BAHAMAS)                          *
*     MODELING OF ADSORPTION COLUMNS, DISTRIBUTED PARAMETER APPROACH        *
*     CASE NO. 3 --- SINGLE SOLUTE COUNTERCURRENT SYSTEM                    *
*     PACKED BED COLUMN WHICH CONTAINS 250 GRAMS OF CARBON                  *
*     DNOSBP IS SOLUTE, CONCENTRATION=100 MICROMOLES PER LITER              *
*     SOLUTION VOLUMETRIC FLOW RATE=60 LITERS PER HOUR                      *
*     ASSUMPTION --- MASS TRANSFER IS RATE LIMITING                        *
*     BED HAS BEEN DIVIDED INTO TEN ELEMENTS                                *
*                                                                          *
INITIAL
***                                                                      ***
***   PRELIMINARY DIMENSIONING FOR CALCOMP PLOTTER OUTPUT FOLLOWS        ***
***                                                                      ***
/     DIMENSION C1P(200),C2P(200),C3P(200),C4P(200),C5P(200),C6P(200),
/    1C7P(200),C8P(200),C9P(200),C10P(200),Q1P(200),Q2P(200),Q3P(200),
/    2Q4P(200),Q5P(200),Q6P(200),Q7P(200),Q8P(200),Q9P(200),Q10P(200),
/    3QTP(200),T(200)
/     DIMENSION BUFR(800)
/     CALL PLOTS (BUFR,800)
      STORAGE OUTIME(200)
      FIXED KEY
      KEY=1
      TABLE OUTIME (1-81)=0.0,0.5,1.0,1.5,2.0,2.5,3.0,3.5,4.0,4.5,...
      5.0,5.5,6.0,6.5,7.0,7.5,8.0,8.5,9.0,9.5,10.0,10.5,11.0,11.5,...
      12.0,12.5,13.0,13.5,14.0,14.5,15.0,15.5,16.0,16.5,17.0,17.5,...
      18.0,18.5,19.0,19.5,20.0,20.5,21.0,21.5,22.0,22.5,23.0,23.5,...
      24.0,24.5,25.0,25.5,26.0,26.5,27.0,27.5,28.0,28.5,29.0,29.5,...
      30.0,30.5,31.0,31.5,32.0,32.5,33.0,33.5,34.0,34.5,35.0,35.5,...
      36.0,36.5,37.0,37.5,38.0,38.5,39.0,39.5,40.0
      CLOCK=OUTIME(KEY)
***                                                                      ***
***   INITIAL CONDITIONS FOR SOLUTION PHASE CONCENTRATION               ***
***                                                                      ***
      INCON ICC1=0.
      INCON ICC2=0.
      INCON ICC3=0.
      INCON ICC4=0.
      INCON ICC5=0.
      INCON ICC6=0.
      INCON ICC7=0.
      INCON ICC8=0.
      INCON ICC9=0.
      INCON ICC10=0.
***                                                                      ***
***   INITIAL CONDITIONS FOR SOLID PHASE CONCENTRATION                  ***
***                                                                      ***
      INCON ICQ1=0.
      INCON ICQ2=0.
      INCON ICQ3=0.
      INCON ICQ4=0.
      INCON ICQ5=0.
      INCON ICQ6=0.
      INCON ICQ7=0.
      INCON ICQ8=0.
      INCON ICQ9=0.
      INCON ICQ10=0.
***                                                                      ***
```

```
***    PARAMETER INPUTS REQUIRED ARE DEFINED AS FOLLOWS                        ***
***                                                                            ***
***    CARBON --- MASS OF CARBON IN BED (GRAMS)                                ***
***    CFLO ----- FLOW OF CARBON THROUGH BED (GRAMS/HOUR)                      ***
***    A --------- EXTERNAL TRANSFER AREA FOR CARBON (SQ CM/GRAM)              ***
***    DP ------- DIAMETER OF CARBON PARTICLES (CM)                            ***
***    RHO ------ PACKED BED DENSITY (GRAMS/LITER)                             ***
***    EPSI ----- PACKED BED POROSITY (DIMENSIONLESS)                          ***
***    QMAX ----- LANGMUIR ULTIMATE UPTAKE CAPACITY (MOLES/GRAM)               ***
***    QO ------- INLET SOLID-PHASE CONCENTRATION OF SOLUTE (MOLES/G)          ***
***    B --------- LANGMUIR ENERGY TERM (LITERS/MOLE)                          ***
***    CO ------- SOLUTION-PHASE CONCENTRATION OF SOLUTE (MOLES/LITER)         ***
***    DL ------- DIFFUSIVITY OF SOLUTE (SQ CM/SEC)                            ***
***    GAMMA ---- KINEMATIC VISCOSITY OF WATER (SQ CM/SEC)                     ***
***    XOR ------ CALCOMP PLOTTING ORIGIN FOR X-DIRECTION                      ***
***    YOR ------ CALCOMP PLOTTING ORIGIN FOR Y-DIRECTION                      ***
***                                                                            ***
***    PARAMETER INPUTS FOLLOW                                                 ***
***                                                                            ***
       PARAM CARBON=250.
       PARAM CFLO=7.0
       PARAM A=150.
       PARAM DP=0.03
       PARAM RHO=381.
       PARAM EPSI=0.45
       PARAM QMAX=8.469E-4
       PARAM QO=0.0
       PARAM B=3.023E5
       PARAM CO=1.E-4
       PARAM DL=6.21E-6
       PARAM AREA=20.25
       PARAM U=60.
       PARAM GAMMA=8.64E-3
       PARAM XOR=0.0
       PARAM YOR=0.0
***                                                                            ***
***    CALCULATED INPUT PARAMETERS ARE DEFINED AS FOLLOWS                      ***
***                                                                            ***
***    V -------- TOTAL VOLUME OF PACKED BED (LITERS)                          ***
***    UBAR ----- SUPERFICIAL VELOCITY OF FLOW THROUGH BED (CM/SEC)            ***
***    NREMOD --- MODIFIED REYNOLDS NUMBER (DIMENSIONLESS)                     ***
***    NSC ------ SCHMIDT NUMBER (DIMENSIONLESS)                               ***
***    JD ------- MASS TRANSFER FACTOR (DIMENSIONLESS)                         ***
***    KF ------- MASS TRANSFER COEFFICIENT (CM/HOUR)                          ***
***                                                                            ***
***    INPUT PARAMETER CALCULATIONS FOLLOW                                     ***
***                                                                            ***
       V=CARBON/(RHO*10.)
       UBAR=(U*0.277777)/AREA
       NREMOD=(UBAR*DP)/(GAMMA*(1.-EPSI))
       NSC=GAMMA/DL
       JD=5.7*NREMOD**(-0.78)
       KF=(JD*UBAR/NSC**(2./3.))*3600.
***                                                                            ***
DYNAMIC
***                                                                            ***
***    DIFFERENTIAL EQUATIONS FOR FIRST ELEMENT                                ***
***                                                                            ***
       C1DOT=(U/(EPSI*V))*(CO-C1)-((KF*A*RHO)/EPSI)*(C1-CSTAR1)*1.E-3
       C1=INTGRL(ICC1,C1DOT)
       Q1DOT=(CFLO/(RHO*V))*(Q2-Q1)+KF*A*(C1-CSTAR1)*1.E-3
       Q1=INTGRL(ICQ1,Q1DOT)
       CSTAR1=Q1/(QMAX*B-Q1*B)
***                                                                            ***
```

```
***      DIFFERENTIAL EQUATIONS FOR SECOND ELEMENT                        ***
***                                                                       ***
         C2DOT=(U/(EPSI*V))*(C1-C2)-((KF*A*RHO)/EPSI)*(C2-CSTAR2)*1.E-3
         C2=INTGRL(ICC2,C2DOT)
         Q2DOT=(CFLO/(RHO*V))*(Q3-Q2)+KF*A*(C2-CSTAR2)*1.E-3
         Q2=INTGRL(ICQ2,Q2DOT)
         CSTAR2=Q2/(QMAX*B-Q2*B)
***                                                                       ***
***      DIFFERENTIAL EQUATIONS FOR THIRD ELEMENT                         ***
***                                                                       ***
         C3DOT=(U/(EPSI*V))*(C2-C3)-((KF*A*RHO)/EPSI)*(C3-CSTAR3)*1.E-3
         C3=INTGRL(ICC3,C3DOT)
         Q3DOT=(CFLO/(RHO*V))*(Q4-Q3)+KF*A*(C3-CSTAR3)*1.E-3
         Q3=INTGRL(ICQ3,Q3DOT)
         CSTAR3=Q3/(QMAX*B-Q3*B)
***                                                                       ***
***      DIFFERENTIAL EQUATIONS FOR FOURTH ELEMENT                        ***
***                                                                       ***
         C4DOT=(U/(EPSI*V))*(C3-C4)-((KF*A*RHO)/EPSI)*(C4-CSTAR4)*1.E-3
         C4=INTGRL(ICC4,C4DOT)
         Q4DOT=(CFLO/(RHO*V))*(Q5-Q4)+KF*A*(C4-CSTAR4)*1.E-3
         Q4=INTGRL(ICQ4,Q4DOT)
         CSTAR4=Q4/(QMAX*B-Q4*B)
***                                                                       ***
***      DIFFERENTIAL EQUATIONS FOR FIFTH ELEMENT                         ***
***                                                                       ***
         C5DOT=(U/(EPSI*V))*(C4-C5)-((KF*A*RHO)/EPSI)*(C5-CSTAR5)*1.E-3
         C5=INTGRL(ICC5,C5DOT)
         Q5DOT=(CFLO/(RHO*V))*(Q6-Q5)+KF*A*(C5-CSTAR5)*1.E-3
         Q5=INTGRL(ICQ5,Q5DOT)
         CSTAR5=Q5/(QMAX*B-Q5*B)
***                                                                       ***
***      DIFFERENTIAL EQUATIONS FOR SIXTH ELEMENT                         ***
***                                                                       ***
         C6DOT=(U/(EPSI*V))*(C5-C6)-((KF*A*RHO)/EPSI)*(C6-CSTAR6)*1.E-3
         C6=INTGRL(ICC6,C6DOT)
         Q6DOT=(CFLO/(RHO*V))*(Q7-Q6)+KF*A*(C6-CSTAR6)*1.E-3
         Q6=INTGRL(ICQ6,Q6DOT)
         CSTAR6=Q6/(QMAX*B-Q6*B)
***                                                                       ***
***      DIFFERENTIAL EQUATIONS FOR SEVENTH ELEMENT                       ***
***                                                                       ***
         C7DOT=(U/(EPSI*V))*(C6-C7)-((KF*A*RHO)/EPSI)*(C7-CSTAR7)*1.E-3
         C7=INTGRL(ICC7,C7DOT)
         Q7DOT=(CFLO/(RHO*V))*(Q8-Q7)+KF*A*(C7-CSTAR7)*1.E-3
         Q7=INTGRL(ICQ7,Q7DOT)
         CSTAR7=Q7/(QMAX*B-Q7*B)
***                                                                       ***
***      DIFFERENTIAL EQUATIONS FOR EIGHTH ELEMENT                        ***
***                                                                       ***
         C8DOT=(U/(EPSI*V))*(C7-C8)-((KF*A*RHO)/EPSI)*(C8-CSTAR8)*1.E-3
         C8=INTGRL(ICC8,C8DOT)
         Q8DOT=(CFLO/(RHO*V))*(Q9-Q8)+KF*A*(C8-CSTAR8)*1.E-3
         Q8=INTGRL(ICQ8,Q8DOT)
         CSTAR8=Q8/(QMAX*B-Q8*B)
***                                                                       ***
***      DIFFERENTIAL EQUATIONS FOR NINTH ELEMENT                         ***
***                                                                       ***
         C9DOT=(U/(EPSI*V))*(C8-C9)-((KF*A*RHO)/EPSI)*(C9-CSTAR9)*1.E-3
         C9=INTGRL(ICC9,C9DOT)
         Q9DOT=(CFLO/(RHO*V))*(Q10-Q9)+KF*A*(C9-CSTAR9)*1.E-3
         Q9=INTGRL(ICQ9,Q9DOT)
         CSTAR9=Q9/(QMAX*B-Q9*B)
***                                                                       ***
```

```
***    DIFFERENTIAL EQUATIONS FOR TENTH ELEMENT                              ***
***                                                                          ***
       C10DOT=(U/(EPSI*V))*(C9-C10)-((KF*A*RHO)/EPSI)*(C10-CSTA10)*1.E-3
       C10=INTGRL(ICC10,C10DOT)
       Q10DOT=(CFLO/(RHO*V))*(Q0-Q10)+KF*A*(C10-CSTA10)*1.E-3
       Q10=INTGRL(ICQ10,Q10DOT)
       CSTA10=Q10/(QMAX*B-Q10*B)
***                                                                          ***
       QT=(Q1+Q2+Q3+Q4+Q5+Q6+Q7+Q8+Q9+Q10)/10.
***                                                                          ***
***    VARIABLES THAT ARE TO BE PLOTTED BY THE CALCOMP PLOTTER               ***
***    ARE READ INTO ARRAYS IN THE FOLLOWING SEGMENT                         ***
***                                                                          ***
       NOSORT
       IF(KEEP.EQ.0) GO TO 100
       IF (TIME.LT.CLOCK) GO TO 100
       C1P(KEY)=(C1*1.E6/20.0)+2.0
       C2P(KEY)=(C2*1.E6/20.0)+2.0
       C3P(KEY)=(C3*1.E6/20.0)+2.0
       C4P(KEY)=(C4*1.E6/20.0)+2.0
       C5P(KEY)=(C5*1.E6/20.0)+2.0
       C6P(KEY)=(C6*1.E6/20.0)+2.0
       C7P(KEY)=(C7*1.E6/20.0)+2.0
       C8P(KEY)=(C8*1.E6/20.0)+2.0
       C9P(KEY)=(C9*1.E6/20.0)+2.0
       C10P(KEY)=(C10*1.E6/20.0)+2.0
       Q1P(KEY)=(Q1*1.E3/0.2)+2.0
       Q2P(KEY)=(Q2*1.E3/0.2)+2.0
       Q3P(KEY)=(Q3*1.E3/0.2)+2.0
       Q4P(KEY)=(Q4*1.E3/0.2)+2.0
       Q5P(KEY)=(Q5*1.E3/0.2)+2.0
       Q6P(KEY)=(Q6*1.E3/0.2)+2.0
       Q7P(KEY)=(Q7*1.E3/0.2)+2.0
       Q8P(KEY)=(Q8*1.E3/0.2)+2.0
       Q9P(KEY)=(Q9*1.E3/0.2)+2.0
       Q10P(KEY)=(Q10*1.E3/0.2)+2.0
       QTP(KEY)=(QT*1.E3/0.2)+2.0
       T(KEY)=(OUTIME(KEY)/5.0)+1.0
       KEY=KEY+1
       CLOCK=OUTIME(KEY)
100    CONTINUE
       SORT
***                                                                          ***
***    EXECUTION CONTROL STATEMENTS                                          ***
***                                                                          ***
METHOD SIMP

TIMER DELT=0.00025,FINTIM=40.,PRDEL=0.5,OUTDEL=0.5
***                                                                          ***
***    OUTPUT CONTROL STATEMENTS                                             ***
***                                                                          ***
       PREPAR C1,C2,C3,C4,C5,C6,C7,C8,C9,C10
       PRINT C1,C6,Q1,Q6,C2,C7,Q2,Q7,C3,C8,Q3,Q8,C4,C9,Q4,Q9,...
       C5,C10,Q5,Q10,C0,Q0
       PRTPLT C1(0.0,1.E-4),C2(0.0,1.E-4),C3(0.0,1.E-4),C4(0.0,1.E-4)
       PRTPLT C5(0.0,1.E-4),C6(0.0,1.E-4),C7(0.0,1.E-4),C8(0.0,1.E-4)
       PRTPLT C9(0.0,1.E-4),C10(0.0,1.E-4)
       PRTPLT Q1(0.0,1.E-3),Q2(0.0,1.E-3),Q3(0.0,1.E-3),Q4(0.0,1.E-3)
       PRTPLT Q5(0.0,1.E-3),Q6(0.0,1.E-3),Q7(0.0,1.E-3),Q8(0.0,1.E-3)
       PRTPLT Q9(0.0,1.E-3),Q10(0.0,1.E-3)
       PRTPLT QT(0.0,1.E-3)
LABEL SIMULATION OF COUNTERCURRENT ADSORPTION COLUMN OPERATION
TITLE SIMULATION OF COUNTERCURRENT ADSORPTION COLUMN OPERATION
***                                                                          ***
```

```
        CALL PLOT1(1.0,7.0,+3)
        CALL PLOT1(9.0,7.0,+2)
        CALL PLOT1(9.0,2.0,+2)
        CALL SYMBOL(4.25,1.25,0.20,'TIME, HOURS',0.0,+11)
        CALL SYMBOL(0.5,2.,0.2,'SOLID-PHASE CONCENTRATION, MM/G',90.,+31)
        RETURN
        END
    ENDJOB

TERMINAL
***                                                                          ***
***     ACTUAL PLOTTING BY THE CALCOMP PLOTTER IS ACCOMPLISHED BY            ***
***     THE FOLLOWING STATEMENTS                                             ***
***                                                                          ***
        CALL GRAPH1(XOR,YOR)
        CALL LINE(T,C1P,81,1,113,5)
        CALL LINE(T,C2P,81,1,114,5)
        CALL LINE(T,C3P,81,1,115,5)
        CALL LINE(T,C4P,81,1,116,5)
        CALL LINE(T,C5P,81,1,117,5)
        CALL LINE(T,C6P,81,1,118,5)
        CALL LINE(T,C7P,81,1,119,5)
        CALL LINE(T,C8P,81,1,120,5)
        CALL LINE(T,C9P,81,1,121,5)
        CALL LINE(T,C10P,81,1,112,5)
        CALL PLOT1(15.,0.,-3)
***                                                                          ***
        CALL GRAPH2(XOR,YOR)
        CALL LINE(T,Q1P,81,1,113,5)
        CALL LINE(T,Q2P,81,1,114,5)
        CALL LINE(T,Q3P,81,1,115,5)
        CALL LINE(T,Q4P,81,1,116,5)
        CALL LINE(T,Q5P,81,1,117,5)
        CALL LINE(T,Q6P,81,1,118,5)
        CALL LINE(T,Q7P,81,1,119,5)
        CALL LINE(T,Q8P,81,1,120,5)
        CALL LINE(T,Q9P,81,1,121,5)
        CALL LINE(T,Q10P,81,1,112,5)
        CALL PLOT1(15.,0.,-3)
***                                                                          ***
        CALL GRAPH2(XOR,YOR)
        CALL LINE(T,QTP,81,1,99,5)
        CALL PLOT1(15.,0.,-3)
***                                                                          ***
    END
    STOP
CCC                                                                          CCC
CCC     SUBROUTINES WHICH GENERATE THE AXES AND LEGENDS FOR EACH             CCC
CCC     OF THE PLOTS ARE DETAILED AS FOLLOWS                                 CCC
CCC                                                                          CCC
CCC     SUBROUTINE FOR CALCOMP OUTPUT OF SOLUTION-PHASE CONCENTRATION        CCC
CCC     VS. TIME FOLLOWS                                                     CCC
CCC                                                                          CCC
        SUBROUTINE GRAPH1(XOR,YOR)
        CALL PLOT1(XOR,YOR,-3)
        CALL AXIS(1.0,2.0,' ',-1,8.0,0.0,0.0,5.0,10.0)
        CALL AXIS(1.0,2.0,' ',+1,5.0,90.0,0.0,20.0,10.0)
        CALL PLOT1(1.0,7.0,+3)
        CALL PLOT1(9.0,7.0,+2)
        CALL PLOT1(9.0,2.0,+2)
        CALL SYMBOL(4.25,1.25,0.20,'TIME, HOURS',0.0,+11)
        CALL SYMBOL(0.5,2.,0.20,'EFFLUENT CONCENTRATION, E-6 M/L',90.,31)
        RETURN
        END
CCC                                                                          CCC
CCC     SUBROUTINE FOR CALCOMP OUTPUT OF SOLID-PHASE CONCENTRATION           CCC
CCC     VS. TIME FOLLOWS                                                     CCC
CCC                                                                          CCC
        SUBROUTINE GRAPH2(XOR,YOR)
        CALL PLOT1(XOR,YOR,-3)
        CALL AXIS(1.0,2.0,' ',-1,8.0,0.0,0.0,5.0,10.0)
        CALL AXIS(1.0,2.0,' ',+1,5.0,90.0,0.0,0.2,10.0)
```

**APPENDIX D**

Program Listing for Case No. 4—
Disolute System

```
*      THOMAS M KEINATH *** AEEP WORKSHOP (BAHAMAS)                          *
*      MODELING OF ADSORPTION COLUMNS, DISTRIBUTED PARAMETER APPROACH        *
*      CASE NO. 4 --- DI-SOLUTE SYSTEM                                       *
*      PACKED BED COLUMN WHICH CONTAINS 250 GRAMS OF CARBON                  *
*      SOLUTE A IS DNOSBP (CONCENTRATION=100 MICROMOLES/LITER)               *
*      SOLUTE B IS PNP (CONCENTRATION=100 MICROMOLES/LITER)                  *
*      SOLUTION VOLUMETRIC FLOW RATE=60 LITERS PER HOUR                      *
*      ASSUMPTION --- MASS TRANSFER IS RATE LIMITING                         *
*      BED HAS BEEN DIVIDED INTO NINE ELEMENTS                               *
*                                                                            *
INITIAL
***                                                                        ***
***    PRELIMINARY DIMENSIONING FOR CALCOMP PLOTTER OUTPUT FOLLOWS         ***
***                                                                        ***
/      DIMENSION CA1P(200),CA2P(200),CA3P(200),CA4P(200),CA5P(200),
/      1CA6P(200),CA7P(200),CA8P(200),CA9P(200),CB1P(200),CB2P(200),
/      2CB3P(200),CB4P(200),CB5P(200),CB6P(200),CB7P(200),CB8P(200),
/      3CB9P(200),QA1P(200),QA2P(200),QA3P(200),QA4P(200),QA5P(200),
/      4QA6P(200),QA7P(200),QA8P(200),QA9P(200),QB1P(200),QB2P(200),
/      5QB3P(200),QB4P(200),QB5P(200),QB6P(200),QB7P(200),QB8P(200),
/      6QB9P(200),T(200)
/      DIMENSION BUFR(800)
/      CALL PLOTS (BUFR,800)
       STORAGE OUTIME(200)
       FIXED KEY
       KEY=1
        TABLE OUTIME (1-81)=0.0,0.5,1.0,1.5,2.0,2.5,3.0,3.5,4.0,4.5,...
       5.0,5.5,6.0,6.5,7.0,7.5,8.0,8.5,9.0,9.5,10.0,10.5,11.0,11.5,...
       12.0,12.5,13.0,13.5,14.0,14.5,15.0,15.5,16.0,16.5,17.0,17.5,...
       18.0,18.5,19.0,19.5,20.0,20.5,21.0,21.5,22.0,22.5,23.0,23.5,...
       24.0,24.5,25.0,25.5,26.0,26.5,27.0,27.5,28.0,28.5,29.0,29.5,...
       30.0,30.5,31.0,31.5,32.0,32.5,33.0,33.5,34.0,34.5,35.0,35.5,...
       36.0,36.5,37.0,37.5,38.0,38.5,39.0,39.5,40.0
       CLOCK=OUTIME(KEY)
***                                                                        ***
***    INITIAL CONDITIONS FOR SOLUTION PHASE CONCENTRATION                 ***
***                                                                        ***
       INCON ICCA1=0.
       INCON ICCA2=0.
       INCON ICCA3=0.
       INCON ICCA4=0.
       INCON ICCA5=0.
       INCON ICCA6=0.
       INCON ICCA7=0.
       INCON ICCA8=0.
       INCON ICCA9=0.
***                                                                        ***
       INCON ICCB1=0.
       INCON ICCB2=0.
       INCON ICCB3=0.
       INCON ICCB4=0.
       INCON ICCB5=0.
       INCON ICCB6=0.
       INCON ICCB7=0.
       INCON ICCB8=0.
       INCON ICCB9=0.
***                                                                        ***
```

```
***    INITIAL CONDITIONS FOR SOLID PHASE CONCENTRATION                      ***
***                                                                          ***
       INCON ICQA1=0.
       INCON ICQA2=0.
       INCON ICQA3=0.
       INCON ICQA4=0.
       INCON ICQA5=0.
       INCON ICQA6=0.
       INCON ICQA7=0.
       INCON ICQA8=0.
       INCON ICQA9=0.
***                                                                          ***
       INCON ICQB1=0.
       INCCN ICQB2=0.
       INCON ICQB3=0.
       INCON ICQB4=0.
       INCON ICQB5=0.
       INCON ICQB6=0.
       INCON ICQB7=0.
       INCON ICQB8=0.
       INCON ICQB9=0.
***                                                                          ***
***    PARAMETER INPUTS REQUIRED ARE DEFINED AS FOLLOWS                       ***
***                                                                          ***
***    CARBON --- MASS OF CARBON IN BED (GRAMS)                               ***
***    A -------- EXTERNAL TRANSFER AREA FOR CARBON (SQ CM/GRAM)              ***
***    DP ------- DIAMETER OF CARBON PARTICLES (CM)                           ***
***    RHO ------ PACKED BED DENSITY (GRAMS/LITER)                            ***
***    EPSI ----- PACKED BED POROSITY (DIMENSIONLESS)                         ***
***    QAMAX ---- LANGMUIR ULTIMATE UPTAKE CAPACITY FOR A (MOLES/GRAM)***
***    QBMAX ---- LANGMUIR ULTIMATE UPTAKE CAPACITY FOR B (MOLES/GRAM)***
***    BA ------- LANGMUIR ENERGY TERM FOR A (LITERS/MOLE)                    ***
***    BB ------- LANGMUIR ENERGY TERM FOR B (LITERS/MOLE)
***    CAO ------ SOLUTION-PHASE CONCENTRATION OF A (MOLES/LITER)             ***
***    CBO ------ SOLUTION-PHASE CONCENTRATION OF B (MOLES/LITER)             ***
***    DLA ------ DIFFUSIVITY OF SOLUTE A (SQ CM/SEC)                         ***
***    DLB ------ DIFFUSIVITY OF SOLUTE B (SQ CM/SEC)                         ***
***    AREA ----- CROSS-SECTIONAL AREA OF COLUMN (SQ CM)                      ***
***    U -------- SOLUTION VOLUMETRIC FLOW RATE (LITERS/HOUR)                 ***
***    GAMMA ---- KINEMATIC VISCOSITY OF WATER (SQ CM/SEC)                    ***
***    XOR ------ CALCOMP PLOTTING ORIGIN FOR X-DIRECTION                     ***
***    YOR ------ CALCOMP PLOTTING ORIGIN FOR Y-DIRECTION                     ***
***                                                                          ***
***    PARAMETER INPUTS FOLLOW                                                ***
***                                                                          ***
       PARAM CARBON=250.
       PARAM A=150.
       PARAM DP=0.03
       PARAM RHO=381.
       PARAM EPSI=0.45
       PARAM QAMAX=8.469E-4
       PARAM QBMAX=9.1E-4
       PARAM BA=3.023E5
       PARAM BB=0.27E4
       PARAM CAO=1.E-4
       PARAM CBO=1.E-4
       PARAM DLA=6.21E-6
       PARAM DLB=1.1696E-5
       PARAM AREA=20.25
       PARAM U=60.
       PARAM GAMMA=8.64E-3
       PARAM XOR=0.0
       PARAM YOR=0.0
***                                                                          ***
***    CALCULATED INPUT PARAMETERS ARE DEFINED AS FOLLOWS                     ***
***                                                                          ***
***    V -------- TOTAL VOLUME OF PACKED BED (LITERS)                         ***
***    UBAR ----- SUPERFICIAL VELOCITY OF FLOW THROUGH BED (CM/SEC)           ***
***    NREMOD --- MODIFIED REYNOLDS NUMBER (DIMENSIONLESS)                    ***
***    NSCA ----- SCHMIDT NUMBER FOR A (DIMENSIONLESS)                        ***
```

```
***    NSCB ----- SCHMIDT NUMBER FOR B (DIMENSIONLESS)                    ***
***    JD ------- MASS TRANSFER FACTOR (DIMENSIONLESS)                    ***
***    KFA ------ MASS TRANSFER COEFFICIENT FOR A (CM/HOUR)               ***
***    KFB ------ MASS TRANSFER COEFFICIENT FOR B (CM/HOUR)               ***
***                                                                       ***
***    INPUT PARAMETER CALCULATIONS FOLLOW                                ***
***                                                                       ***
       V=CARBON/(RHO*9.)
       UBAR=(U*0.277777)/AREA
       NREMOD=(UBAR*DP)/(GAMMA*(1.-EPSI))
       NSCA=GAMMA/DLA
       NSCB=GAMMA/DLB
       JD=5.7*NREMOD**(-0.78)
       KFA=(JD*UBAR/NSCA**(2./3.))*3600.
       KFB=(JD*UBAR/NSCB**(2./3.))*3600.
***                                                                       ***
DYNAMIC
***                                                                       ***
***    DIFFERENTIAL EQUATIONS FOR FIRST ELEMENT                           ***
***                                                                       ***
       CA1DOT=(U/(EPSI*V))*(CA0-CA1)-((KFA*A*RHO)/EPSI)*(CA1-CSTA1)*1.E-3
       CA1=INTGRL(ICCA1,CA1DOT)
       CB1DOT=(U/(EPSI*V))*(CB0-CB1)-((KFB*A*RHO)/EPSI)*(CB1-CSTB1)*1.E-3
       CB1=INTGRL(ICCB1,CB1DOT)
       QA1DOT=KFA*A*(CA1-CSTA1)*1.E-3
       QA1=INTGRL(ICQA1,QA1DOT)
       QB1DOT=KFB*A*(CB1-CSTB1)*1.E-3
       QB1=INTGRL(ICQB1,QB1DOT)
       CSTA1=QBMAX*BB*QA1/(QAMAX*QBMAX*BA*BB-QAMAX*QB1*BA*BB-QBMAX*QA1...
       *BA*BB)
       CSTB1=QAMAX*BA*QB1/(QAMAX*QBMAX*BA*BB-QAMAX*QB1*BA*BB-QBMAX*QA1...
       *BA*BB)
***                                                                       ***
***    DIFFERENTIAL EQUATIONS FOR SECOND ELEMENT                          ***
***                                                                       ***
       CA2DOT=(U/(EPSI*V))*(CA1-CA2)-((KFA*A*RHO)/EPSI)*(CA2-CSTA2)*1.E-3
       CA2=INTGRL(ICCA2,CA2DOT)
       CB2DOT=(U/(EPSI*V))*(CB1-CB2)-((KFB*A*RHO)/EPSI)*(CB2-CSTB2)*1.E-3
       CB2=INTGRL(ICCB2,CB2DOT)
       QA2DOT=KFA*A*(CA2-CSTA2)*1.E-3
       QA2=INTGRL(ICQA2,QA2DOT)
       QB2DOT=KFB*A*(CB2-CSTB2)*1.E-3
       QB2=INTGRL(ICQB2,QB2DOT)
       CSTA2=QBMAX*BB*QA2/(QAMAX*QBMAX*BA*BB-QAMAX*QB2*BA*BB-QBMAX*QA2...
       *BA*BB)
       CSTB2=QAMAX*BA*QB2/(QAMAX*QBMAX*BA*BB-QAMAX*QB2*BA*BB-QBMAX*QA2...
       *BA*BB)
***                                                                       ***
***    DIFFERENTIAL EQUATIONS FOR THIRD ELEMENT                           ***
***                                                                       ***
       CA3DOT=(U/(EPSI*V))*(CA2-CA3)-((KFA*A*RHO)/EPSI)*(CA3-CSTA3)*1.E-3
       CA3=INTGRL(ICCA3,CA3DOT)
       CB3DOT=(U/(EPSI*V))*(CB2-CB3)-((KFB*A*RHO)/EPSI)*(CB3-CSTB3)*1.E-3
       CB3=INTGRL(ICCB3,CB3DOT)
       QA3DOT=KFA*A*(CA3-CSTA3)*1.E-3
       QA3=INTGRL(ICQA3,QA3DOT)
       QB3DOT=KFB*A*(CB3-CSTB3)*1.E-3
       QB3=INTGRL(ICQB3,QB3DOT)
       CSTA3=QBMAX*BB*QA3/(QAMAX*QBMAX*BA*BB-QAMAX*QB3*BA*BB-QBMAX*QA3...
       *BA*BB)
       CSTB3=QAMAX*BA*QB3/(QAMAX*QBMAX*BA*BB-QAMAX*QB3*BA*BB-QBMAX*QA3...
       *BA*BB)
***                                                                       ***
***    DIFFERENTIAL EQUATIONS FOR FOURTH ELEMENT                          ***
***                                                                       ***
       CA4DOT=(U/(EPSI*V))*(CA3-CA4)-((KFA*A*RHO)/EPSI)*(CA4-CSTA4)*1.E-3
       CA4=INTGRL(ICCA4,CA4DOT)
       CB4DOT=(U/(EPSI*V))*(CB3-CB4)-((KFB*A*RHO)/EPSI)*(CB4-CSTB4)*1.E-3
       CB4=INTGRL(ICCB4,CB4DOT)
       QA4DOT=KFA*A*(CA4-CSTA4)*1.E-3
```

```
QA4=INTGRL(ICQA4,QA4DOT)
QB4DOT=KFB*A*(CB4-CSTB4)*1.E-3
QB4=INTGRL(ICQB4,QB4DOT)
CSTA4=QBMAX*BB*QA4/(QAMAX*QBMAX*BA*BB-QAMAX*QB4*BA*BB-QBMAX*QA4...
*BA*BB)
CSTB4=QAMAX*BA*QB4/(QAMAX*QBMAX*BA*BB-QAMAX*QB4*BA*BB-QBMAX*QA4...
*BA*BB)
```
***        ***
***    DIFFERENTIAL EQUATIONS FOR FIFTH ELEMENT        ***
***        ***
```
CA5DOT=(U/(EPSI*V))*(CA4-CA5)-((KFA*A*RHO)/EPSI)*(CA5-CSTA5)*1.E-3
CA5=INTGRL(ICCA5,CA5DOT)
CB5DOT=(U/(EPSI*V))*(CB4-CB5)-((KFB*A*RHO)/EPSI)*(CB5-CSTB5)*1.E-3
CB5=INTGRL(ICCB5,CB5DOT)
QA5DOT=KFA*A*(CA5-CSTA5)*1.E-3
QA5=INTGRL(ICQA5,QA5DCT)
QB5DOT=KFB*A*(CB5-CSTB5)*1.E-3
QB5=INTGRL(ICQB5,QB5DOT)
CSTA5=QBMAX*BB*QA5/(QAMAX*QBMAX*BA*BB-QAMAX*QB5*BA*BB-QBMAX*QA5...
*BA*BB)
CSTB5=QAMAX*BA*QB5/(QAMAX*QBMAX*BA*BB-QAMAX*QB5*BA*BB-QBMAX*QA5...
*BA*BB)
```
***        ***
***    DIFFERENTIAL EQUATIONS FOR SIXTH ELEMENT        ***
***        ***
```
CA6DOT=(U/(EPSI*V))*(CA5-CA6)-((KFA*A*RHO)/EPSI)*(CA6-CSTA6)*1.E-3
CA6=INTGRL(ICCA6,CA6DOT)
CB6DOT=(U/(EPSI*V))*(CB5-CB6)-((KFB*A*RHO)/EPSI)*(CB6-CSTB6)*1.E-3
CB6=INTGRL(ICCB6,CB6DOT)
QA6DOT=KFA*A*(CA6-CSTA6)*1.E-3
QA6=INTGRL(ICQA6,QA6DCT)
QB6DOT=KFB*A*(CB6-CSTB6)*1.E-3
QB6=INTGRL(ICQB6,QB6DOT)
CSTA6=QBMAX*BB*QA6/(QAMAX*QBMAX*BA*BB-QAMAX*QB6*BA*BB-QBMAX*QA6...
*BA*BB)
CSTB6=QAMAX*BA*QB6/(QAMAX*QBMAX*BA*BB-QAMAX*QB6*BA*BB-QBMAX*QA6...

*BA*BB)
```
***        ***
***    DIFFERENTIAL EQUATIONS FOR SEVENTH ELEMENT        ***
***        ***
```
CA7DOT=(U/(EPSI*V))*(CA6-CA7)-((KFA*A*RHO)/EPSI)*(CA7-CSTA7)*1.E-3
CA7=INTGRL(ICCA7,CA7DOT)
CB7DOT=(U/(EPSI*V))*(CB6-CB7)-((KFB*A*RHO)/EPSI)*(CB7-CSTB7)*1.E-3
CB7=INTGRL(ICCB7,CB7DCT)
QA7DOT=KFA*A*(CA7-CSTA7)*1.E-3
QA7=INTGRL(ICQA7,QA7DOT)
QB7DOT=KFB*A*(CB7-CSTB7)*1.E-3
QB7=INTGRL(ICQB7,QB7DOT)
CSTA7=QBMAX*BB*QA7/(QAMAX*QBMAX*BA*BB-QAMAX*QB7*BA*BB-QBMAX*QA7...
*BA*BB)
CSTB7=QAMAX*BA*QB7/(QAMAX*QBMAX*BA*BB-QAMAX*QB7*BA*BB-QBMAX*QA7...
*BA*BB)
```
***        ***
***    DIFFERENTIAL EQUATIONS FOR EIGHTH ELEMENT        ***
***        ***
```
CA8DOT=(U/(EPSI*V))*(CA7-CA8)-((KFA*A*RHO)/EPSI)*(CA8-CSTA8)*1.E-3
CA8=INTGRL(ICCA8,CA8DOT)
CB8DOT=(U/(EPSI*V))*(CB7-CB8)-((KFB*A*RHO)/EPSI)*(CB8-CSTB8)*1.E-3
CB8=INTGRL(ICCB8,CB8DOT)
QA8DOT=KFA*A*(CA8-CSTA8)*1.E-3
QA8=INTGRL(ICQA8,QA8DOT)
QB8DOT=KFB*A*(CB8-CSTB8)*1.E-3
QB8=INTGRL(ICQB8,QB8DOT)
CSTA8=QBMAX*BB*QA8/(QAMAX*QBMAX*BA*BB-QAMAX*QB8*BA*BB-QBMAX*QA8...
*BA*BB)
CSTB8=QAMAX*BA*QB8/(QAMAX*QBMAX*BA*BB-QAMAX*QB8*BA*BB-QBMAX*QA8...
*BA*BB)
```
***        ***

```
***     DIFFERENTIAL EQUATIONS FOR NINTH ELEMENT                    ***
***                                                                 ***
        CA9DOT=(U/(EPSI*V))*(CA8-CA9)-((KFA*A*RHO)/EPSI)*(CA9-CSTA9)*1.E-3
        CA9=INTGRL(ICCA9,CA9DOT)
        CB9DOT=(U/(EPSI*V))*(CB8-CB9)-((KFB*A*RHO)/EPSI)*(CB9-CSTB9)*1.E-3
        CB9=INTGRL(ICCB9,CB9DOT)
        QA9DOT=KFA*A*(CA9-CSTA9)*1.E-3
        QA9=INTGRL(ICQA9,QA9DOT)
        QB9DOT=KFB*A*(CB9-CSTB9)*1.E-3
        QB9=INTGRL(ICQB9,QB9DOT)
        CSTA9=QBMAX*BB*QA9/(QAMAX*QBMAX*BA*BB-QAMAX*QB9*BA*BB-QBMAX*QA9...
        *BA*BB)
        CSTB9=QAMAX*BA*QB9/(QAMAX*QBMAX*BA*BB-QAMAX*QB9*BA*BB-QBMAX*QA9...
        *BA*BB)
***                                                                 ***
        QAT=(QA1+QA2+QA3+QA4+QA5+QA6+QA7+QA8+QA9)/9.
        QBT=(QB1+QB2+QB3+QB4+QB5+QB6+QB7+QB8+QB9)/9.
***                                                                 ***
***     VARIABLES THAT ARE TO BE PLOTTED BY THE CALCOMP PLOTTER     ***
***     ARE READ INTO ARRAYS IN THE FOLLOWING SEGMENT              ***
***                                                                 ***
        NOSORT
        IF(KEEP.EQ.0) GO TO 100
        IF (TIME.LT.CLOCK) GO TO 100
        CA1P(KEY)=(CA1*1.E6/20.0)+2.0
        CA2P(KEY)=(CA2*1.E6/20.0)+2.0
        CA3P(KEY)=(CA3*1.E6/20.0)+2.0
        CA4P(KEY)=(CA4*1.E6/20.0)+2.0
        CA5P(KEY)=(CA5*1.E6/20.0)+2.0
        CA6P(KEY)=(CA6*1.E6/20.0)+2.0
        CA7P(KEY)=(CA7*1.E6/20.0)+2.0
        CA8P(KEY)=(CA8*1.E6/20.0)+2.0
        CA9P(KEY)=(CA9*1.E6/20.0)+2.0
        QA1P(KEY)=(QA1*1.E3/0.2)+2.0
        QA2P(KEY)=(QA2*1.E3/0.2)+2.0
        QA3P(KEY)=(QA3*1.E3/0.2)+2.0
        QA4P(KEY)=(QA4*1.E3/0.2)+2.0
        QA5P(KEY)=(QA5*1.E3/0.2)+2.0
        QA6P(KEY)=(QA6*1.E3/0.2)+2.0
        QA7P(KEY)=(QA7*1.E3/0.2)+2.0
        QA8P(KEY)=(QA8*1.E3/0.2)+2.0
        QA9P(KEY)=(QA9*1.E3/0.2)+2.0
        CB1P(KEY)=(CB1*1.E6/30.0)+2.0
        CB2P(KEY)=(CB2*1.E6/30.0)+2.0
        CB3P(KEY)=(CB3*1.E6/30.0)+2.0
        CB4P(KEY)=(CB4*1.E6/30.0)+2.0
        CB5P(KEY)=(CB5*1.E6/30.0)+2.0
        CB6P(KEY)=(CB6*1.E6/30.0)+2.0
        CB7P(KEY)=(CB7*1.E6/30.0)+2.0
        CB8P(KEY)=(CB8*1.E6/30.0)+2.0
        CB9P(KEY)=(CB9*1.E6/30.0)+2.0
        QB1P(KEY)=(QB1*1.E3/0.05)+2.0
        QB2P(KEY)=(QB2*1.E3/0.05)+2.0
        QB3P(KEY)=(QB3*1.E3/0.05)+2.0
        QB4P(KEY)=(QB4*1.E3/0.05)+2.0
        QB5P(KEY)=(QB5*1.E3/0.05)+2.0
        QB6P(KEY)=(QB6*1.E3/0.05)+2.0
        QB7P(KEY)=(QB7*1.E3/0.05)+2.0
        QB8P(KEY)=(QB8*1.E3/0.05)+2.0
        QB9P(KEY)=(QB9*1.E3/0.05)+2.0
        T(KEY)=(OUTIME(KEY)/5.0)+1.0
        KEY=KEY+1
        CLOCK=OUTIME(KEY)
100     CONTINUE
        SORT
***                                                                 ***
***     EXECUTION CONTROL STATEMENTS                                ***
***                                                                 ***
METHOD SIMP
TIMER DELT=0.00025,FINTIM=40.,PRDEL=0.5,OUTDEL=0.5
***                                                                 ***
```

```
***    OUTPUT CONTROL STATEMENTS                                         ***
***                                                                      ***
       PREPAR CB1,CB2,CB3,CB4,CB5,CB6,CB7,CB8,CB9
       PRINT CA1,CB1,QA1,QB1,CA2,CB2,QA2,QB2,CA3,CB3,QA3,QB3,...
       CA4,CB4,QA4,QB4,CA5,CB5,QA5,QB5,CA6,CB6,QA6,QB6,CA7,CB7,QA7,...
       QB7,CA8,CB8,QA8,QB8,CA9,CB9,QA9,QB9,CA0,CB0
       PRTPLT CA1(0.0,1.E-4),CA2(0.0,1.E-4),CA3(0.0,1.E-4),CA4(0.0,1.E-4)
       PRTPLT CA5(0.0,1.E-4),CA6(0.0,1.E-4),CA7(0.0,1.E-4),CA8(0.0,1.E-4)
       PRTPLT CA9(0.0,1.E-4),CB1(0.0,2.E-4),CB2(0.0,2.E-4),CB3(0.0,2.E-4)
       PRTPLT CB4(0.0,2.E-4),CB5(0.0,2.E-4),CB6(0.0,2.E-4),CB7(0.0,2.E-4)
       PRTPLT CB8(0.0,2.E-4),CB9(0.0,2.E-4),QA1(0.0,1.E-3),QA2(0.0,1.E-3)
       PRTPLT QA3(0.0,1.E-3),QA4(0.0,1.E-3),QA5(0.0,1.E-3),QA6(0.0,1.E-3)
       PRTPLT QA7(0.0,1.E-3),QA8(0.0,1.E-3),QA9(0.0,1.E-3),QB1(0.0,3.E-4)
       PRTPLT QB2(0.0,3.E-4),QB3(0.0,3.E-4),QB4(0.0,3.E-4),QB5(0.0,3.E-4)
       PRTPLT QB6(0.0,3.E-4),QB7(0.0,3.E-4),QB8(0.0,3.E-4),QB9(0.0,3.E-4)
LABEL SIMULATION OF ADSORPTION BREAKTHROUGH PROFILE
TITLE SIMULATION OF ADSORPTION BREAKTHROUGH PROFILE
***                                                                      ***
TERMINAL
***                                                                      ***
***    ACTUAL PLOTTING BY THE CALCOMP PLOTTER IS ACCOMPLISHED BY         ***
***    THE FOLLOWING STATEMENTS                                          ***
***                                                                      ***
       CALL GRAPH1(XOR,YOR)
       CALL LINE(T,CA1P,81,1,113,5)
       CALL LINE(T,CA2P,81,1,114,5)
       CALL LINE(T,CA3P,81,1,115,5)
       CALL LINE(T,CA4P,81,1,116,5)
       CALL LINE(T,CA5P,81,1,117,5)
       CALL LINE(T,CA6P,81,1,118,5)
       CALL LINE(T,CA7P,81,1,119,5)
       CALL LINE(T,CA8P,81,1,120,5)
       CALL LINE(T,CA9P,81,1,121,5)
       CALL PLOT1(15.,0.,-3)
***                                                                      ***
       CALL GRAPH2(XOR,YOR)
       CALL LINE(T,QA1P,81,1,113,5)
       CALL LINE(T,QA2P,81,1,114,5)
       CALL LINE(T,QA3P,81,1,115,5)
       CALL LINE(T,QA4P,81,1,116,5)
       CALL LINE(T,QA5P,81,1,117,5)
       CALL LINE(T,QA6P,81,1,118,5)
       CALL LINE(T,QA7P,81,1,119,5)
       CALL LINE(T,QA8P,81,1,120,5)
       CALL LINE(T,QA9P,81,1,121,5)
       CALL PLOT1(15.,0.,-3)
***                                                                      ***
       CALL GRAPH3(XOR,YOR)
       CALL LINE(T,CB1P,81,1,113,5)
       CALL LINE(T,CB2P,81,1,114,5)
       CALL LINE(T,CB3P,81,1,115,5)
       CALL LINE(T,CB4P,81,1,116,5)
       CALL LINE(T,CB5P,81,1,117,5)
       CALL LINE(T,CB6P,81,1,118,5)
       CALL LINE(T,CB7P,81,1,119,5)
       CALL LINE(T,CB8P,81,1,120,5)
       CALL LINE(T,CB9P,81,1,121,5)
       CALL PLOT1(15.,0.,-3)
***                                                                      ***
       CALL GRAPH4(XOR,YOR)
       CALL LINE(T,QB1P,81,1,113,5)
       CALL LINE(T,QB2P,81,1,114,5)
       CALL LINE(T,QB3P,81,1,115,5)
       CALL LINE(T,QB4P,81,1,116,5)
       CALL LINE(T,QB5P,81,1,117,5)
       CALL LINE(T,QB6P,81,1,118,5)
       CALL LINE(T,QB7P,81,1,119,5)
       CALL LINE(T,QB8P,81,1,120,5)
       CALL LINE(T,QB9P,81,1,121,5)
       CALL PLOT1(15.,0.,-3)
```

```
***                                                                          ***
END
STOP
CCC                                                                          CCC
CCC     SUBROUTINES WHICH GENERATE THE AXES AND LEGENDS FOR EACH             CCC
CCC     OF THE PLOTS ARE DETAILED AS FOLLOWS                                 CCC
CCC                                                                          CCC
CCC     SUBROUTINE FOR CALCOMP OUTPUT OF SOLUTION-PHASE CONCENTRATION        CCC
CCC     OF COMPONENT A VS. TIME FOLLOWS                                      CCC
CCC                                                                          CCC
        SUBROUTINE GRAPH1(XCR,YOR)
        CALL PLOT1(XOR,YOR,-3)
        CALL AXIS(1.0,2.0,' ',-1,8.0,0.0,0.0,5.0,10.0)
        CALL AXIS(1.0,2.0,' ',+1,5.0,90.0,0.0,20.0,10.0)
        CALL PLOT1(1.0,7.0,+3)
        CALL PLOT1(9.0,7.0,+2)
        CALL PLOT1(9.0,2.0,+2)
        CALL SYMBOL(4.25,1.25,0.20,'TIME, HOURS',0.0,+11)
        CALL SYMBOL(0.5,2.,0.20,'EFFLUENT CONCENTRATION, E-6 M/L',90.,31)
        RETURN
        END
CCC                                                                          CCC
CCC     SUBROUTINE FOR CALCOMP OUTPUT OF SOLID-PHASE CONCENTRATION           CCC
CCC     OF COMPONENT A VS. TIME FOLLOWS                                      CCC
CCC                                                                          CCC
        SUBROUTINE GRAPH2(XOR,YOR)
        CALL PLOT1(XOR,YOR,-3)
        CALL AXIS(1.0,2.0,' ',-1,8.0,0.0,0.0,5.0,10.0)
        CALL AXIS(1.0,2.0,' ',+1,5.0,90.0,0.0,0.2,10.0)
        CALL PLOT1(1.0,7.0,+3)
        CALL PLOT1(9.0,7.0,+2)
        CALL PLOT1(9.0,2.0,+2)
        CALL SYMBOL(4.25,1.25,0.20,'TIME, HOURS',0.0,+11)
        CALL SYMBOL(0.5,2.,0.2,'SOLID-PHASE CONCENTRATION, MM/G',90.,+31)
        RETURN
        END
CCC                                                                          CCC
CCC     SUBROUTINE FOR CALCOMP OUTPUT OF SOLUTION-PHASE CONCENTRATION        CCC
CCC     OF COMPONENT B VS. TIME FOLLOWS                                      CCC
CCC                                                                          CCC
        SUBROUTINE GRAPH3(XOR,YOR)
        CALL PLOT1(XOR,YOR,-3)
        CALL AXIS(1.0,2.0,' ',-1,8.0,0.0,0.0,5.0,10.0)
        CALL AXIS(1.0,2.0,' ',+1,5.0,90.0,0.0,30.0,10.0)
        CALL PLOT1(1.0,7.0,+3)
        CALL PLOT1(9.0,7.0,+2)
        CALL PLOT1(9.0,2.0,+2)
        CALL SYMBOL(4.25,1.25,0.20,'TIME, HOURS',0.0,+11)
        CALL SYMBOL(0.5,2.,0.20,'EFFLUENT CONCENTRATION, E-6 M/L',90.,31)
        RETURN
        END
CCC                                                                          CCC
CCC     SUBROUTINE FOR CALCOMP OUTPUT OF SOLID-PHASE CONCENTRATION           CCC
CCC     OF COMPONENT B VS. TIME FOLLOWS                                      CCC
CCC                                                                          CCC
        SUBROUTINE GRAPH4(XOR,YOR)
        CALL PLOT1(XOR,YOR,-3)
        CALL AXIS(1.0,2.0,' ',-1,8.0,0.0,0.0,5.0,10.0)
        CALL AXIS(1.0,2.0,' ',+1,5.0,90.0,0.0,0.05,10.0)
        CALL PLOT1(1.0,7.0,+3)
        CALL PLOT1(9.0,7.0,+2)
        CALL PLOT1(9.0,2.0,+2)
        CALL SYMBOL(4.25,1.25,0.20,'TIME, HOURS',0.0,+11)
        CALL SYMBOL(0.5,2.,0.2,'SOLID-PHASE CONCENTRATION, MM/G',90.,+31)
        RETURN
        END
ENDJOB
```

# CHAPTER 2

# MODELING AND SIMULATION OF THE
# AGGREGATION OF SUSPENSIONS

Robert S. Gemmell

Department of Civil Engineering
Northwestern University
Evanston, Illinois

## INTRODUCTION

The aggregation of colloidal or pseudocolloidal particles into larger flocs is commonly undertaken as a unit process in water and wastewater treatment. Its purpose is to facilitate the removal of such particles from the water by sedimentation and/or filtration. The particles must normally be destabilized, at least partially, through the use of chemical coagulants such as alum or polyelectrolytes before successful aggregation can occur. The coagulants serve to reduce or shield the effective charge on the colloidal particles thereby reducing the mutual repulsion of these particles, and provide a binding agent or bridging mechanism between particles. Adsorption or precipitation on the surface of the colloidal particle might be involved. The gelatinous nature of the precipitated coagulant may also aid in the entrapment of particles. While considerable work has been done in studying the destabilization of colloidal particles, the joint modeling of destabilization and aggregation has not progressed appreciably beyond the incorporation of a collision efficiency factor sometimes called a "stickiness" ratio in the kinetic expressions for particle contacting and agglomeration. For this reason, the modeling of particle destabilization is not discussed further except in the context of reported collision efficiency factors.

## Kinetics of Particle Aggregation

The first models of the kinetics of particle aggregation were proposed by Smoluchowski.[1] He identified two particle transport mechanisms that give rise to interparticle collisions and thereby provide aggregation or contact opportunities: Brownian diffusion and fluid motion.

## Perikinetic Flocculation

In a colloidal suspension, the number of particles diffusing radially inward through the surface of a sphere centered on a stationary particle is proportional to the surface area of the sphere, the Brownian diffusion coefficient of the particles, and the particle concentration gradient in the radial direction. As the diffusion is radial, this number must also equal the number of collisions with the central particle. Integration over the proper limits for the particle concentration gradient, combined with recognition that the central particle is also subject to diffusion, gives

$$I_{ij} = 4\pi D_{ij} R_{ij} n_i n_j \qquad (1)$$

where $I_{ij}$ is the number of contacts per unit of time and volume between particles of size i and size j; $D_{ij}$ is the mutual diffusion coefficient of particles i and j (approximately $D_i + D_j$); $R_{ij}$ is the radius of an interaction of the two particles, the distance between centers of two particles forming a lasting contact (commonly taken as the sum of radii of the particles); and $n_i$ and $n_j$ are the respective number concentrations of i and j particles. The rate of change in number concentration of k particles ($k = i + j$) can then be written as:

$$\frac{dn_k}{dt} = \frac{1}{2} \sum_{\substack{i=1 \\ j=k-i}}^{k-1} 4\pi D_{ij} R_{ij} n_i n_j - n_k \sum_{i=1}^{\infty} 4\pi D_{ik} R_{ik} n_i \qquad (2)$$

Here the first summation represents aggregation into size k while the second represents growth out of size k.

The Brownian diffusion coefficient is inversely proportional to the size of the particle, and when all particles are about the same size, $D_{ij} R_{ij}$ may be approximated as $2DR$, where D and R are values characteristic of the common particle size. Use of this approximation permits an analytic solution to the rate equations for the number concentration all particles (irrespective of size) as a function of time:

$$\frac{N_t}{N_0} = \frac{1}{1 + t/T} \qquad (3)$$

where $N_t$ is the total number concentration of particles of all sizes at time t; $N_o$ is the initial number concentration of particles, and

$$T = \frac{1}{4\pi DRN_o}$$

the half-time of flocculation. For water at room temperature,

$$T \cong \frac{2 \times 10^{11}}{N_o}$$

If not all collisions result in lasting combinations, a collision efficiency factor expressing the proportion of collisions that result in aggregation can be inserted in the denominator of the half-time equation.

The number concentration of particles of size k at time t can also be found:

$$\frac{n_k}{N_o} = \frac{(t/T)^{k-1}}{(1 + t/T)^k} \tag{4}$$

Equations 3 and 4 are shown graphically as Figure 2.1.

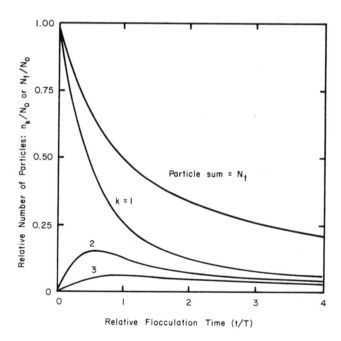

**Figure 2.1.** Change in number of particles of stated size during perikinetic flocculation (after Overbeek).

Substantial heterodispersity of the suspension may make the approximation $D_{ij}R_{ij} = 2DR$ unacceptable. A more exact relationship is

$$D_{ij}R_{ij} = D_1 r_i \left\{ 4 + \left[ \left( \frac{r_i}{r_j} \right)^2 - \frac{r_j}{r_i} \right)^2 \right]^{1/2} \right\}$$ (5)

but the use of this prevents direct analytic solution to the rate equations and requires, instead, simulation or numerical solutions.

**Orthokinetic Flocculation**

When the liquid suspending the particles is in motion, this motion can transport the particles and thereby give rise to aggregation. The model proposed by Smoluchowski for this aggregation presumed that, on the scale of the particles' radius of interaction, the liquid motion could be characterized by a constant local velocity gradient.[1] The flow was basically undirectional, so that the velocity of one particle relative to another could be considered as the product of the local velocity gradient and the distance separating the particles measured normal to the direction of flow (Figure 2.2). For $n_i$-particles per unit volume, the number of contacts of i particles with a central j particle per unit time is

$$J_i = 2n_i \int_{Z=0}^{R_{ij}} Z \frac{du}{dz} \left[ 2\sqrt{R_{ij} - Z^2} \right] dz = \frac{4}{3} n_i \frac{du}{dz} R_{ij}^3$$

and for $n_j$ such central particles per unit volume,

$$J_{ij} = \frac{4}{3} n_i n_j R_{ij}^3 \frac{du}{dz}$$ (6)

The rate of change of the number of k-particles is

$$\frac{dn_k}{dt} = \frac{1}{2} \sum_{\substack{i=1 \\ j=k-i}}^{k-1} \frac{4}{3} n_i n_j (r_i + r_j)^3 \frac{du}{dz} - n_k \sum_{i=1}^{\infty} \frac{4}{3} n_i (r_i + r_k)^3 \frac{du}{dz}$$ (7)

Here the radius of interaction $R_{ij}$ has been replaced by the sum of the particle radii.

If the suspension is monodisperse, $R_{ij}$ equals the particle diameter and $n_i$ and $n_j$ equal n, the number concentration of particles. Since n $d^3$ equals 6 $\phi/\pi$, where $\phi$ is the floc volume fraction, this substitution is occasionally made in Equation 6. Its use in Equation 7 is less than satisfactory. However, without the use of such a substitution, Equation 7 can only be solved numerically.

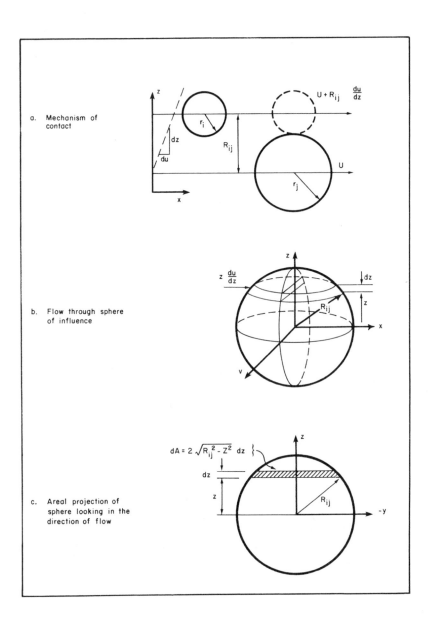

**Figure 2.2.** Definition sketches for orthokinetic flocculation.

The significant role played by particle size in the orthokinetic flocculation equation can be seen from the ratio of Equation 6 to Equation 1. For a velocity gradient of 1 sec$^{-1}$ and water temperature of 20°C,

$$\frac{J_{ij}}{I_{ij}} = 1.25 \times 10^{11} \, (R_{ij})^3 \tag{8}$$

and the orthokinetic and perikinetic flocculation contribute equally when the radius of interaction $R_{ij}$ is about $2\mu$. In heterodisperse systems, this condition is satisfied when the larger particle is of this size. For larger velocity gradients or larger particles, orthokinetic flocculation dominates. Since the local velocity gradient $du/dz$ is not commonly known, an average value G is often calculated from the rate of energy dissipation in the fluid. Camp[2] proposed that

$$G = \left( \frac{P}{\mu \, V} \right)^{\frac{1}{2}} \tag{9}$$

where P is the power dissipated in the liquid as a result of fluid motion or mixing, $V$ is the volume of the liquid in which the power is dissipated, and $\mu$ is the viscosity of the liquid.

The aggregation of particles according to the velocity gradient equation applies wherever the velocity gradients are approximately linear over a distance of a particle diameter. It could apply to cases of turbulent flow near a solid boundary. In turbulent flow, however, a more important mechanism is likely to be turbulent diffusion. Levich[3] has suggested that the number of particle contacts per unit time and volume resulting from turbulent diffusion is:

$$H_{ij} = 12\pi n_i n_j R_{ij}^{\,3} \, \beta\sqrt{\epsilon_0/\nu} \tag{10}$$

where $\epsilon_0$ is the rate of energy dissipation per unit volume of liquid and $\beta$ is a constant. Equation 9 applies to particles larger than about $0.1\mu$ in diameter. It is similar in many respects to the velocity gradient flocculation equation when the local velocity gradient is replaced by its estimated mean value G. Argaman and Kaufmann[4] developed a similar equation. For particles smaller than about $0.1\mu$, the perikinetic flocculation equation applies.

## SIMULATION

To employ the aggregation rate equations in numerical simulations, some operating rules must be established. One involves establishing the relationship between the size (radius or diameter) of a particle and the number of primary or elemental particles it contains. Nearly all simulation models of aggregation kinetics utilize a coalescence scheme in which a spherical particle of volume i combines with a spherical particle of volume j to form a spherical particle of volume i + j = k. Thus the index i, j or k designates the particle volume and the particle radius is proportional to the cube root of the index. This is a convenient but not especially realistic approach to particle agglomeration.

Another rule is occasioned by the need to replace the infinite bounds on summations with finite bounds. This effectively establishes a maximum size for stable flocs, and is in general accord with experience for orthokinetic flocculation. In most simulations of flocculation, this limit has been arbitrarily established, but recent studies by Parker, *et al.*[5] suggest that the maximum stable floc size can be expressed as

$$d_S = CG^{-n} \tag{11}$$

Here, $d_S$ is the maximum stable floc size; G is the mean velocity gradient; C is a floc strength coefficient and n an exponent ranging from 2 to ½ depending on the hydraulic regime and type of floc involved. The floc strength coefficient also varies according to type of floc and the hydraulic conditions imposed.

The establishment of a maximum size for stable flocs has further repercussions on the modeling process, since a methodology must be devised to prevent the permanent formation of oversized flocs. One simple mechanism that has been used is to assert that the collision efficiency factor is zero for all possible aggregations of flocs i + j greater than the limiting size. In Equation 7, then, the upper limit on the second summation would be L-k, where L is the volume index of the maximum stable particle. A further requirement that k ≤ L is imposed on both summations.

A more plausible approach, perhaps, is to view oversized flocs as unstable and subject to immediate breakup from the hydraulic stresses placed on the floc by the fluid motion. If the floc breaks into smaller ones of the size that just combined to form it, the result is equivalent to the zero collision efficiency approach previously described. Alternatively, the oversized floc could break up into any arbitrarily selected number and size of smaller, stable flocs. Little experimental evidence exists to support the selection of a particular breakup mechanism. Bartok and Mason[6] studied the breakup of droplets suspended in a liquid under shear flow and found that two alternative breakup mechanisms resulted depending on the ratio of the viscosities of the droplet liquid and the suspending liquid. When the viscosity of the droplet was smaller than that of the suspending fluid the droplet took on an ellipsoidal configuration and continued to break down at the center until it split into two nearly equal sized droplets (Figure 2.3). When the viscosity of the droplet was greater than that of the suspending

a. Droplet viscosity < suspending liquid viscosity

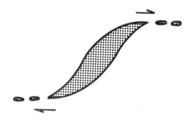

b. Droplet viscosity > suspending liquid viscosity

**Figure 2.3.** Breakup of droplets in shear flow.

fluid, the droplet took on a sigmoidal shape and broke up by the loss of small droplets at each end.

Argaman and Kaufmann suggested that the breakup of chemical flocs in turbulent flows is by the erosion of primary particles from the surface of the floc.[4] The rate of erosion is proportional to the mean velocity gradient and the surface area of the floc. Thomas suggested that floc breakup was principally the result of dynamic pressure differences on opposite sides of the floc, resulting in bulgy deformation and floc splitting.[7] The frequency of breakup was held to be proportional to the size of the floc and the eddy frequency. Parker and his co-workers[5] reviewed both of these approaches and concluded that a surface erosion phenomenon was more plausible. They proposed that the erosion may be expressed as:

$$\frac{dn_1}{dt} = K_\beta X G^m \tag{12}$$

where $K_\beta$ is a floc breakup rate coefficient, X is the suspended solids concentration, and m is a breakup rate exponent where value is 2 or 4 depending on the hydraulic regime of the flocculation.

## SIMULATION RESULTS

In this section, selected results obtained with several different simulation models are presented by way of illustration.

### Perikinetic Flocculation

A model based on Equation 2 utilizing a maximum floc size of 40 units and the zero collision efficiency factor method of controlling floc growth beyond the maximum size was developed by Fair and Gemmell.[8] The value of $D_{ij}R_{ij}$ was computed by Equation 5 and by the approximation 2DR and the resulting changes in the total number of particles during flocculation obtained by the two approaches were compared with each other and with the analytic solution (Equation 3). These results are shown in Figure 2.4.

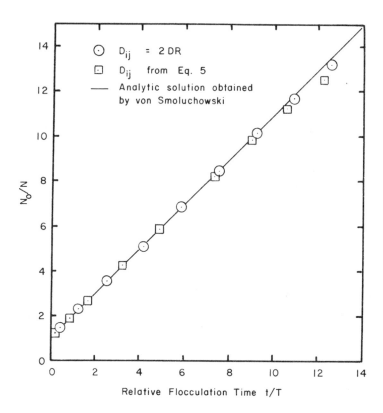

**Figure 2.4.** Perikinetic flocculation: linearizing plot for second-order decrease in total number of particles.

The increase in mean floc size for these simulations is shown in Figure 2.5. The results indicate that the analytic solution is a suitable representation of the perikinetic flocculation process.

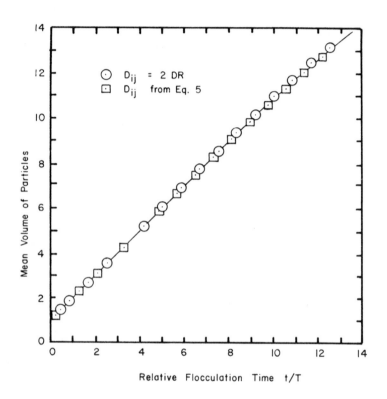

**Figure 2.5.** Perikinetic flocculation: increase in mean size.

**Orthokinetic Flocculation**

The first major numerical simulation of orthokinetic flocculation was performed by Gemmell.[9] The simulation model was based on Equation 2 with a maximum floc size of MAXVOL. It employed four alternative breakup mechanisms: BRAKUP 1 employed the zero collision efficiency factor approach; BRAKUP 2, 3, and 4 arbitrarily broke an oversized particle into 2, 3, or 4 particles of essentially equal size. The results for several simulations are shown in Figures 2.6 and 2.7. These were batch-type flocculations in which the initial suspension was monodisperse containing $N_0$ size 1 particles per unit volume. The parameter C referred to

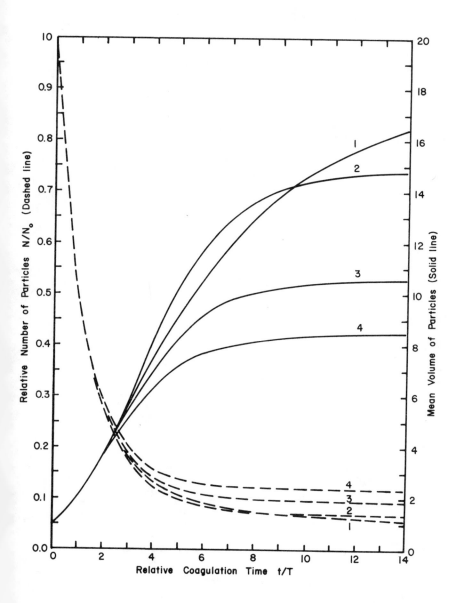

**Figure 2.6.** Effects of conjunction and breakup on floc growth and total number of particles in orthokinetic coagulation. C = 0.01, Maxvol = 20, T = 17.7; 1, Brakup 1 (Run 21); 2, Brakup 2 (Run 29); 3, Brakup 3 (Run 13); 4, Brakup 4 (Run 30).

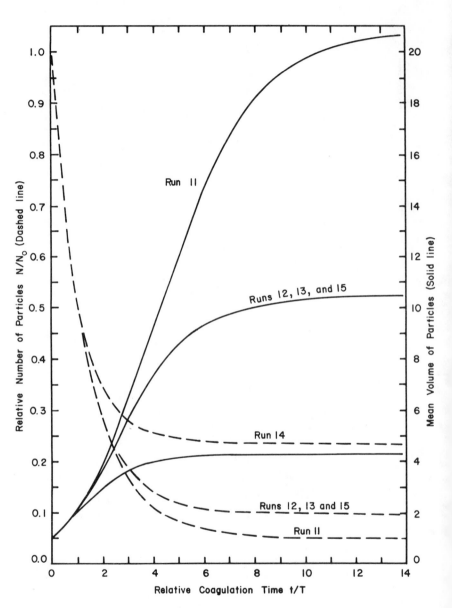

**Figure 2.7.** Effects of other input variables on floc growth and total number of particles in orthokinetic coagulation. All runs used Brakup 3. Run 11, C = 0.02, Maxvol = 40, T = 8.7; Run 12, C = 0.04, Maxvol = 20, T = 40; Run 13, C = 0.01, Maxvol = 20, T = 17.7; Run 14, C = 0.01, Maxvol = 8, T = 18.6; Run 15, C = 0.02; Maxvol = 20, T = 8.7.

on the figures is the product $GN_0$. Notice that floc breakup provides a feedback mechanism that permits the systems to attain a steady-state equilibrium within a reasonably short time. The observed flocculation half-time T for each simulation was of such a value that the product $GTN_0$ was essentially constant.

The batch-type models just described were also converted to continuous flow models for completely mixed flocculation tanks by removing, after each time step of simulation, a fixed proportion of the particles of each size and replacing them with an equivalent volume of primary particles. The proportion of particles removed at each time step is equal to $Q\Delta t$, the product of flow rate through the flocculation and the magnitude of the simulation time-step.

Harris *et al.*[10] pursued these studies further utilizing the floc volume fraction $\phi$ and a particle size distribution function to replace the $R_{ij}$ term in the basic equation. They succeeded in developing a general equation for the number of primary particles in the effluent of the $m^{th}$ compartment of a flocculation basin as

$$(n_0/n_1)_m = 1 + Kd_1G \frac{\phi\tau}{m} \tag{13}$$

where $n_0$ and $n_1$ are the influent and effluent concentrations of primary particles, respectively; $\tau$ is the retention of the series of reactors, $\phi$ is the floc volume fraction; $d_1$ is the diameter of a primary particle and m is the number of completely mixed tanks in the series through which the flow has passed. Size distributions of flocs obtained from the simulations are shown in Figure 2.8. For these simulations the maximum floc size was 100 and oversized particles were broken into two equalized parts.

Argaman and Kaufman,[4] in studying flocculation in turbulently mixed vessels, incorporated the surface erosion concept of floc breakup and the general model developed by Harris to conclude that

$$\frac{n_1^0}{n_1^m} = \frac{\left(1 + K_F K_S G \frac{T}{m}\right)^m}{1 + K_B G^2 \frac{T}{m} \sum\limits_{i=1}^{m-1} \left(1 + K_1 - K_S K_P G \frac{T}{m}\right)^i} \tag{14}$$

Here $n_1^0$ is the number of primary particles inflowing to the first of m turbulently mixed tanks in series; $n_1^m$ is the number of primary particles in the effluent from the $m^{th}$ tank; and $K_F$, $K_S$, $K_B$, and $K_P$ are constants. The relationship is shown graphically in Figure 2.9.

Chang developed a simulation model of flocculation in quiescent settling.[11] The basic equations for the model are Equations 6 and 7, but the velocity

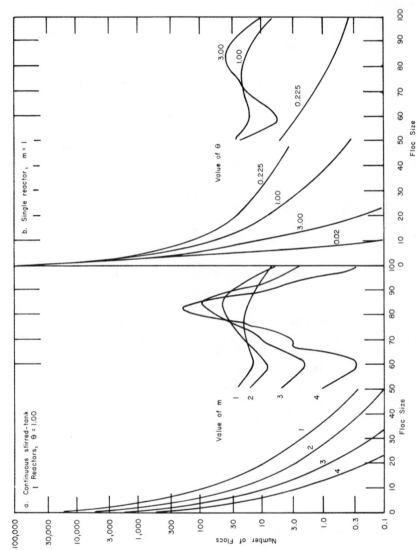

**Figure 2.8.** Simulated floc size distributions (after Harris).

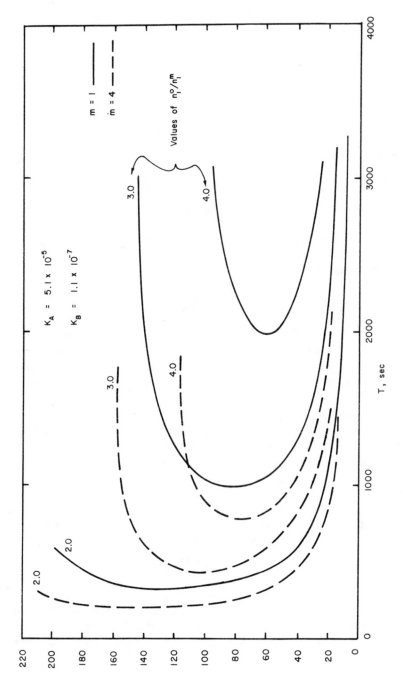

**Figure 2.9.** Relationship of performance to G, T, and m (after Argaman).

gradient term is replaced by the difference in particle settling velocities divided by the radius of interaction.

$$\frac{du}{dz} = \frac{V_i - V_j}{R_{ij}}$$

The flocculation and settling is then considered to take place in a tall column subdivided into a number of strata and the settling and flocculation for any time step are for each strata sequentially moving upward from the bottom most strata. The initial suspension must be heterodisperse. The effect of hindered settling was also included. A sample of the results is shown in Figure 2.10.

Furthermore, Chang has adapted the model to represent a continuous flow settling basin with turbulent mixing in a vertical direction.[11] This mixing is accomplished by the interchanging of strata in the tall column according to a Monte Carlo process based on the Prandtl mixing length and the eddy frequency.

## OPPORTUNITIES FOR FURTHER STUDIES

A variety of simulation models for particle aggregation have been developed, each having its own characteristic strengths and weaknesses. Verification of any simulation model of flocculation is seriously impeded by deficiencies in existing methods for determining floc size distributions experimentally.

Floc breakup is a dominating characteristic of particle aggregation, and certainly warrants additional study experimentally as well as numerically. The use of the floc volume fraction as a surrogate for the radius of interaction can only be successful if it is *adequately* modified by an adequate size distribution characteristic function.

Vold studied, by computer simulation, growth of a single floc composed of unit-sized spheres joined together by surface adhesion without coalescence.[] She found that the effective floc radius, r, was proportional to the number, n, of primary particles comprising the floc raised to the 0.43 power.

$$r = K_1 n^{0.43} \tag{15}$$

This indicates that the floc size increases more rapidly than with a coalescence approach, and the density of a floc decreases accordingly. The Vold model has been verified empirically by Lagvankar and Gemmell for chemical flocs.[13] No simulation models for aggregation have been developed as yet utilizing this type of floc growth.

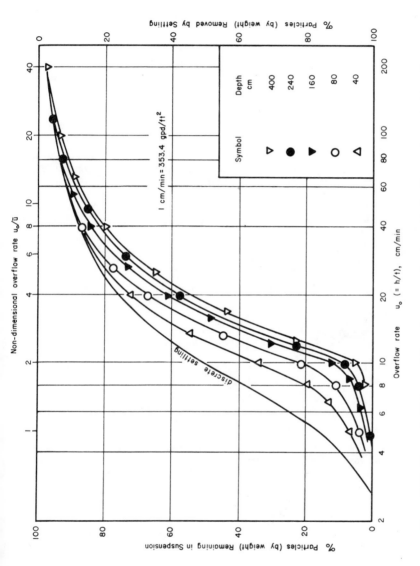

**Figure 2.10.** Removal efficiency of flocculent suspension in an ideal sedimentation tank (after Chang).

The substantial if not dominant role of perikinetic flocculation in the aggregation of submicron particles has been largely ignored. This type of flocculation may have an important effect on the performance of filters in terms of effluent turbidity, since filter collection efficiencies have been shown to be lowest for particles in the one micron size range.

In general, simulation models should focus more clearly on the effect of particle agglomeration on subsequent treatment processes. The time is approaching when the modeling of process sequences such as flocculation, sedimentation, and filtration can be gainfully employed to better our insights on the roles played by these processes in concert.

## REFERENCES

1. Smoluchowski, M. "Drei Vortrage uber Diffusion, Brownsche Molekulerbewegung und Koagulatim von Kolloidteilchen," *Physik. Z.* **17**, 557 (1916). Also, "Versuch einer Mathematischen Theorie der Koagulationskinetic Losungen," *Z. Physik. Chem.* **92**, 129 (1917).
2. Camp, T. R. and P. C. Stein. "Velocity Gradients and Internal Work in Fluid Motion," *J. Bos. Soc. Civil Eng.* **30**, 0.219 (1943).
3. Levich, V. G. *Physico Chemical Hydrodynamics*, Chapter 5 (Englewood Cliffs, New Jersey: Prentice-Hall, Inc., 1962).
4. Argaman, Y. and W. J. Kaufmann. "Turbulence and Flocculation," *J. San. Eng. Div. ASCE* **96**, 0.223 (1970).
5. Parker, D. S., W. J. Kaufmann and D. Jenkins. "Floc Breakup in Turbulent Flocculation Processes," *J. San. Eng. Div. ASCE* **98**, 79 (1972).
6. Bartok, W. and S. G. Mason. "Particle Motions in Sheared Suspensions. VIII. Singlets and Doublets of Fluid Spheres," *J. Colloid. Sci.* **14**, 13 (1959).
7. Thomas, G. D. "Turbulent Disruption of Flocs in Small Particle Size Suspensions," *A. I. Ch.E. J.* **10**, 517 (1964).
8. Fair, G. M. and R. S. Gemmell. "A Mathematical Model of Coagulation," *J. Colloid. Sci.* **19**, 360 (1964).
9. Gemmell, R. S. "Some Aspects of Orthokinetic Flocculation," Ph.D. Thesis, Harvard University, Cambridge, Mass. (1963).
10. Harris, H. S., W. J. Kaufmann and R. B. Krine. "Orthokinetic Flocculation in Water Purification," *J. San. Eng. Div. ASCE* **92**, 95 (1966).
11. Chang, S. C. "Computer Simulation of Flocculent Settling," Ph.D. Thesis, Northwestern University (June, 1972).
12. Vold, M. J. "Computer Simulation of Floc Formation in a Colloid Suspension," *J. Colloid Sci.* **18**, 684 (1963).
13. Lagvankar, A. L. and R. S. Gemmell. "A Size-Density Relationship for Flocs," *J. Amer. Water Works Assoc.* **60**, 1040 (1972).

# CHAPTER 3

## CHEMICAL PRECIPITATION MODELING IN SANITARY ENGINEERING

John F. Ferguson

Department of Civil Engineering
University of Washington
Seattle, Washington 98105

### INTRODUCTION

While chemical precipitation processes have been widely used in sanitary engineering for suspended solids removal, hardness, and phosphate removal, modeling of the processes for teaching, research or design has remained at a rudimentary level. The purpose of this chapter is to take some of the data now available and construct models based on the best present understanding of the chemical processes taking place. These models are intended to be useful to students in consolidating their knowledge of chemical principles and precipitation processes. These same models might, with further development, be useful to engineers in making preliminary decisions concerning alternative phosphate removal processes. Models for describing chemical aspects of chemical precipitation processes might conceptually be divided into three categories—stoichiometric, equilibrium, and kinetic. These categories are hierarchical in that certain systems can be adequately described by stoichiometry alone, but equilibrium models incorporate stoichiometry, and kinetic models incorporate both stoichiometric and equilibrium relations.

Stoichiometry in a model of a precipitation reaction implies that reactants combine to form precipitates in fixed, constant proportions and that the precipitation reaction goes to completion. Water softening is an example of a process in sanitary engineering that can be modeled successfully with stoichiometric relations. The process is described by several reactions, which are assumed to go rapidly to completion. Equilibrium solubilities and kinetic effects are not explicitly considered but are usually compensated for by assuming arbitrary residuals of calcium and magnesium ions. The model is often expressed as a bar diagram method for determining chemical dosages and product water characteristics. Phosphate removal up to 80-90% with iron III [Fe(III)] or aluminum (Al) can often be treated as simple stoichiometric reactions.

Chemical equilibrium in precipitation models is included when the pH changes greatly during the process or when the extent of the precipitation reaction is controlled by a solubility product expression rather than by a stoichiometric relation. Stability-constant relationships are incorporated, and several chemical components and many soluble species usually must be considered.

Such a model is the description of phosphate precipitation with alum or ferric salts as the dissolution equilibrium of aluminum or ferric phosphate with varying pH.[1] In this widely cited model, the residual phosphate concentration is a strong function of pH, with minimum values at pH 6 for Al and 5 for Fe. Other examples of chemical equilibria models include precipitation calculations for the $CaO$-$P_2O_5$-$H_2O$-HCl system,[2] and the solubility calculations for $Ca_5(PO_4)_3OH$ with varying pH and $(Ca^{2+})$.[3] None of these models predicts chemical dosages for phosphate removal. However, in principle, chemical equilibrium models can be used to predict the extent of phosphate removal, the final pH, concentrations of all species and the chemical dosage needed.

The fact that most chemical reactions in sanitary engineering processes do not reach chemical equilibrium has hindered the application of these models to engineering systems. The predictions of the models, while qualitatively correct, have usually been numerically incorrect, sometimes by more than one order of magnitude. Other factors that tend to make the numerical predictions unrealistic include competing reactions with components not included in the model, such as condensed and organic phosphates, nonideality of the solution, kinetically hindered reactions, temperature effects, failure of the system to be either completely closed or open. The models are very complex, though conceptually simple. Although they have been useful in understanding the systems, they have yielded no useful numerical results. In this chapter, the assumption of chemical equilibrium is relaxed in order to consider conditions of steady-state of metastable equilibrium. Models are developed for precipitations where the solubility of the solids is represented by concentration or activity products rather than thermodynamic solubility products. This procedure applies empiricism to the equilibrium models and is a simple way to compensate for the factors preventing the attainment of equilibrium and for factors that nullify the simplifying assumptions about the behavior of the chemical system.

The third class of precipitation models consists of those that include a mathematical description of the rate of the precipitation reaction. Precipitation processes currently used in sanitary engineering are rapid compared to the reaction times available in the reactors used. Thus the kinetics of precipitation reactions have been neglected entirely in modeling phosphate removal and water softening reactions. It is the author's belief that at least one kinetically slow reaction has potential application in phosphate removal.[4] The precipitation of calcium phosphate at pH values close to 8 is slow in a batch reaction, but phosphate removal can be increased in a continuous process by recycle of precipitated solids. Study of the reaction rate has progressed so that the effects of several factors on the reaction rate are now understood. The rate equation for the reaction has been combined with reactor characteristics for several reactor configurations, and the results are presented in this chapter. The model can be used to predict the residual phosphate, pH, and solution concentrations and to compare the trade-offs between various reactor conformations and sizes and chemical dosages.

The examples developed in this paper fit the second and third categories, equilibrium and kinetic models. All deal with phosphate precipitation processes. Water softening is elegantly modeled by simple stoichiometry, and the use of bar diagrams to apply the stoichiometric relations to determine chemical dosages and residual solutions concentrations is well known. Principles of equilibrium have been applied to this system by Langelier[5] and Caldwell and Lawrence;[6] further elaboration would be useless.

There is a set of assumptions common to all the precipitation models, which should be stated explicitly. First, the complex wastewater system is assumed to be adequately represented by just a few chemical components. The presence of many inorganic and usually all organic species and microorganisms is neglected. Second, the system is assumed to be either isolated from a gas phase or to be in equilibrium with a gas phase. In general, mass transfer to, from, or within the solution is assumed not to be rate-limiting. Third, surface chemistry and flocculation kinetics are neglected so that models do not describe solids separation and produce estimates that describe insolubilization but not solids separation. The assumptions restrict the applicability to real systems; however, to some extent relaxing the assumptions of strict chemical equilibria can bring the model results more in line with engineering systems.

There are other assumptions and simplifications that have been made for convenience that could be relaxed. These include the assumption of ideal solution behavior; activity corrections could readily be made. Reaction temperature is assumed to be $25°C$; other temperatures could be modeled after some effort was expended to estimate stability constants. Many solution species, especially ion pair complexes, have been ignored if present in insignificant concentrations. Most estimates of pH change due to reaction are based on simple assumptions concerning stoichiometric precipitations; the results of model calculations could be used to refine the estimates by using successive approximations.

Three precipitation models are considered in this chapter. First, a model is described that is strictly analogous to an acid-base titration computation; it calculates the final pH of a solution for chemical dosages of alum, lime, or ferric chloride. Such a model might have application in phosphate removal modeling, if a functional relationship between final solution pH and residual phosphate concentration were known. Such relations have been presented by Menar and Jenkins for calcium phosphate precipitation in activated sludge[7] and by Seiden and Patel for tertiary calcium phosphate precipitation (Figure 3.1).[8] Iron, aluminum and calcium phosphate solubility relationships have also been presented as a function of pH.[9] In the

(a)

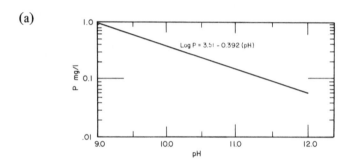

(b)

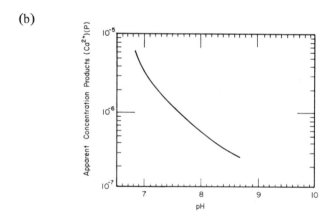

Figure **3.1.** Relationships between phosphate residual and pH. (a) Tertiary lime precipitation,[8] and (b) Calcium and phosphate in activated sludge.[7]

context of this chapter, the pH model is mathematically similar to the equilibrium precipitation models. It serves to introduce the second class of models, modified chemical equilibrium models, which are used to predict residual phosphate concentration, final solution concentrations, and the quantities of solids produced. These models are not completely worked out and only partially verified. With further development they should adequately describe the chemical reactions taking place in rapid precipitations used to remove phosphate from wastewater. The final set of models describe calcium phosphate precipitation at slightly alkaline pH values. Using a general rate expression, the effects of solution parameters are considered for several reactor types. These models, too, are only partially verified by laboratory data and as yet are not substantiated with data from engineering systems.

All the models fit into the calculation scheme shown in Figure 3.2. Initial conditions are used to calculate initial species concentrations using

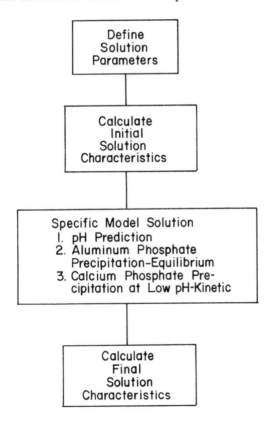

**Figure 3.2.** General computation pattern for the models.

the chemical dosage, calculations specific to the particular problem are performed to find the final pH and/or the final phosphate concentration. Final solution species and amounts of solids are then calculated and with the computer program are printed. The calculations performed by the computer are rather elementary; BASIC language was utilized on a time-sharing commercial system (Leasco-Response, Bethesda, Maryland).

## pH PREDICTION MODEL

The final pH is calculated for a solution with known initial orthophosphate, ammonia nitrogen, total alkalinity and pH and dosed with alum, ferric chloride or lime. It is assumed that the added chemicals contribute known amounts of acid or base, that the system has no gas phase and rapidly comes to equilibrium. The computation is identical to that of calculating the final pH of a mixture of weak acids and bases titrated with a strong acid or base. In Table 3.1 the chemical components, species, and stability constants ($25°C$, 1 atm) used in the calculations are presented.

**Table 3.1. Chemical Species and Stability Constants ($25°C$, 1 atm) Used in Calculations**

| Reaction | Stability Constant | Log K |
|---|---|---|
| $H_2O = H^+ + OH^-$ | $K_w$ | -14.0 |
| $H_2CO_3 = H^+ + HCO_3^-$ | $K_1$ | - 6.2 |
| $HCO_3^- = H^+ + CO_3^{2-}$ | $K_2$ | -10.2 |
| $H_3PO_4 = H^+ + H_2PO_4^-$ | $K_1$ | - 2.2 |
| $H_2PO_4^- = H^+ + HPO_4^{2-}$ | $K_2$ | - 7.2 |
| $HPO_4^{2-} = H^+ + PO_4^{3-}$ | $K_3$ | -12.2 |
| $NH_4OH + H^+ = NH_4^+ + H_2O$ | $K$ | 9.2 |
| $Al^{3+} + OH^- = Al(OH)^{2+}$ | $K(AlOH_2^+)$ | 9.0 |
| $Al^{3+} + 4(OH^-) = Al(OH)_4^-$ | $K(Al(OH)_4^-)$ | 32.5 |
| $Ca^{2+} + CO_3^{2-} = CaCO_3°(aq)$ | $K(CaCO_3°)$ | 3.2 |
| $Ca^{2+} + HPO_4^{2-} = CaHPO_4°(aq)$ | $K(CaHPO_4°)$ | 2.7 |
| $Ca^{2+} + PO_4^{3-} = CaPO_4^-$ | $K(CaPO_4^-)$ | 6.5 |
| $Ca_5(PO_4)_3OH = 5Ca^{2+} + 3PO_4^{3-} + OH^-$ | * | -49 |
| $CaCO_3 = Ca^{2+} + CO_3^{2-}$ | * | - 8 |
| $Al(OH)_3 = Al^{3+} + 3OH^-$ | * | -30.4 |
| $Al_{1.4}PO_4(OH)_{1.2} = 1.4Al^{3+} + PO_4^{3-} + 1.2OH^-$ | * | -32.2 |

*Denotes concentration products.

The computations are made in several steps which are outlined in the following sections. These steps are very similar to ones used for the chemical equilibrium models:

1. The initial analytical description of the wastewater is taken, and the initial species distribution for phosphoric acid, ammonia and carbonic acid species is calculated,

2. The chemical dosage is taken and the equivalent acid or base addition computed,

3. The electroneutrality (charge balance) equation is solved for the final pH (in this case, by a sequential search procedure ending with a pH value arbitrarily near the actual value),

4. The final distribution of species and analytical concentrations are calculated.

To compute the carbonate and bicarbonate concentration from the initial alkalinity, corrections must be made for the influence of phosphate and ammonia on the alkalinity. For pH values between 6 and 9, the following relationships are valid and can be solved for the bicarbonate concentration:

$$Alk = (HCO_3^-) + (HPO_4^{2-}) + (NH_4OH) \qquad (1)$$

$$HPO_4{}^{2-} = P_T/(1+H^+/K_2) \qquad (2)$$

$$NH_4OH = N_T/(1+H^+ \cdot K) \qquad (3)$$

All carbonic and phosphoric acid species concentrations can be calculated from stability products and then summed in an electroneutrality balance to compute the initial concentration of cations associated with the anionic species.

$$M_i^+ = 2(HPO_4{}^{2-}) + (H_2PO_4^-) + 2(CO_3{}^{2-}) + (HCO_3^-) + (OH^-) - (H^+) - (NH_4^+) \qquad (4)$$

The value, $M_i^+$, is characteristic of the initial conditions and is used in subsequent calculations to compute final pH values.

In the calculations to predict final pH, the acid ($C_A$) associated with hydrolysis of $Al^{3+}$ or $Fe^{3+}$ and base ($C_B$) from lime dissolution is determined from the chemical dosages.

$$Al^{3+} + 3H_2O = Al(OH)_3 + 3H^+ \qquad (5)$$

$$C_A \text{ (Alum)} = \text{Alum (mg/l)} \times 10^{-5}$$

$$Fe^{3+} + 3H_2O = Fe(OH)_3 + 3H^+ \qquad (6)$$

$$C_A \text{ (ferric chloride)} = FeCl_3 \text{ (mg/l)} \times 1.86 \times 10^{-5}$$

$$CaO + H_2O = Ca^{2+} + 2(OH^-) \tag{7}$$

$$C_B \text{ (lime)} = \text{Lime (mg/l)} \times 3.57 \times 10^{-5}$$

This acid or base addition shifts the dissociation of the weak acids and bases in the solution. Implicitly, all aluminum or iron is assumed to precipitate and none of the added calcium. The acid-base titration calculation is made by computing the hydrogen ion concentration necessary to balance the new electroneutrality condition.

$$C_B - C_A + M_i^+ + (H^+) + (NH_4^+) = 3(PO_4^{3-}) + 2(HPO_4^{2-}) + (H_2PO_4^-)$$

$$+ 2(CO_3^{2-}) + (HCO_3^-) + (OH^-) \tag{8}$$

$C_B$, $C_A$, and $M_i^+$ are concentrations in equivalents/l; all others are molar.

For a closed system, the equation can be written as a function of the component concentrations, $C_B$, $C_A$, $M_i^+$, stability constants and $H^+$.

$$C_B - C_A + M_i^+ + (H^+) + \frac{N_T(H^+)K}{1 + (H^+)K} - \frac{C_T \left(1 + 2\dfrac{K_2}{(H^+)}\right)}{\left(\dfrac{(H^+)}{K_1} + 1 + \dfrac{K_2}{(H^+)}\right)}$$

$$- \frac{P_T \left(1 + 2\dfrac{(H^+)}{K_2} + 3\dfrac{(H^+)^2}{K_2K_3}\right)}{\left(1 + \dfrac{K_2}{(H^+)} + \dfrac{K_2K_3}{(H^+)^2}\right)} - \frac{Kw}{(H^+)} = 0 \tag{9}$$

The equation is a complex polynomial that for specific cases could be simplified and solved by hand. Alternatively, the calculation method can be generalized and performed on a computer. A scheme (Figure 3.3) for a successive search for the correct values of $H^+$ to satisfy Equation 9 is used in the appended program. This procedure took an average of about 15 iterations to compute $(H^+)$ within 3% of the correct value. The species concentrations and the final alkalinity are then calculated using stability constants.

The procedure is too simplified to give accurate answers. With alum and iron salt addition, $CO_2$ will be supersaturated and tend to be lost to the atmosphere in most engineering reactors, and the pH will rise above values predicted for a closed system. With lime, precipitation of calcium carbonate or phosphate and $CO_2$ absorption from the atmosphere all tend to decrease the pH below the value computed by the acid-base titration calculation.

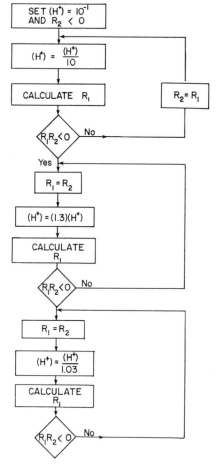

**Figure 3.3.** Calculation scheme to find solution pH by successive approximations.

Predicted pH values for several wastewater alkalinities for alum and lime addition are shown in Figure 3.4. The pH predictions have been used with Seiden and Patel's relation[8] and Stumm's solubility relation[1] to predict phosphate residuals as a function of precipitant dose. These computations can be used with laboratory jar tests of the coagulants to demonstrate the utility of chemical equilibria calculations as well as to emphasize the importance of wastewater alkalinity and final pH as variables in phosphate precipitation along with precipitant dose and phosphate concentration.

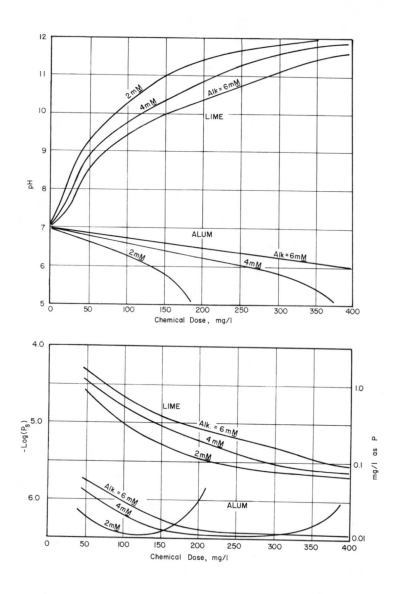

**Figure 3.4.** Application of the pH prediction model. (a) Predicted pH for alum and lime addition for varying initial alkalinities; (b) Predicted phosphate residuals using Figure 3.1a for lime,[8] and aluminum phosphate solubility[1] for alum.

## MODIFIED CHEMICAL EQUILIBRIUM MODELS

### Aluminum Phosphate (AlPO$_4$) Precipitation

As alum is added to a wastewater solution, aluminum phosphate is precipitated with a stoichiometry of 1.4 mol Al to 1 mol P.[10]  Aluminum [Al(III)] is essentially insoluble at neutral pH values, and the precipitation is stoichiometric until about 80-90% of the phosphate is precipitated. Simultaneous with phosphate precipitation, the pH of the solution drops due to protons released from hydroxide-phosphate precipitation. As the alum dose is increased further, the phosphate concentration is controlled by the concentration product describing the aluminum phosphate solubility. With further alum addition, aluminum hydroxide precipitates and establishes a maximum $Al^{3+}$ concentration at any pH, which consequently establishes a minimum soluble phosphate concentration at any pH. Further additions of alum change the pH but do not further reduce the phosphate residual. There consequently is a minimum phosphate solubility for alum precipitation that very nearly corresponds to the solubility of an aluminum phosphate solid where the pH may be independently adjusted with acid or base.[1]  This model of the phosphate removal process ignores coprecipitation and adsorption as mechanisms of removal. In the proposed model, concentration products derived from rapid precipitation experiments are used to describe aluminum phosphate solubility. The model predicts the final solution pH and the amounts of precipitates as well as giving a refined estimate of the residual phosphate compared to earlier models.

Studies by Recht and Ghassemi determined the ratio of aluminum and phosphate in precipitation to be 1.4:1, with excess aluminum precipitated as an amorphous aluminum hydroxide [Al(OH)$_3$].[10]  From their data a concentration product of $10^{-32.2}$ for a $Al_{1.4}(PO_4)(OH)_{1.2}$ precipitate was calculated. This product was obtained for pH between 6 and 8 at a Al:P molar ratio of 2 and an initial phosphate concentration of 12 mg/l P. A concentration product of $10^{-30.4}$ was used to describe the solubility of amorphous Al(OH)$_3$. This combination of concentration products was chosen so that predictions of orthophosphate solubility as a function of alum dosage and final pH would be reasonably consistent with data from several prior studies.

The predictions were obtained from a model of the precipitation process that has the following characteristics. Equilibrium with the metastable, even fictitious, solids was assumed. Initial concentrations for the wastewater and the alum dosage were used to compute final solution concentrations and amounts of solid species.

Two cases are of interest: (1) where the alum dosage results in precipitation of aluminum phosphate alone, or (2) where alum dosage results in precipitation of both aluminum phosphate and aluminum hydroxide.

In the first case where $1.4\ P_T > Al_T$, the precipitation is described by an equilibrium calculation in which the final pH is determined by the protons released from the precipitation and hydrolysis of $Al^{3+}$. Mass balances for aluminum phosphate precipitation are shown.

$$Al_T = 1.4Al_{1.4}\ PO_4(OH)_{1.2} + (Al^{3+}) + (AlOH^{2+}) + [Al(OH_4)^-] \qquad (10)$$

$$P_T = Al_{1.4}\ PO_4(OH)_{1.2} + (H_2PO_4^-) + (HPO_4^{2-}) \qquad (11)$$

The quantity of hydrogen ions, $C_A$, added due to alum addition may be calculated from the assumed precipitation reaction where soluble aluminum species are negligible.

$$1.4Al^{3+} + H_2PO_4^- + H_2O = Al_{1.4}\ PO_4(OH)_{1.2} + 3.2\ H^+ \qquad (12)$$

$$C_A = \frac{3.2}{1.4}\ Al_T \qquad (13)$$

An electroneutrality condition is written for the solution after precipitation.

$$C_A + (HCO_3^-) + (H_2PO_4^-) + 2(HPO_4^{2-}) + (OH^-) = (H^+) + (NH_4^+) + M_i^+ \qquad (14)$$

$(CO_3^{2-})$, $(PO_4^{3-})$, and soluble aluminum species have been neglected in Equation 14, since they are not significant in the pH range of interest. These equations may be solved exactly by substituting the concentration product for $Al_{1.4}PO_4(OH)_{1.2}$ (Equation 15) and the stability relations (Table 3.1) into Equations 10, 11 and 14.

$$(Al^{3+})^{1.4}(PO_4^{3-})(OH^-)^{1.2} = 10^{-32.2} \qquad (15)$$

Alternatively, since the soluble phosphate, $P_S$, is nearly equal to $P_T - Al_T/1.4$, the charge balance (Equation 14) can be solved directly for $(H^+)$  The $(PO_4^{3-})$ concentration can then be calculated from the mass balances (Equations 10 and 11) and the concentration product (Equation 15). This second procedure has been used in the program listed in the Appendix. It produces a pH estimate that is slightly incorrect, but the error is less than the precision ($\pm 3\%$) used for finding $(H^+)$ and is insignificant in comparison to all the uncertainties in the model.

When $Al_T/1.4$ exceeds $P_T$ the solution will become supersaturated with respect to $Al(OH)_3$ and both precipitates may form. The $(Al^{3+})$ concentration and the $Al(OH)_3$ concentration product are computed to determine if both solids are predicted to exist. If it does, the problem must be solved again. Mass balances for aluminum and phosphate are shown for the case of two solids present.

$$Al_T = Al(OH)_3 + 1.4 \, Al_{1.4} \, PO_4(OH)_{1.2} + (Al^{3+}) + (AlOH^{2+}) + [Al(OH)_4^-] \quad (16)$$

$$P_T = Al_{1.4} \, PO_4(OH)_{1.2} + (H_2PO_4^-) + (HPO_4^{2-}) \quad (17)$$

Two solubility relations must be satisfied:

$$(Al^{3+})(OH^-)^3 = 10^{-30.4} \quad (18)$$

$$(Al^{3+})^{1.4}(PO_4^{3-})(OH^-)^{1.2} = 10^{-32.2} \quad (19)$$

These determine unique values for $(Al^{3+})$ and $(PO_4^{3-})$ for any pH value. The acidity due to precipitation of $Al^{3+}$ is a function of the amounts of $Al(OH)_3$ and aluminum phosphate formed. The value of $C_A$ ranges from $3Al_T$ for $Al(OH)_3$ to $(3.2/1.4)Al_T$ for $Al_{1.4}PO_4(OH)_{1.2}$. The latter value has been used in the calculations, though the answer could be refined by estimating $C_A$ from the predicted amounts of the precipitates and repeating the computations.

$$C_A = \frac{3.2}{1.4} Al_T \quad (20)$$

The charge balance is unchanged from Equation 14.

The solubility relationships may be used to eliminate $(PO_4^{3-})$ from the electroneutrality equation, leaving a polynomial in $(H^+)$ that can be solved as before. By substituting $(H^+)$, $(Al^{3+})$ and $(PO_4^{3-})$ in the mass balances the amount of the precipitates can be computed. Predictions for both precipitates present are particularly interesting since many engineering applications involve use of excess alum to precipitate phosphate to low residual levels. The phosphate residual is strongly a function of pH; however, phosphate removal might also be a function of the colloidal properties of the precipitate. These properties are not modeled. Predictions from the model are reasonably consistent with both laboratory and field data for pH values above 6. As an example, data from our laboratory are reproduced in Figure 3.5 along with predictions from the precipitation and pH models. Predictions of residual phosphate, pH and alkalinity are all close to the experimental values.

The model in its present version predicts steadily decreasing phosphate as the pH drops below 6. This result is obviously incorrect and will eventually be corrected. The qualitative difference between the model and experience could be corrected or explained by either polymeric hydroxo-aluminum species or by aluminum phosphate ion pair complexes. The hydroxometal polymer would predominate over $Al^{3+}$ at pH values below 6 and hence increase the solubility of phosphate, which is controlled by

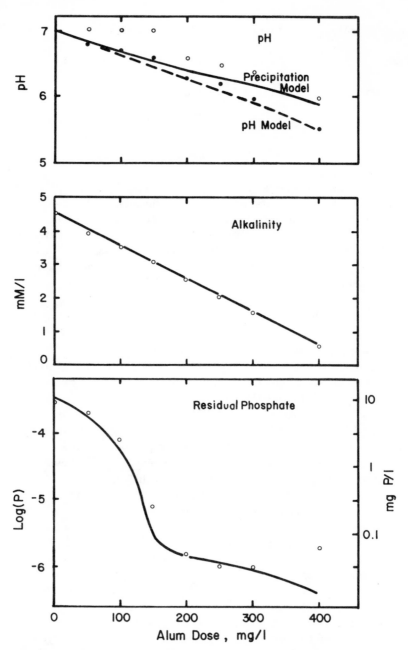

**Figure 3.5.** Comparison of alum precipitation model with Jar Test Data: (a) pH values from pH prediction model (- - -) and precipitation model (——) compared to 1-min measured values (●) and 15-min values (◉), (b) Final alkalinity values, and (c) Residual phosphate values.

($Al^{3+}$). There is experimental evidence for polymers such as in Equation 21 both as intermediates in precipitation and perhaps as stable species.

$$13 \ Al^{3+} + 34 \ OH^- = Al_{13}(OH)_{34}^{5+} \tag{21}$$

The phosphate ion pair complex would add to the soluble phosphate and clearly would have to be the predominant phosphate species for pH values below 6. The model could be modified by including additional terms in the mass balances (Equations 10, 11, 16 and 17) with appropriate values for stability constants for the new species. This unsolved problem with the pseudo-equilibria model has demonstrated, at least to the author, that formal models can effectively demonstrate when simplifying assumptions are wrong. Judgment is required at all stages in making and using engineering models. Use of such models in teaching forces the student and instructor to understand the chemistry of the precipitation process.

This precipitation model can be used to demonstrate the effect of varying alum dosages on a wastewater. By adding terms for independent acid or base addition, the cost of pH adjustment to obtain an optimum final pH can be compared to the cost of adding more alum in order to reduce the pH. An example is shown in Figure 3.6 where a high alkalinity (300 mg/l) wastewater is treated with (1) progressive dosages of alum or (2) 150 mg/l alum (a stoichiometric amount) and progressive dosages of $H_2SO_4$. Typical costs for alum ($60/ton) and $H_2SO_4$ ($50/ton) indicate that the cheapest chemical dosage to reach a residual 0.1 mg/l is predicted to be combination

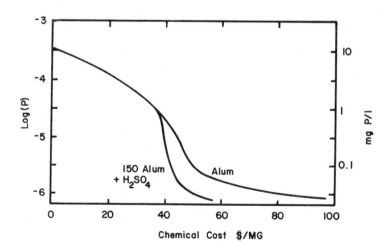

**Figure 3.6.** Comparison of chemical cost for alum or alum plus sulfuric acid to achieve various phosphate residuals for a high alkalinity wastewater.

alum and acid ($41/MG for 150 mg/l alum and 19 mg/l $H_2SO_4$ compared with $49/MG for 200 mg/l alum alone). To reach a residual of 1 mg/l, alum alone (150 mg/l) is sufficient. As the alkalinity increases, the advantages of acid addition become even greater. Using alum as an acid to reduce the pH is shown to be very costly.

These conclusions were reached without the aid of a model for alum precipitation of oxidation pond effluent at Lancaster, California.[11] The model might have been used in other studies to reduce the number of tests and pilot studies needed to determine suitable conditions for precipitation.

### Ferric Phosphate (FePO$_4$) Precipitation

Procedures analogous to those for aluminum phosphate may be used to predict phosphate removal by precipitation with Fe(III) salts. Data on stoichiometry and solubility of precipitates can be used to select concentration products that reproduce the behavior of the system.[10] Mass balances, solubility relations and the electroneutrality conditions can then be combined to calculate residual phosphate, pH, and amounts of solids precipitated. The details of these calculations have not been worked out but are similar in all respects to aluminum phosphate precipitation.

### Ferrous Phosphate [Fe$_3$(PO$_4$)$_2$] Precipitation

Knowledge of $Fe_3(PO_4)_2$ precipitation is poorly developed. Recently solubility data for vivianite have been reported.[12,13] However, the kinetics of precipitation and of competing reactions such as oxidation of Fe(II) to Fe(III) or precipitation of siderite ($FeCO_3$) are not understood sufficiently to justify any quantitative modeling.

### Calcium Phosphate [Ca$_3$(PO$_4$)$_2$] Precipitation

When lime is added to wastewater to raise the pH above 9, both calcium phosphate and calcium carbonate ($CaCO_3$) precipitate. This precipitation is rapid and can be treated essentially in the same way as alum or ferric salt precipitation. At a pH below 8.5, calcium phosphate (apatite) precipitates slowly and with predictable kinetics. Calcium carbonate does not precipitate unless the alkalinity exceeds $\cong 300$ mg/l. The slow precipitation can be used to effect phosphate removal by recycling solids in a reactor with a large plug flow component such as used in the conventional activated sludge process.

## Calcium Phosphate-Calcium Carbonate Precipitation at High pH Values

For the system $CaO-P_2O_5-CO_2-H_2O-HCl$ without a gas phase, the mass balances are complicated by the fact that there are ion pair complexes, including $CaCO_3^{\circ}(aq)$, $CaHPO_4^{\circ}(aq)$, and $CaPO_4^-$, that are significant fractions of the soluble calcium and phosphate. The $Ca_2(PO_4)_2$ that forms is a poorly crystalline apatite not unlike the mineral phase of bone or tooth enamel. Calcium carbonate precipitates as small crystals of calcite.

The solubility of these precipitates have been characterized by concentration products on the order of $10^{-49}$ for $Ca_5(PO_4)_3OH$ but depending on precipitation time and the solution alkalinity. Calcium carbonate solubility is described by a concentration product of about $10^{-8}$, a value slightly greater than $K_{sp} = 10^{-8.35}$.

Magnesium concentration also affects the system.[14] The precipitation model should, but does not yet include both the solubility of brucite $[Mg(OH)_2]$ and the effect of Mg on the concentration products for calcium phosphate and carbonate.

For precipitation at pH greater than 9, either, neither, or both precipitates might form. Only one case is of practical interest—that of both precipitates. The calcium concentration after lime addition will always exceed $P_T$ in a wastewater. At all reasonable alkalinities the solubility of $CaCO_3$ will be exceeded and both precipitates will be expected to form rapidly. A pseudo-equilibrium calculation can be used to predict the phosphate residual for various pH, Ca, $P_T$ concentrations and lime dosages.

The applicable mass balances are given.

$$Ca_T = CaCO_3 + 5Ca_5(PO_4)_3OH + (Ca^{2+}) + (CaHPO_4^{\circ}) + (CaPO_4^-) + (CaCO_3^{\circ}) \quad (22)$$

$$P_T = 3Ca_5(PO_4)_3OH + (HPO_4^{2-}) + (PO_4^{3-}) + (CaHPO_4^{\circ}) + (CaPO_4^-) \quad (23)$$

$$CO_{3T} = CaCO_3 + (CaCO_3^{\circ}) + (HCO_3^-) + (CO_3^{2-}) \quad (24)$$

Two solubility relations must be satisfied:

$$(Ca^{2+})^5(PO_4^{3-})^3(OH^-) = 10^{-49} \quad (25)$$

$$(Ca^{2+})(CO_3^{2-}) = 10^{-8} \quad (26)$$

The lime dose results in base addition:

$$C_B = \frac{CaO \ (mg/l)}{56 \times 10^3} \times 2 \quad (27)$$

The precipitation reactions cause the release of protons:

$$5(Ca^{2+}) + 3(HPO_4^{2-}) + H_2O = Ca_5(PO_4)_3\,OH + 4H^+ \qquad (28)$$

$$Ca^{2+} + HCO_3^- = CaCO_3 + H^+ \qquad (29)$$

The electroneutrality condition then is as follows:

$$C_A + (HCO_3^-) + 2(CO_3^{2-}) + 2(HPO_4^{2-}) + 3(PO_4^{3-}) + (CaPO_4^-) + (OH^-)$$

$$= C_B + M_i^+ + (H^+) + (NH_4^+) \qquad (30)$$

Using stability relationships, the set of six equations (22-26, 30) can be used to solve for six unknown quantities [*e.g.*, $(CO_3^{2-})$, $(PO_4^{3-})$, $(Ca^{2+})$, $(H^+)$, $CaCO_3$, and $Ca_5(PO_4)_3OH$].

The algebra is complex and has not been worked out in detail. General computer programs to solve such chemical systems are available.[15,16] Solution of the equations should result in much better predictability of phosphate removal than attained using stoichiometry or the empirical correlation between pH and phosphate (Figures 3.1a and 3.4a).

### Calcium Phosphate Precipitation at pH 7.5-8.5

Precipitation of calcium phosphate at pH 8 has been described by the following expression for phosphate removal after an initial induction period[4] where P is soluble orthophosphate and $C_T$ is bicarbonate alkalinity.

$$\frac{dP}{dt} = -\frac{k(P)^{2.8}}{(C_T)} \qquad (31)$$

Removal is proportional to the phosphate concentration raised to 2.8 power and inversely proportional to the bicarbonate concentration. The expression was derived for batch precipitation tests in which the pH was maintained at $8 \pm 0.1$, and the initial $Ca^{2+}$ concentration at 2.0 m$M$. The rate constant k is proportional to the available surface area for crystal growth, a quantity assumed to remain constant during precipitation. Further tests to be described elsewhere have indicated that the kinetics of the precipitation reaction (after the induction period) can be approximated by the rate expression

$$\frac{dP}{dt} = -k'\,(Ca^{2+})^5(PO_4^{3-})^3(H^+)^{-1}(C_T)^{-1} \qquad (32)$$

This reaction, which has a large temperature coefficient, is described by the above equation for calcium concentrations between 1 and 2.5 m$M$, pH between 7.6 and 8.4, $P_T$ between 0 and 1 m$M$, and alkalinity 0.5 and 7 m$M$. If $(PO_4^{3-})$ is expressed as a function of P for this pH range, the rate expression is

$$\frac{dP}{dt} = -k''(Ca^{2+})^5 (P)^3 [H^+]^{-4} (C_T)^{-1} \tag{33}$$

For 25°C, $[H^+] = 10^{-pH}$, other concentrations in mmol/l, k has a value of $1.3 \times 10^{16}$ hr$^{-1}$. This expression can be used to predict steady-rate phosphate removal in various reactors such as completely stirred tank reactors (CSTR) and tubular reactors with or without solids recycle. Solids recycling reactors are assumed to increase the solids concentration by the ratio $\theta_s/\theta$, where $\theta_s$ is solids retention time and $\theta$, hydraulic retention time. Since precipitated solids are agglomerates of very small particles ($<0.05\mu$), the rate constant is assumed to increase in proportion to $\theta_s/\theta$. Predictions from such a model are not likely to be very accurate, although the precipitation model has predicted phosphate removals within 10% of observed values for a single CSTR and for four CSTRs with solid recycle. The model is useful as an example of the effect of reactor configuration on a process. The reaction order is greater than one for phosphate, thus, plug flow reactors are much more effective than CSTRs of similar detention time. The process, too, is potentially attractive economically since only modest lime doses are required to reach pH 8. The only equally economical phosphate removal process would be luxury uptake in activated sludge, which in some cases might be due to calcium phosphate precipitation rather than to a biological mechanism.

The model is presented in the following sections by considering four reactor types: CSTR, CSTR with solids recycle, series of CSTRs with solids recycle from last to first, and tubular reactor with solid recycle. Equations can be developed to be incorporated into the general precipitation model to calculate the extent of the precipitation, phosphate removal, and the calcium and hydroxide dose to achieve assumed pH values and calcium concentrations. The general model is used to compute the initial and final solution characteristics.

## CSTR

The CSTR (Figure 3.7a) is not a practical reactor configuration; however, phosphate precipitation can readily be described. A mass balance on soluble phosphate results in the following equation at steady state.

$$QP_T = QP + V\frac{dP}{dt} = 0 \tag{34}$$

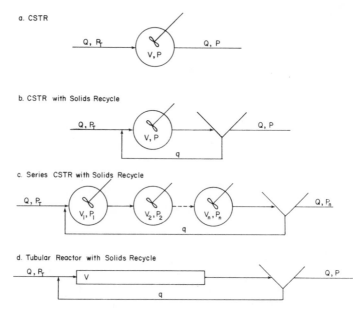

**Figure 3.7.** Reactor types considered with low pH calcium phosphate precipitation.

Substituting the rate expression (Equation 33) and noting that the hydraulic detention time $\theta$ is $V/Q$

$$P_T - P = k'' \, \theta \, (P)^3 (Ca^{2+})^5 (H^+)^{-4} (C_T)^{-1} \tag{35}$$

The equation models phosphate removal as a function of phosphate, calcium, bicarbonate and pH values and reactor detention time. To a reasonable approximation for waste treatment condition, the initial calcium, bicarbonate, and pH can be assumed equal to final concentrations. For initial conditions $P_T = 0.3$ m$M$, $Ca^{2+} = 2$ m$M$, $C_T$ alk 4 m$M$, $(H^+) = 10^{-8}$, reaction to precipitate 80% of the phosphate would result in $Ca^{2+} = 1.6$ m$M$ and alk = 3.7 m$M$; the pH would be reduced to about 7.4, but as $CO_2$ diffuses out of the solution the pH would increase back very close to 8.

The model could be refined by making successive approximations of the extent of the reaction and correcting values for $Ca^{2+}$, pH and $C_T$ between approximations. The usefulness of the numerical results does not merit this false precision, and the model calculations can readily be performed by hand.

## CSTR with Solids Recycle

For the reactor type of Figure 3.7b, the precipitation reaction rate constant, K, is obtained by multiplying $k''$ by $\theta_s/\theta$, the ratio of solids retention time to hydraulic retention time, which is also the measure of the increase in concentration in solids over what would be found in a once-through reactor.

$$K = k'' \frac{\theta_s}{\theta} \qquad (36)$$

$\theta_s$ is defined as the total quantity of solids in the system divided by the solids lost from the system per day. For a precipitation process operated in a conventional activated sludge system, $\theta_s$ is on the order of several days while $\theta$ in the aeration tank is several hours.

At steady state solids lost per day would be equal to solids produced per day. With efficient solids capture

$$\theta_s = \frac{Vc}{QFP_T R} \qquad (37)$$

where F is the coefficient relating weight of phosphate removed to weight of solids produced; R, the phosphate removal efficiency; c, the concentration of solids in the reactor; and V, the volume of the reactor. $\theta_s$ can be manipulated by controlling c by wasting solids or by changing the hydraulic detention time $\theta$.

The removal equation can readily be derived by considering the phosphate mass balance (Equation 34).

$$P_T - P = K \; \theta (P)^3 (Ca^{2+})^5 (H^+)^{-4} (C_T)^{-1} \qquad (38)$$

The calculation could readily be incorporated into the general precipitation model (Figure 3.2) by using known or assumed reactor characteristics (Q, V, $\theta_s$) and chemical characteristics ($C_T$, $Ca^{2+}$, pH, $P_T$) to compute the extent of the reaction. The quantity of protons ($C_A$) released due to the precipitation reactions can be computed from the stoichiometry of the reaction.

$$5Ca^{2+} + 3HPO_4^{2-} + H_2O \rightarrow Ca_5(PO_4)_3OH + 4H^+ \qquad (39)$$

The electroneutrality equation can be written to compute $C_B$, the base needed to attain the assumed pH value. The calcium needed can be calculated from a mass balance on calcium for the system, where Ca (to be added) equals ($Ca^{2+}$) plus $5Ca_5(PO_4)_3OH$ minus Ca (initial).

### CSTRs in Series with Solids Recycle

Reactors not intended to be completely mixed can often be approximated as a series of CSTRs. For activated sludge aeration tanks the number of CSTRs best representing the mixing in the aeration reactor ranges from 2 to 5 depending on the nearness to plug flow or completely mixed conditions. Removal of P in each reactor can be derived from a mass balance on phosphate. The hydraulic residence times are equal.

$$\theta_1 = \theta_2 = \theta_n = \frac{\theta}{n} \tag{40}$$

The rate constant K applies in each reactor and calcium, bicarbonate and pH are assumed to remain approximately constant for all reactors.

$$P_T - P_1 = K\,\theta_1 (P_1)^3 (Ca^{2+})^5 (H^+)^{-4} (C_T)^{-1} \tag{41}$$

$$P_1 - P_2 = K\,\theta_2 (P_2)^3 (Ca^{2+})^5 (H^+)^{-4} (C_T)^{-1} \tag{42}$$

$$P_{n-1} - P_n = K\,\theta_n (P_n)^3 (Ca^{2+})^5 (H^+)^{-4} (C_T)^{-1} \tag{43}$$

These equations can be conveniently solved by hand for $n \leqslant 5$. Alternatively a computer may be used in which case changes in all concentrations may be incorporated into the removal expression.

The solids recycle flow, q, dilutes the influent waste to some extent.

$$P_T = \frac{QP_T + qP_n}{Q + q} \tag{44}$$

Successive approximations can be used to reach an answer of any desired precition. Results for calculations for 1, 2, and 4 CSTRs are presented for typical conditions ($P_T = 0.3$ m$M$, $Ca^{2+} = 2$ m$M$, $C_T = 3.4$ m$M$, pH = 8, T = 25°C, $\theta_s/\theta \doteq 30$) in Figure 3.8 to show the expected phosphate residual as a function of $\theta$. The calculations were made with assistance of the computer to solve polynomials by the Newton-Raphson method, but corrections were not made for changes in reactant concentrations or detention time due to recycle.

Since the reaction order is greater than one, there are great economies in required tank volume for a poorly mixed reactor to accomplish a level of phosphate removal—or great increase in removal for the same tank volume—compared to a single CSTR with solids recycle.

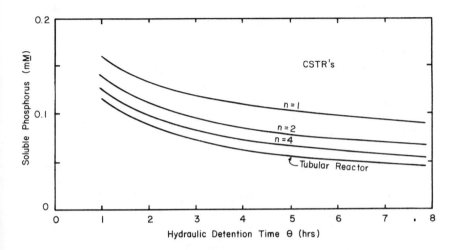

**Figure 3.8.**   Predicted phosphate residuals for various reactor types.
Low pH calcium phosphate precipitation: $P_T$ = 0.3 m$M$,
Ca$^{2+}$ = 2 m$M$, $C_T$ = 3.4 m$M$, pH 8, T = 25°C, $\theta_s/\theta \doteq$ 30.

## Tubular Reactor with Solids Recycle

As the number of CSTRs in series increases, the reaction approaches
that predicted for a tubular reactor with solids recycle. This limiting reactor configuration (Figure 3.7d) has a single solution obtained by integrating
the rate equations from 0 to $\theta$, where the initial concentration is given by
Equation 44.

$$\tfrac{1}{2}\left(\frac{QP_T + qP}{Q + q}\right)^{-2} - \tfrac{1}{2}P^{-2} = \frac{-K\theta}{C_T(1 + q/Q)} \tag{45}$$

The integrated equation may be solved using the Newton-Raphson
method yielding a relation between $\theta$, q/Q and phosphate removal. For
q/Q = 0.2 and other parameters unchanged, phosphate residual versus $\theta$
predictions are shown in Figure 3.8. The increase in effectiveness from
four CSTRs to ideal plug flow is smaller than the increase from two to
four CSTRs.

The chemical dosage to achieve a desired phosphate removal for a given
reactor can be calculated as discussed previously. Typical calcium and base
requirements are on the order of 1 m$M$. The process should use very

modest quantities of chemicals. If the precipitation process is used with activated sludge, biological production of $CO_2$ may depress the pH and cause higher requirements of base. Alternatively, high rates of aeration may be used to strip $CO_2$ and keep the pH high enough for rapid calcium phosphate crystal growth.

The process is very sensitive to the initial calcium concentration and the pH with slight increases markedly increasing the rate of removal. With both parameters, increases result in supersaturation of calcium carbonate so that only moderate levels Ca (2.5 m$M$) and pH (8.5) can be attained before the model fails due to competitive calcium carbonate precipitation.

The model describes a process that has not yet been shown to be applicable to wastewaters. The process does provide a good example of the interrelationship between reaction kinetics and reactor type and can readily be used as an example when reactor types and characteristics are being discussed. When presented with other kinetically controlled reactions (*e.g.*, disinfection approximated by Chick's law as first-order, and biological substrate removal approximated by Monod's relation as mixed-order), the third-order precipitation makes an excellent contrasting example of a process where the flow regime greatly affects the treatment effectiveness.

## CONCLUSIONS

Several models for phosphate precipitation have been presented incorporating concepts of stoichiometry, chemical equilibrium and reaction kinetics. The models are not verified in any meaningful sense, thus their use in engineering is only potential. The models are useful in formalizing our understanding of the chemistry of the precipitation processes and do have teaching applications. The models especially allow the rapid formulation of alternatives among chemicals and dosages for $AlPO_4$ precipitation and among chemical dosage and reactor type and volume for $Ca_3(PO_4)_2$ precipitation at slightly alkaline pH.

## REFERENCES

1.  Stumm, W. In *Advances in Water Pollution Research*, Vol. 2 (New York, Macmillan, 1964), p. 216.
2.  Ferguson, J. F. "The Precipitation of Calcium Phosphate from Fresh Waters and Waste Waters," Ph.D. Dissertation, Stanford University, Stanford, California (1969).
3.  Leckie, J. O. and W. Stumm. "Phosphate Precipitation," In *Water Quality Improvement by Physical and Chemical Processes,*" E. F. Gloyna and W. Eckenfelder, Eds. (Austin: University of Texas, 1970), p. 237.

4.  Ferguson, J. F., D. Jenkins, and J. Eastman. "Calcium Phosphate Precipitation at Slightly Alkaline pH Values," *J. Water Poll. Control Fed.* **45**, 620 (1973).
5.  Langelier, W. F. "Chemical Equilibrium in Water Treatment," *J. Amer. Water Works Assoc.* **38**, 169 (1946).
6.  Caldwell, D. H. and W. B. Lawrence. "Water Softening and Conditioning Problems," *Ind. Eng. Chem.* **45**, 535 (1953).
7.  Menar, A. B. and D. Jenkins. "The Fate of Phosphorus in Sewage Treatment Processes, Part II—Mechanism of Enhanced Phosphate Removal by Activated Sludge," SERL Report 68-6, University of California, Berkeley (1968).
8.  Seiden, L. and K. Patel. *Mathematical Model of Tertiary Treatment by Lime Addition*, Report TWRC-14 (Washington, D. C.: Federal Water Pollution Control Administration, U.S. Department of the Interior, 1969).
9.  Stumm, W. and J. J. Morgan. *Aquatic Chemistry* (New York: Wiley-Interscience, 1970).
10. Recht, H. L. and M. Ghassemi. *Kinetics and Mechanism of Precipitation and Nature of the Precipitate Obtained in Phosphate Removal from Wastewater Using Aluminum (III) and Iron (III) Salts*, Report 17010EK1 04/70 (Washington, D.C.: Federal Water Quality Administration, 1970).
11. Dryden, F. D. and G. Stern. "Renovated Wastewater Creates Recreational Lake," *Environ. Sci. Tech.* **2**, 268 (1968).
12. Singer, P. C. "Anaerobic Control of Phosphate by Ferrous Iron," *J. Water Poll. Cont. Fed.* **44**, 663 (1972).
13. Nriagu, J. "Stability of Vivianite and Ion Pair Formation in the System $Fe_3(PO_4)_2 - H_3PO_4 - H_2O$," *Geochim. Cosmochim. Acta* **36**, 459 (1972).
14. Ferguson, J. F. and P. L. McCarty. "Effects of Carbonate and Magnesium on Calcium Phosphate Precipitation," *Environ. Sci. Tech.* **5**, 534 (1971).
15. Ingri, N., W. Kakolowicz, L. G. Sillen, and B. Warnquist. "High Speed Computers as a Supplement to Graphical Methods," *Talanta* **14**, 1261 (1967).
16. Morel, F. M. and J. J. Morgan. "A Numerical Method for Computing Equilibria in Aqueous Chemical Systems," *Environ. Sci. Tech.* **6**, 58 (1972).

## APPENDIX

## Computer Programs in Basic

CHEM

```
10   REM PROGRAM IN BASIC
20   REM PROGRAM TO CALCULATE ESTIMATED FINAL PH FOR
30   REM PRECIPITANTS ALUM, FERRIC CHLORIDE OR LIME
40   REM NAME CHEMICAL SPECIES VARIABLES AS WELL AS COMBINATIONS
50   REM OF SPECIES
60   REM H0 = OH,       H1 = H+,       H9 = PH.
70   REM C0 = H2CO3,  C1 = HCO3,    C2 = CO3,    C9 = T. CARB.
80   REM N0 = NH3,    N1 = NH4,     N9 = T. AMM.
90   REM P1 = H2PO4,  P2 = HPO4,    P3 = PO4,    P9 = T.O-PHOS
100  REM A9 = ACID,   B9 = BASE,    L9 = ALK,    M9 = CATION
110  REM STABILITY CONSTANTS WRITTEN FOR LIGAND ASSOCIATING
120  REM WITH PROTON
130  REM K1 = H2O,     K2 = HCO3,   K3 = CO3,    K4 = NH4
140  REM K5 = HPO4,   K6 = PO4.
150  REM DEFINE STABILITY CONSTANT VALUES
160  LET K1=1/10↑14
170  LET K2=1/10↑6.35
180  LET K3=1/10↑10.33
190  LET K4=10↑9.25
200  LET K5=1/10↑7.2
210  LET K6=1/10↑12.2
220  REM READ ALKALINITY, T.O-PHOS, T.AMM-N, AND PH
230  REM READ CONCENTRATIONS IN MILLIMOLES/LITER
240  REM N IS NUMBER OF DIFFERENT INITIAL WATER CONDITIONS
250  READ N
260  READ L9,P9,N9,H9
270  DATA 1,6,.3,2,8
280  FOR J=1 TO N
290  PRINT USING 300,L9,P9,N9,H9
300  :ALK = ##.#, TOP = ##.#, TAM = ##.#, PH = ##.#
310  REM CONVERT TO MOLAR CONCENTRATIONS
320  LET L9=L9/1000
330  LET P9=P9/1000
340  LET N9=N9/1000
350  LET H1=1/10↑H9
360  REM CALCULATE HCO3 FROM INITIAL CONCENTRATIONS
370  LET P2=P9/(1+H1/K5)
380  LET N0=N9/(1+H1*K4)
390  LET C1=L9-P2-N0
400  REM ALSO COMPUTE TOTAL CARBONATE SO SUBROUTINE TO COMPUTE
410  REM ALL SPECIES CAN BE GENERALIZED
420  LET C9=C1*(1+H1/K2+K3/H1)
430  REM SUBROUTINE TO COMPUTE ALL SPECIES NEEDED IN INITIAL CHRG BAL
440  GOSUB 1020
450  REM SET A9 AND B9 ZERO THEN USE PRINT SUBROUTINE TO SHOW INITIAL
460  REM SPECIES CONCENTRATIONS
470  LET A9=0
480  LET B9=0
490  GOSUB 1240
500  REM KEEP MI+ AS M8
510  LET M8=M9
520  REM ADD PRECIPITANTS AND COMPUTE ACID OR BASE
530  REM M IS # OF COAGULANT DOSES
540  READ M
550  DATA 3
560  FOR I=1 TO M
570  READ D1,D2,D3
580  REM D1 = LIME, D2 = ALUM, D3 = FERRIC CHLORIDE IN MG/L
590  DATA 0,100,0,0,300,0,0,500,0
600  PRINT USING 610,D1,D2,D3
610  :LIME = ###.#, ALUM = ###.#, FERRIC CHLORIDE = ###.# MG/L
620  LET A9=0
```

```
630   LET B9=0
640   REM COMPUTE ACID OR BASE FROM PRECIPITANT DOSES
650   LET B9=B9+(D1/56)*2/1000
660   LET A9=A9+(D2/300)*3/1000
670   LET A9=A9+(D3/162)*3/1000
680   REM FIND FINAL PH BY SOLVING CHARGE BALANCE THAT
690   REM INCLUDES ACID OR BASE FROM PRECIPITANT
700   LET J=0
710   REM SUCCESSIVE SEARCH FOR H+ FROM PH 2, BY /10,*1.3,/1.03
720   LET H1=.1
730   REM NOTE THAT R9 MUST BE SET INITIALLY EITHER < OR > ZERO
740   REM THERE IS NO EASY WAY TO TELL WHICH WILL CONVERGE
750   LET R9=-.01
760   LET H1=H1/10
770   REM CALCULATE SPECIES AND CHARGE BALANCE (M9)
780   GOSUB 1020
790   REM COMPUTE AND COMPARE REMAINDERS FOR CHARGE BALANCE
800   GOSUB 1160
810   REM IF REMAINDERS FOR SUCCESSIVE APPROXIMATIONS
820   REM HAVE SAME SIGN, THEN LET PH =PH + 1
830   IF S9>0 THEN 760
840   LET H1=H1*1.3
850   GOSUB 1020
860   GOSUB 1160
870   IF S9>0 THEN 840
880   LET H1=H1/1.03
890   GOSUB 1020
900   GOSUB 1160
910   IF S9>0 THEN 880
920   REM PRINT CYCLES AND REMAINDER AFTER CONVERGENCE TO APPROX H+
930   PRINT "J = ",J,"R = ",R8
940   REM CALCULATE FINAL PH AND ALKALINITY
950   LET H9=-LOG(H1)/2.303
960   REM L9 IS FINAL ALKALINITY
970   LET L9=C1+2*C2+N0+P2+2*P3+H0-H1
980   GOSUB 1240
990   NEXT I
1000  NEXT J
1010  STOP
1020  REM SUBROUTINE TO COMPUTE SPECIES FOR KNOWN TOTAL COMPONENT
1030  REM CONCENTRATIONS AND PH. ALSO COMPUTE MI+ FROM CHARGE BALANCE
1040  LET P2=P9/(1+H1/K5+K6/H1)
1050  LET P1=H1*P2/K5
1060  LET P3=K6*P2/H1
1070  LET C1=C9/(1+H1/K2+K3/H1)
1080  LET C0=H1*C1/K2
1090  LET C2=K3*C1/H1
1100  LET N0=N9/(1+H1*K4)
1110  LET N1=N0*H1*K4
1120  LET H0=K1/H1
1130  REM USE CHARGE BALANCE TO CALCULATE MI+(M9)
1140  LET M9=H0-H1+C1+2*C2-N1+P1+2*P2+3*P3
1150  RETURN
1160  REM SUBROUTINE TO COMPUTE AND COMPARE CHARGE BALANCE REMAINDER
1170  LET R8=R9
1180  LET R9=A9+M9-B9-M8
1190  REM CHECK TO AVOID NON-CONVERGENCE
1200  LET J=J+1
1210  IF J>100 THEN 980
1220  LET S9=R8*R9
1230  RETURN
1240  REM SUBROUTINE TO PRINT CONCENTRATIONS OF SPECIES AND COMBINATIONS
1250  PRINT "STATUS OF H+,CO3,NH3,PO4,ACID AND ALK"
1260  PRINT H0,H1,H9
1270  PRINT C0,C1,C2,C9
1280  PRINT N0,N1,N9
1290  PRINT P1,P2,P3,P9
1300  PRINT A9,B9,L9,M9
1310  RETURN
1320  END
```

```
RUN
CHEM

ALK =  6.0, TOP =    .3, TAM =  2.0, PH =  8.0
STATUS OF H+,CO3,NH3,PO4,ACID AND ALK
  .000001        1.00000E-08     8
  1.26142E-04    5.63456E-03     2.63547E-05     5.78706E-03
  1.06480E-04    1.89352E-03     .002
  4.10399E-05    2.58944E-04     1.63382E-08     .0003
  0              0               .006            4.35372E-03
LIME =      .0, ALUM = 100.0, FERRIC CHLORIDE =    .0 MG/L
J =         16              R =              -1.24797E-05
STATUS OF H+,CO3,NH3,PO4,ACID AND ALK
  1.26267E-07    7.91971E-08     7.10001
  8.71090E-04    4.91307E-03     2.90163E-06     5.78706E-03
  1.41009E-05    1.98590E-03     .002
  1.66973E-04    1.33026E-04     1.05980E-09     .0003
  .001           0               5.06605E-03     3.36605E- 3
LIME =      .0, ALUM = 300.0, FERRIC CHLORIDE =    .0 MG/L
J =         18              R =              -1.62590E-05
STATUS OF H+,CO3,NH3,PO4,ACID AND ALK
  2.47379E-08    4.04238E-07     6.39221
  2.74903E-03    3.03768E-03     3.51483E-07     5.78706E-03
  2.77836E-06    1.99722E-03     .002
  2.59496E-04    4.05036E-05     6.32202E-11     .0003
  .003           0               3.08128E-03     1.38128E-03

STOP




SUNDAY

10    REM PROGRAM IN BASIC
20    REM PROGRAM TO CALCULATE PHOSPHATE, PH, AND SOLUTION COMPOSITION
30    REM FOR ALUM PRECIPITATION
40    REM NAME CHEMICAL SPECIES VARIABLES AS WELL AS COMBINATIONS
50    REM OF SPECIES
60    REM H0 = OH,      H1 = H+,      H9 = PH.
70    REM C0 = H2CO3,   C1 = HCO3,    C2 = CO3,    C9 = T. CARB.
80    REM N0 = NH3,     N1 = NH4,     N9 = 1. AMM.
90    REM P1 = H2PO4,   P2 = HPO4,    P3 = PO4,    P9 = T.O-PHOS
100   REM A9 = ACID,    B9 = BASE,    L9 = ALK,    M9 = CATION
110   REM STABILITY CONSTANTS WRITTEN FOR LIGAND ASSOCIATING
120   REM WITH PROTON
130   REM K1 = H2O,     K2 = HCO3,    K3 = CO3,    K4 = NH4
140   REM K5 = HPO4,    K6 = PO4.
150   REM DEFINE STABILITY CONSTANT VALUES
160   LET K1=1/10↑14
170   LET K2=1/10↑6.35
180   LET K3=1/10↑10.33
190   LET K4=10↑9.25
200   LET K5=1/10↑7.2
210   LET K6=1/10↑12.2
220   REM SOLUBLE ALUMINUM SPECIES AL3+, ALOH2+, AL(OH)4-
230   REM ARE INCLUDED IN PRECIPITATION CALCULATIONS
240   REM BUT NOT IN INITIAL SOLUTION CALCULATIONS
250   REM READ ALKALINITY, T.O-PHOS, T.AMM-N, AND PH
260   REM READ CONCENTRATIONS IN MILLIMOLES/LITER
270   REM N IS NUMBER OF DIFFERENT INITIAL WATER CONDITIONS
280   READ N
290   READ L9,P9,N9,H9
300   DATA 1,4.5,.3,0,7
310   FOR J=1 TO N
320   PRINT USING 330,L9,P9,N9,H9
330   :ALK = ##.#, TOP = ##.#, TAM = ##.#, PH = ##.#
```

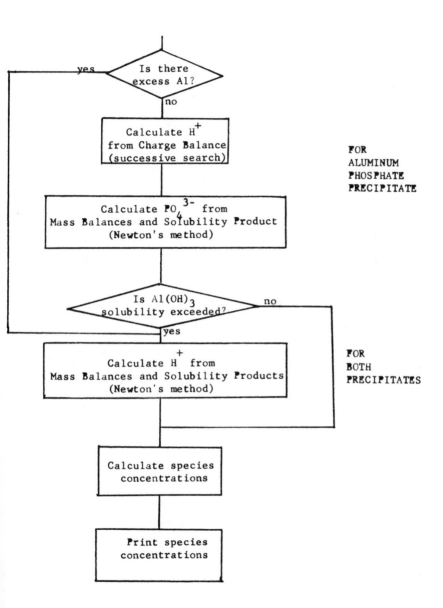

**Figure 3.A.** Specific solution procedure used for aluminum phosphate precipitation.

```
340  REM CONVERT TO MOLAR CONCENTRATIONS
350  LET L9=L9/1000
360  LET P9=P9/1000
370  LET N9=N9/1000
380  LET H1=1/10↑H9
390  REM CALCULATE HCO3 FROM INITIAL CONCENTRATIONS
400  LET P2=P9/(1+H1/K5)
410  LET N0=N9/(1+H1*K4)
420  LET C1=L9-P2-N0
430  REM ALSO COMPUTE TOTAL CARBONATE SO SUBROUTINE TO COMPUTE
440  REM ALL SPECIES CAN BE GENERALIZED
450  LET C9=C1*(1+H1/K2+K3/H1)
460  REM SUBROUTINE TO COMPUTE ALL SPECIES NEEDED IN INITIAL CHRG BAL
470  GOSUB 1500
480  REM SET A9 AND B9 ZERO THEN USE PRINT SUBROUTINE TO SHOW INITIAL
490  REM SPECIES CONCENTRATIONS
500  LET A9=0
510  LET B9=0
520  REM TAG Q =0 TO INDICATE NO PRECIPITATES FORMED
530  LET Q=0
540  GOSUB 1900
550  REM KEEP M1+ AS M8 AND TOTAL PHOSPHATE AS P7
560  LET M8=M9
570  LET P7=P9
580  LET P7=P9
590  REM ADD ALUM AND CALCULATE APPROXIMATE ACID
600  REM M IS # OF COAGULANT DOSES
610  READ M
620  DATA 7
630  FOR I=1 TO M
640  READ D2
650  REM D2 IS ALUM DOSE IN MG/L
660  DATA 50,100,150,200,250,300,400
670  PRINT USING 680,D2
680  :ALUM DOSAGE = ###.# MG/L
690  LET A9=0
700  REM CALCULATE A8 MOLAR ALUMINUM DOSE
710  LET A8=D2/300000.
720  REM FOR ACID ASSUME ALUMINUM PHOSPHATE PRECIPITATE ONLY
730  REM STOICHIOMETRY 1.4 AL : PO4 : 1.2 OH
740  LET A9=A9+A8*3.2/1.4
750  REM IS EXCESS THEN BOTH PRECIPITATES WILL FORM
760  IF A8/1.4>P7 THEN 1150
770  REM CONSIDER CASE FOR ALUMINUM PHOSPHATE PRECIPITATE ONLY
780  REM SOLUBLE P = TOTAL P - ALUMINUM/1.4
790  LET P9=P7-A8/1.4
800  LET J=0
810  LET H1=1/10↑4
820  LET R9=-1
830  LET H1=H1/10
840  REM CALCULATE H+ FROM CHARGE BALANCE BY SUCCESSIVE APPROXIMATIONS
850  GOSUB 1500
860  GOSUB 1640
870  IF S9>0 THEN 830
880  LET H1=H1*1.3
890  GOSUB 1500
900  GOSUB 1640
910  IF S9>0 THEN 880
920  LET H1=H1/1.03
930  GOSUB 1500
940  GOSUB 1640
950  IF S9>0 THEN 920
960  REM HAVE H+ FROM CHARGE BALANCE,
970  REM USE MASS BALANCE TO CALCULATE (NEWTON-RAPHSON) AL3+ AND PO4
980  GOSUB 2000
990  LET P8=E3*(H1/K6+(H1↑2)/(K5*K6))
1000 REM IF SOLUBLE PO4 IS MUCH LESS THAN ASSUMED, THEN REPEAT CYCLE
1010 IF (P7-A8/1.4)/P8>1.5 THEN 1030
1020 GOTO 1060
1030 LET P9=(P9+P8)/2
```

```
1040    GOTO 800
1050    REM COMPUTE AL+3 TO CHECK AL(OH)3 SOLUBILITY
1060    LET A7=H1↑.857/(((10↑11)*(E3↑.7143)))
1070    REM COMPUTE Q/K
1080    LET A5=A7/((10↑11.6)*(H1↑3))
1090    REM IF Q/K > 1.2 , GO TO CASE FOR BOTH SOLIDS
1100    IF A5>1.2 THEN 1170
1110    REM TAG Q = L TO INDICATE ALUMINUM PHOSPHATE PRECIPITATE ONLY
1120    LET Q=1
1130    REM GO TO FINAL CONCENTRATION COMPUTATIONS AND PRINT RESULTS
1140    GOTO 1310
1150    REM CONSIDER CASE OF ALUMINUM PHOSPHATE AND ALUMINUM HYDROXIDE
1160    REM TAG Q = 2 TO INDICATE BOTH PRECIPITATES
1170    LET Q=2
1180    LET H1=.001
1190    LET R9=-1
1200    LET H1=H1/10
1210    REM CALCULATE H+ FROM CHARGE BALANCE WITH PO4 AND AL3+
1220    REM FROM SOLUBILITY RELATIONS USING SUCCESSIVE APPROXIMATIONS
1230    GOSUB 2300
1240    IF S9>0 THEN 1200
1250    LET H1=H1*1.3
1260    GOSUB 2300
1270    IF S9>0 THEN 1250
1280    LET H1=H1/1.03
1290    GOSUB 2300
1300    IF S9>0 THEN 1280
1310    REM CALCULATE FINAL PH AND ALKALINITY
1320    LET P9=E3*(H1/K6+H1↑2/(K5*K6))
1330    LET H9=-LOG(H1)/2.303
1340    GOSUB 1500
1350    REM L9 IS FINAL ALKALINITY
1360    LET L9=C1+2*C2+NO+P2+2*P3+HO-H1
1370    GOSUB 1900
1380    NEXT I
1390    NEXT J
1400    STOP


1500    REM SUBROUTINE TO COMPUTE SPECIES FOR KNOWN COMPONENT
1510    REM CONCENTRATIONS AND PH. ALSO COMPUTE MI+ FROM CHARGE BALANCE
1520    LET P2=P9/(1+H1/K5+K6/H1)
1530    LET P1=H1*P2/K5
1540    LET P3=K6*P2/H1
1550    LET C1=C9/(1+H1/K2+K3/H1)
1560    LET CO=H1*C1/K2
1570    LET C2=K3*C1/H1
1580    LET NO=N9/(1+H1*K4)
1590    LET N1=NO*H1*K4
1600    LET HO=K1/H1
1610    REM USE CHARGE BALANCE TO CALCULATE MI+(M9)


SUNDAY

1620    LET M9=HO-H1+C1+2*C2-N1+P1+2*P2+3*P3
1630    RETURN
1640    REM SUBROUTINE TO COMPUTE CHARGE BALANCE FOR ONE PRECIPITATE
1650    LET R8=R9
1660    LET R9=A9+HO+C1+2*C2+P1+2*P2-H1-M8-N1
1670    LET J=J+1
1680    IF J>100 THEN 1700
1690    LET S9=R8*R9
1700    RETURN
```

```
1800 REM SUBROUTINE TO COMPUTE CHARGE BALANCE FOR NO PRECIPITATES
1810 LET R8=R9
1820 LET R9=A9+M9-B9-M8
1830 REM CHECK TO AVOID NON-CONVERGENCE
1840 LET J=J+1
1850 IF J>100 THEN 1370
1860 LET S9=R8*R9
1870 RETURN


1900 REM SUBROUTINE TO PRINT CONCENTRATIONS OF SPECIES AND COMBINATIONS
1910 PRINT "STATUS OF H+,CO3,NH3,PO4,ACID AND ALK"
1920 PRINT H0,H1,H9,Q
1930 PRINT C0,C1,C2,C9
1940 PRINT N0,N1,N9
1950 PRINT P1,P2,P3,P9
1960 PRINT A9,B9,L9,M9
1970 RETURN
1980 REM
1990 REM
2000 REM SUBROUTINE TO FIND PO4 FROM MASS BALANCES FOR ALUMINUM
2010 REM PHOSPHATE SOLID AND KNOWN H+ USING NEWTON-RAPHSON METHOD
2020 LET J=0
2030 LET E3=1/10↑9
2040 LET G1=A8-1.4*P7
2050 LET G2=-(H1↑.857)*(1+1/(H1*(10↑5))+1/((H1↑4)*(10↑23.5)))
2060 LET G3=G2/((10↑11)*(E3↑.7143))
2070 LET G4=1.4*E3*(H1/K6+H1↑2/(K5*K6))
2080 LET G5=G1+G3+G4
2090 LET D1=-.7143*G3/E3
2100 LET D2=G4/E3
2110 LET D3=D1+D2
2120 REM MODIFY NEWTON APPROX BY 1/2 TO CONVERGE MORE SLOWLY
2130 LET E4=E3-G5/(2*D3)
2140 REM IF PO4 IS COMPUTED < 0, METHOD HAS FAILED
2150 IF E4<0 THEN 2240
2160 REM GET SUCCESSIVE APPROXIMATIONS WITHIN FACTOR OF 1.02
2170 IF E4/E3<.98 THEN 2200
2180 IF E4/E3>1.02 THEN 2200
2190 GOTO 2240
2200 LET E3=E4
2210 LET J=J+1
2220 IF J>100 THEN 2240
2230 GOTO 2040
2240 RETURN
2250 REM


2300 REM SUBROUTINE TO CALCULATE REMAINDER FOR CHARGE BALANCE
2310 REM FOR BOTH PRECIPITATES IN ORDER TO FIND H+
2320 LET J=0
2330 LET R8=R9
2340 LET E3=1/((10↑31.64)*(H1↑3))
2350 LET L1=A9+K1/H1-N9*H1*K4/(1+H1*K4)-H1-M8
2360 LET L2=C9*(1+2*K3/H1)/(1+H1/K2+K3/H1)
2370 LET L3=(2*H1/K6+H1↑2/(K5*K6))*E3
2380 LET L4=-(10↑11.6)*(3+2/(H1*(10↑5))-1/((H1↑4)*(10↑23.5)))
2390 LET L5=L4*(H1↑3)
2400 LET R9=L1+L2+L3+L5
2410 LET J=J+1
2420 IF J>100 THEN 2440
2430 LET S9=R8*R9
2440 RETURN
2450 END
```

```
RUN
SUNDAY  1240

ALK =   4.5, TOP =   .3, TAM =   .0, PH =   7.0
STATUS  OF  H+,CO3,NH3,PO4,ACID AND ALK
 1.00000E-07     1.00000E-07     7                0
 9.81444E-04     4.38394E-03     2.05052E-06      5.36744E-03
 0               0               0
 1.83941E-04     1.16059E-04     7.32278E-10      .0003
 0               0               .0045            4.80410E-03
ALUM DOSAGE =   50.0 MG/L
STATUS  OF  H+,CO3,NH3,PO4,ACID AND ALK
 7.92308E-08     1.26214E-07     6.89765          1
 1.18215E-03     4.18374E-03     1.55045E-06      5.36744E-03
 0               0               0
 1.25451E-04     6.27143E-05     3.13516E-10      1.88165E-04
 3.80952E-04     0               4.24951E-03      4.43767E-03
ALUM DOSAGE =   100.0 MG/L
STATUS  OF  H+,CO3,NH3,PO4,ACID AND ALK
 6.46584E-08     1.54659E-07     6.8094           1
 1.38014E-03     3.98609E-03     1.20551E-06      5.36744E-03
 0               0               0
 4.57476E-05     1.86635E-05     7.61406E-11      6.44112E-05
 7.61905E-04     0               4.00708E-03      4.07149E-03
ALUM DOSAGE =   150.0 MG/L
STATUS  OF  H+,CO3,NH3,PO4,ACID AND ALK
 4.82886E-08     2.07088E-07     6.68264          2
 1.69992E-03     3.66668E-03     8.28164E-07      5.36744E-03
 0               0               0
 2.77873E-06     8.46623E-07     2.57949E-12      3.62535E-06
 1.14286E-03     0               3.66903E-03      3.67265E-03
ALUM DOSAGE =   200.0 MG/L
STATUS  OF  H+,CO3,NH3,PO4,ACID AND ALK
 3.51413E-08     2.84565E-07     6.54464          2
 2.08853E-03     3.27837E-03     5.38859E-07      5.36744E-03
 0               0               0
 2.02218E-06     4.48371E-07     9.94157E-13      2.47055E-06
 1.52381E-03     0               3.27965E-03      3.28212E-03
ALUM DOSAGE =   250.0 MG/L
STATUS  OF  H+,CO3,NH3,PO4,ACID AND ALK
 2.62445E-08     3.81033E-07     6.41788          2
 2.47070E-03     2.89638E-03     3.55543E-07      5.36744E-03
 0               0               0
 1.51022E-06     2.50079E-07     4.14108E-13      1.76030E-06
 1.90476E-03     0               2.89699E-03      2.89875E-03
ALUM DOSAGE =   300.0 MG/L
STATUS  OF  H+,CO3,NH3,PO4,ACID AND ALK
 2.01881E-08     4.95342E-07     6.30396          2
 2.82222E-03     2.54498E-03     2.40313E-07      5.36744E-03
 0               0               0
 1.16171E-06     1.47976E-07     1.88488E-13      1.30968E-06
 2.28571E-03     0               2.54513E-03      2.54644E-03
ALUM DOSAGE =   400.0 MG/L
STATUS  OF  H+,CO3,NH3,PO4,ACID AND ALK
 1.12599E-08     8.88110E-07     6.05044          2
 3.57118E-03     1.79616E-03     9.45968E-08      5.36744E-03
 0               0               0
 6.47941E-07     4.60328E-08     3.27039E-14      6.93974E-07
 3.04762E-03     0               1.79552E-03      1.79621E-03

DONE AT 1244
```

# CHAPTER 4

# OPTIMAL PROCESS DESIGN FOR IRON(II) OXIDATION

Donald T. Lauria and Charles R. O'Melia

Department of Environmental Sciences and Engineering
University of North Carolina
Chapel Hill, North Carolina 27514

## INTRODUCTION

A process consists of one or more activities that accomplish a specific purpose. Definition of the purpose does not usually determine the process to be employed. The removal of organic material from sewage, for example, can be accomplished by plain sedimentation, biological filtration, sedimentation preceded by chemical coagulation, or a combination of these and other activities. The first step in process design, therefore, is process selection, which consists of specifying the appropriate activities in proper sequence. While in many cases process selection is straightforward, this is not always true, particularly when the wastewater to be treated has unique characteristics.

Once the process is determined, its overall performance must be identified. This can be done most completely and usefully by developing a mathematical equation or model that defines the quality of water or wastewater effluent from the process as a function of influent quality, process parameters, and process decision variables over which the designer has control. Such a model is called a technological function and its development constitutes one of the most important but difficult phases of process design. It is so difficult, in fact, that the technological function is often only roughly approximated and, in many cases, development of a mathematical statement of process performance is simply not achieved.

If the technological function is known, process design is not difficult. Given a target efficiency or target effluent quality for the process, the designer can easily select levels of the decision variables that will achieve this. An example will illustrate. Suppose laboratory or other studies have shown that the efficiency of a particular process depends on only two variables: x, which is the size of the tank in which the process is carried out, and y, which is the amount of a required chemical. Assume that the technological function has been determined to be $E = xy/(1 + xy)$, where E is efficiency expressed as a decimal. If we select a target efficiency of 0.9 (*i.e.,* 90%) and decide x should be 1, y would have to be 9. Clearly, there is an infinite number of designs that will produce 90% efficiency.

While many combinations of the decision variables satisfy the process efficiency requirement, some combinations are better than others from the economic standpoint. For example, the x, y combinations (20, .45) might cost 100 while the (1, 9) combination might only cost 70. *Optimal* design therefore implies selecting the decision variables to minimize the costs associated with them. Designers, of course, take this into account by only proposing values for the decision variables that are thought to be "economical." Such values are usually associated with design standards. After a few trial proposals and calculation of associated costs, the best combination is selected.

Instead of relying on judgment or standards for selection of "economical" decision variables, an optimization model can be developed and solved using calculus or some other mathematical technique to determine the best design. This is the approach taken herein. Optimization models include two components: (a) the technological function, and (b) a mathematical expression of total cost as a function of the decision variables. The latter is called the objective function; it represents the desires of the designer to be "economical" while the technological function represents his desire to be "efficient."

In this chapter, the process of concern is oxidation of Fe(II) to Fe(III) by oxygen. The rate of oxidation of ferrous iron by oxygen is dependent on pH. With increasing pH, the time required to achieve a given oxidation efficiency is reduced. pH adjustment can be achieved by neutralization using lime, by aeration to remove carbon dioxide, and by a combination of these processes. Therefore, the design variables for this problem include the volume of the reactor, lime dosage and aerator height, assuming use of a tray type aerator.

Three different cases are analyzed. In Case 1 it is assumed that only lime addition is available for pH adjustment. The appropriate technological and objective functions are developed and the resulting optimization model is solved by the calculus to identify the best design for given values of economic and process parameters. A sensitivity analysis is then made to determine the variation in optimal design for changes in the parameters.

In Case 2 it is assumed that only aeration is available for pH adjustment. Technological and objective functions are developed as before followed by identification of the optimal design and sensitivity analysis.

In Case 3 lime for neutralization and aeration for $CO_2$ stripping are assumed to be simultaneously available. The optimization model is developed, solved and analyzed.

The objectives herein are twofold: first, presentation of a methodology for the optimal design of iron oxidation processes based on pertinent chemical and economic concepts, and second, identification of optimal designs for specific values of the chemical and economic parameters. Underlying the presentation is the hope that designers will see some advantages in the approach suggested for process design making.

## CASE 1–NEUTRALIZATION PLUS DETENTION

### Technological Function

The stoichiometry of the oxidation of Fe(II) to Fe(III) may be represented as follows:[1]

$$Fe^{2+} + \tfrac{1}{4} O_2 + 2\tfrac{1}{2} H_2O = Fe(OH)_3(s) + 2H^+ \tag{1}$$

Equation 1 indicates that the oxidation of 1 mg/l of Fe(II) will consume 0.14 mg/l of dissolved oxygen; for most real systems, therefore, the consumption of oxygen will be negligible.

The kinetics of the oxygenation of Fe(II) by oxygen in bicarbonate solu-solutions have been described by Stumm and Lee[1] as follows:

$$\frac{d[Fe(II)]}{dt} = -k\,[P_{O_2}]\,[OH^-]^2\,[Fe(II)] \tag{2}$$

where k is a constant, $[P_{O_2}]$ is the partial pressure of oxygen in the gas phase at equilibrium with dissolved oxygen in the water to be treated, and $[OH^-]$ is the concentration of hydroxide ions. $k = 8 \times 10^{13}$ atm$^{-1}$ mole$^{-2}$ liter$^2$ min$^{-1}$ at 20°C; $[P_{O_2}] = 0.21$ atm for water at equilibrium with air; and $[OH^-]$ has units (mol/l).     Equation 2 indicates that the oxidation of ferrous iron is a fourth-order reaction, but first-order with respect to the concentration of ferrous iron. It can therefore be rewritten as a pseudo first-order reaction:

$$\frac{d[Fe(II)]}{dt} = -k'\,[Fe(II)] \tag{3}$$

where

$$k' = k[P_{O_2}]\,[OH^-]^2 \tag{4}$$

Use of Equation 3 for designing a reactor implies that pH and dissolved oxygen are constant throughout the reactor contents. Use of Equation 2 implies that reaction is not catalyzed or retarded by substances present in natural waters. In some cases this can be significant, since there is evidence that $Cu^{2+}$ accelerates the reaction while certain soluble organic substances can slow the reaction rate.

Now we must relate the reaction kinetics of Equation 3 to performance of a continuous-flow water treatment facility. The oxidation efficiency of a well-mixed continuous flow reactor within which a first-order reaction occurs, is as follows:[2,3]

$$E = \frac{k't}{1 + k't} \tag{5}$$

where E is oxidation efficiency expressed as a decimal and t is tank detention time in minutes. By substituting Equation 4 into Equation 5, we obtain an expression for efficiency in terms of t and [OH⁻]. While t is one of our decision variables, [OH⁻] is not. Hence, it is necessary to describe [OH⁻] as a function of lime dosage and substitute the result in Equation 4.

Let m be the initial alkalinity of the raw water to be treated (equivalents per liter) and j be the initial concentration of carbon dioxide (mol/l). For pH levels below 8.3, [OH⁻] is related to B, the dosage of lime; m, the initial alkalinity; and j, the initial $CO_2$ concentration by the following equation:

$$[OH] = 2 \times 10^{-8} \left[ \frac{m + 2B}{j - 2B} \right] \tag{6}$$

This equation is based on the conceptual definition of alkalinity, m = $[HCO_3^-]$ + $2[CO_3^{2-}]$ + $[OH^-]$ - $[H^+]$.[4] Values of $10^{-14}$ and $5 \times 10^{-7}$ are used for the hydrolysis product of water and the first acidity constant of carbonic acid, respectively. It is also assumed that calcium or ferrous carbonate precipitates do not form.

We can now substitute Equation 6 into Equation 4 and the result into Equation 5 to obtain an expression of efficiency as a function of decision variables t and B, detention time and lime dosage:

$$E = \frac{(6.72 \times 10^{-3})t[(m + 2B)/(j - 2B)]^2}{1 + (6.72 \times 10^{-3}) \, t[(m + 2B)/(j - 2B)]^2} \tag{7}$$

Equation 7 assumes that the acidity of the      which is produced in the oxidation reaction has a negligible effect on the alkalinity of the water, that the water is saturated with oxygen ($P_{O_2}$ = 0.21 atm), and that the temperature is 20°C (k = $8 \times 10^{13}$ $atm^{-1}$ $mol^{-2}$ $liter^2$ $min^{-1}$). In Equation 7, initial alkalinity (m) and carbon dioxide concentration (j) are chemical characteristics of the water source to be treated. They are variables outside the control of the designer and are called parameters.

## Objective Function

The objective function to be minimized is total present-value cost, which is the sum of reactor cost and the present-value cost of lime for neutralization. Reactor cost is assumed to vary linearly with size, in which case it is simply equal to the product of design flow, unit cost (*i.e.,* price) and detention time. Assuming a design flow of 1 mgd and a unit cost of S dollars per gallon, the cost of the reactor is 695 St.

The present-value cost of lime is in principle obtained by developing expressions for the annual costs of lime, discounting each annual cost to present value and summing over the economic life of the facility. In practice, the present-value cost is the product of design flow, lime dosage, lime price and a present-value factor. To facilitate computation, continuous discounting is assumed in which case the present-value factor is $(1 - e^{-rn})/r$, where r is the discount rate and n is the economic life or time horizon. For a design flow of 1 mgd and lime price of P dollars per pound, the present-value cost of lime is $1.71 \times 10^8$ PB $(1 - e^{-rn})/r$. Reactor and lime costs may now be summed to form the total present-value cost function:

$$695 \text{ St} + 1.71 \times 10^8 \text{ P} \left[ \frac{1 - e^{-rn}}{r} \right] \text{B} \qquad (8)$$

This expression of the design objective includes the decision variables B and t, and in addition includes four economic parameters outside the designer's control: S, the unit tank cost; P, the price of lime; r, the discount rate; and n, the economic life of the facility. The cost of chemical facilities has been ignored. Assuming this cost is independent of dosage, it can be handled as a fixed charge which does not affect optimal process design. Alternatively, it can be included in the price of lime.

## Optimization Model

The design problem is to minimize the objective function, Equation 8, while satisfying the technological function, Equation 7. Note that the latter is a perfectly general expression of process efficiency. Let us specify a target level of efficiency; say E = 0.9. Now the design problem is to find the optimal amounts of detention time (t*) and lime dosage (B*) that minimize total present-value cost while achieving an iron oxidation efficiency of 90%.

Mathematically, there are two approaches to solution. One is to solve Equation 7 for one of the decision variables and substitute into Equation 8. In this case, B can be solved in terms of t; substitution into Equation 8 results in a single expression in terms of a single unknown and parameters. The value of t* for which the expression is minimal is found by setting the derivative with respect to t equal to zero and solving. The optimal value of B* is then obtained by back-substituting t* into Equation 7.

Alternatively, the optimization problem can be solved by Lagrangian analysis. The approach is to (1) expression Equation 7 as an implicit function by subtracting 0.9 (the value of E) from both sides of the equation; (2) multiply this implicit function by an undetermined Lagrangian multiplier, $\lambda$; (3) subtract this expression from Equation 8 to obtain a "Lagrangian" function in terms of B, t, $\lambda$ and parameters; and (4) solve the partial derivatives with respect to B, t and $\lambda$ set equal to zero for the unknowns. An example of the application of Lagrangian analysis to a problem of this type is presented elsewhere.[3]

While either approach is valid, each has its own merits. The substitution method reduces the problem to one in which there is a single unknown and, consequently, need for only a single derivative. Clearly, mathematical manipulation is simplified. Lagrangian analysis, on the other hand, introduces an additional unknown, $\lambda$, which complicates the mathematics. Lambda, however, has important economic implications; its value reflects the marginal present-value cost of meeting the efficiency constraint of 90%.

In our problem, the substitution method was used simply to facilitate computation. The resulting optimality expressions are

$$(t^*)^{3/2} + 73.4(t^*) + 1345(t^*)^{1/2} - 2.26 \times 10^6 \left(\frac{P}{S}\right)\left(\frac{1 - e^{-rn}}{r}\right)(m + j) = 0 \qquad (9)$$

$$B^* = \frac{36.9j - m\, t^{*\frac{1}{2}}}{2t^{*\frac{1}{2}} + 73.4} \qquad (10)$$

It should be noted that although Equation 9 is a function of only a single decision variable, it is difficult to solve for t* explicitly. Solution, however, can readily be obtained by numerical methods; in this case Newton's method was used. The value for t* is then substituted into Equation 10 to determine B*. It should be noted that Equation 10 is identical to Equation 7, the technological function, with 0.9 substituted for E.

A final observation on formulation of the design problem is needed. The approach here has been to minimize total present-value cost subject to meeting an efficiency constraint. Alternatively, if the annual amount of available funds were limited, the design problem would be to maximize the efficiency of removal, subject to meeting a budgetary constraint. The technological function in this case would become the objective function and the (equivalent annual) cost expression would be the constraint. A third alternative would be to minimize total present-value cost subject to meeting a specified target concentration of Fe(II) in the reactor effluent, say 0.1 mg/l. The approach to each of these alternative design problems is similar to that presented above.

### Results

The effects of economic life or time horizon (n) and the ratio of capital to operating prices (S/P) on the optimal design are presented in Figure 4.1. These results are obtained by solving Equations 9 and 10 for $4 = 5\%$, $m = 10^{-3}$ eq/l (50 mg/l as $CaCO_3$), and initial pH = 6.3 ($j = 10^{-3}$ mol/l or 44 mg $CO_2/l$). The effects of the discount rate (r) and (S/P) are presented in Figure 4.2. Here n is assumed to be 20 years, with m and j remaining at $10^{-3}$ eq/l and $10^{-3}$ mol/l, respectively.

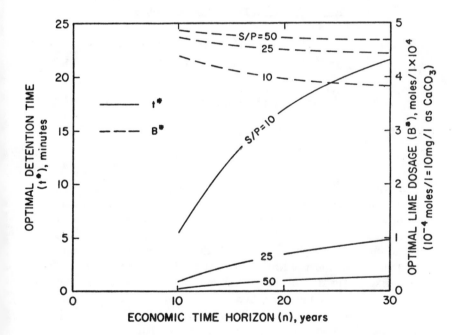

**Figure 4.1.** Effect of economic time horizon on optimal design using neutralization. Alkalinity = $10^{-3}$ eq/l, initial pH = 6.3, r = 5%, oxidation efficiency = 90%.

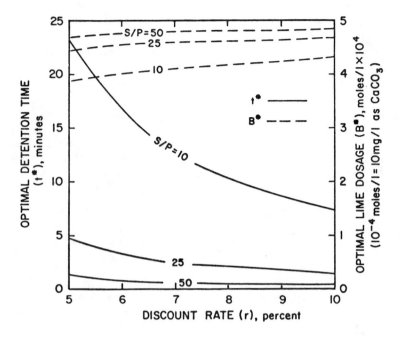

**Figure 4.2.** Effect of discount rate (r) on optimal design using neutralization. Alkalinity = $10^{-3}$ eq/l, initial pH = 6.3, n = 20 yrs, oxidation efficiency = 90%.

The effects of initial alkalinity on optimal design are presented in Figure 4.3. Optimal detention time (t*), optimal lime addition (B*), and minimum total present-value cost are plotted as functions of initial alkalinity (m). Results are presented for two pH levels corresponding to two initial $CO_2$ concentrations. Given the variables pH, alkalinity, and $CO_2$ concentration, only, two may be independent in any water. When two are known, the third is fixed. The data in Figures 4.3 and 4.4 are expressed in terms of m and pH, rather than m and j, because pH is measured more frequently. In these calculations, r = 5%, n = 20 years, S = \$0.20/gal, and P = \$0.02/lb (S/P = 10). These and similar calculations are alternatively presented in Figure 4.4 where the optimal design is plotted as a function of initial pH (or initial j) for an initial alkalinity of $10^{-2}$ eq/l (500 mg/l as $CaCO_3$); the economic parameters are as before.

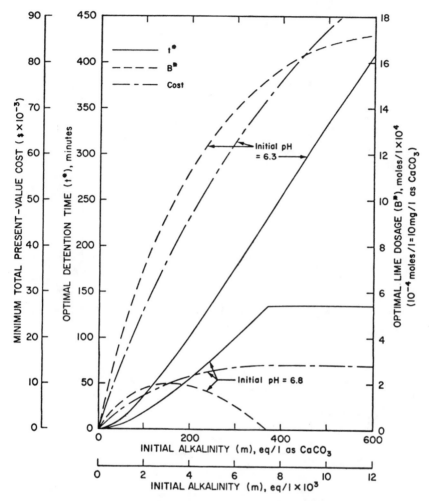

**Figure 4.3.** Effect of initial alkalinity on optimal design using neutralization. r = 50%, n = 20 yrs, S = $0.20/gal, p = $0.02/lb, oxidation efficiency = 90%.

## Discussion: Economic Parameters

Notice in Equation 9 that parameters increasing the magnitude of the last term on the left hand side will require an increase in the optimal detention time. Hence, increases in lime costs (P) and the economic life or time horizon of the facility (n) will require a larger reactor. Similarly, an increase in tank cost (S) or the discount rate (r) will result in a smaller

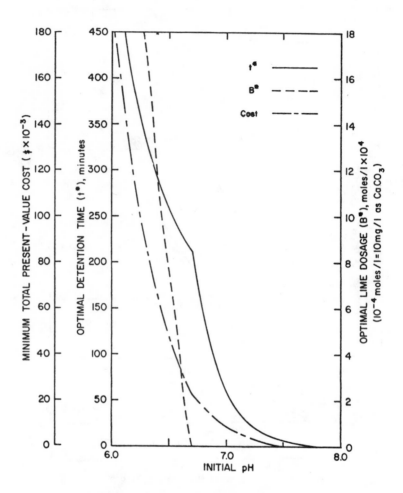

**Figure 4.4.** Effect of initial pH on optimal design using neutralization. r = 5%, n = 20 yrs, S = $0.20/gal, p = $0.02/lb, oxidation efficiency = 90%. Initial alkalinity = $10^{-2}$ eq/l (500 mg/l as CaCO₃).

facility. These effects are shown more clearly in Figures 4.1 and 4.2 from which the following observations can be made.

1. The particular values of S and P are not as important as their ratio. Hence, for given values of the discount rate and the economic time horizon, identical designs are obtained with S = $0.50/gal and P = $0.05/lb as with S = $0.10/gal and P = $0.01/lb. While the designs are identical, their total costs will differ considerably.

2. Increasing values of S/P indicate that tankage (*i.e.,* capital) is relatively more expensive than chemicals (*i.e.,* operation). As S/P increases, the optimal reactor size decreases and the required chemical dosage increases.

3. As r increases, the optimal detention time decreases; correspondingly, chemical requirements increase. Similarly, optimal detention time increases and chemical dosage decreases as n is increased. These effects are most pronounced at low values of S/P.

Although these observations are based on a specific example, the following more general statements are implied for systems where capital can be substituted for operation and vice versa. Clearly, both capital and operating costs must be considered simultaneously for optimal design. In cases where capital prices are high in relation to operating prices, investments should be relatively small. This is also true where the opportunity cost of capital (r) is high. Such conditions (high value of S/P and/or high r) are common in developing countries. For situations where technology is changing rapidly so that equipment and processes become obsolete (n is small), operation should be substituted for capital investment. A similar situation arises when equipment service life is short.

Design standards that specify a definite resourse use (*e.g.,* a specific detention time) imply a specific relationship among economic parameters, particularly the ratio of capital to operating costs. This implied relationship may be quite different than the actual S/P ratio and hence may not provide for desired environmental quality at minimum cost. Even if standards were in perfect agreement with the price levels, discount rate and economic time horizon of a particular country, they may be far from optimal in another country with a different economy. Standards for an industrial state are in general not appropriate for an agrarian society.

### Discussion: Chemical Parameters

Referring again to Equation 9 we see that high values of m, the initial alkalinity or j, the initial $CO_2$ content of the water, cause optimal detention time to increase. These effects and the corresponding changes in optimal lime dosage are shown in Figures 4.3 and 4.4 from which the following observations can be made:

1. Comparing waters with the same initial pH but with different alkalinities, it is evident that waters with high initial alkalinities require longer detention times and higher costs than those with lower alkalinities. For constant initial pH, increasing the alkalinity corresponds to increasing the $CO_2$ concentration. As the initial alkalinity increases the optimal lime dosage increases at first, then passes through a maximum, and finally decreases to zero. This decrease occurs because at high alkalinities the buffer capacity of the water is so great that costs for lime to increase pH become excessive and investment is

properly placed in larger tanks. When B* is found to equal zero, the buffer capacity of the water is so high that iron oxidation is best accomplished by reaction time alone, without pH adjustment with lime. It follows that optimal detention time and minimum cost are independent of alkalinity when the optimal lime addition is zero (*e.g.*, for initial alkalinity greater than 370 mg/l as $CaCO_3$ and initial pH = 6.8; see Figure 4.3).

2. Comparing waters with the same initial alkalinity but with different initial concentrations of hydrogen ions (pH), it is evident that high initial pH levels lead to reduced optimal detention times, lime dosages, and costs. For constant initial alkalinity, a high pH corresponds to a low $CO_2$ concentration. Stated another way, waters with high $CO_2$ (low pH) require large lime dosages, long reaction times, and high costs.

3. For every alkalinity, there exists a pH below which lime addition is economical and above which detention alone provides the least cost solution. For an alkalinity of 500 mg/l, the pH is about 6.7 (Figure 4.4). At pH levels below 6.7, carbon dioxide should be neutralized with lime, while above pH 6.7 oxidation of Fe(II) should be achieved without chemical neutralization.

Here again, design standards cannot be applied without some rather considerable excessive costs. Minimum detention times based on experience in treating waters with high alkalinity (*e.g.*, iron-bearing hypolimnetic waters in North Carolina). Design standards developed by experience with one type of water are in general not appropriate for waters with different characteristics.

The costs of neutralization for ground waters in Illinois are seen to be considerably greater than costs for hypolimnetic waters in North Carolina. In such cases, it is useful to evaluate alternate treatment procedures in an effort to reduce costs. Following is a consideration of one such procedure, aeration to provide $CO_2$ stripping.

## CASE 2–AERATION PLUS DETENTION

### Technological Function

In this model, we consider the use of aeration to raise pH by stripping carbon dioxide from solution; detention time, of course, is required to provide for oxidation of Fe(II). Lime for pH adjustment is not available. As in the previous case, the first task is to express oxidation efficiency as a function of decision variables and parameters. This implies replacing $k'$ in Equation 5 with an expression that defines the effect of aeration on $CO_2$ and in turn on pH.

Assume that aeration is accomplished in a gravity-type tray aerator. The relationship between j, the initial $CO_2$ concentration (mol/l) and p, the

$CO_2$ remaining after aeration (mol/l)  is as follows:[5]

$$\frac{C_s - P}{C_s - j} = e^{-k_La(2\,h/g)^{1/2}} \tag{11}$$

Here $C_s$ is the saturation concentration of $CO_2$ in water at equilibrium with air, $k_La$ is the overall mass transfer coefficient for aeration, g is gravity acceleration, and h is the head provided for the $CO_2$ stripping process. $C_s = 1.1 \times 10^{-5}$  mol $/l^{-1}$ at $20°C$, $g = 32.2$ ft, $sec^{-2}$; the units of $k_La$ and h are $sec^{-1}$ and feet, respectively. The effects of multiple trays in series in the aerator are not considered. By solving Equation 11 for p, the dependence of final $CO_2$ concentration on decision variable h and process parameters j and $k_La$ is seen to be

$$p = 1.1 \times 10^{-5} - (1.1 \times 10^{-5} - j)e^{-0.249\,k_La\,h^{1/2}} \tag{12}$$

The alkalinity of a water remains unchanged when $CO_2$ is added or withdrawn. At a pH less than 8.3, the concentration of hydroxide ions in the water leaving the aerator is related to the initial alkalinity and the $CO_2$ concentration after aeration as follows:

$$[OH^-] = 2 \times 10^{-8}\,(m/p) \tag{13}$$

Note that Equation 13 is identical to Equation 6 with $B = 0$. Substituting Equation 12 into Equation 13, the $[OH^-]$ in the water leaving the aerator can be expressed as a function of (h), the head provided during aeration (our decision variable), the transfer coefficient ($k_La$), initial $CO_2$ concentration (j), and alkalinity (m).

$$[OH^-] = \frac{2 \times 10^{-8}\,m}{1.1 \times 10^{-5} - (1.1 \times 10^{-5} - j)\,e^{-0.249\,k_La\,h^{1/2}}} \tag{14}$$

It is important to note that this same $[OH^-]$ concentration will exist in the contents of the reactor and also in the reactor effluent since the unit is assumed to be well mixed. Equation 14 can be substituted into Equation 4 to obtain $k'$, the pseudo first-order rate constant for the iron oxidation reaction. The resulting equation for $k'$ must in turn be substituted into Equation 5 to obtain an expression for reactor efficiency in terms of decision variables t and h and parameters m, j and $k_La$. The final mathematical expression for E is cumbersome.

$$E = \frac{6.72 \times 10^{-3}\,t[m/(1.1 \times 10^{-5} - (1.1 \times 10^{-5} - j)\,e^{-0.249\,k_La\,h^{1/2}})]^2}{1 + 6.72 \times 10^{-3}\,t[m/(1.1 \times 10^{-5} - (1.1 \times 10^{-5} - j)\,e^{-0.249\,k_La\,h^{1/2}})]^2}$$

To simplify the mathematics, the efficiency equation will be left in terms of p, a surrogate decision variable. Once p*, the optimal value, is found, it will be necessary to use Equation 12 to determine the corresponding value of h*, the actual decision variable. The resulting technological function that expresses efficiency in terms of decision variables t and p and parameter m is

$$E = \frac{6.72 \times 10^{-3} \ t(m/p)^2}{1 + 6.72 \times 10^{-3} \ t(m/p)^2} \qquad (15)$$

## Objective Function

The objective function to be minimized is the sum of present-value reactor and aeration costs. For a 1 mgd plant, reactor cost remains unchanged from before (695 St). Present-value aeration costs include two components, the cost of the aerator and the present-value cost of power for pumping raw water to the aerator inlet. In this model, the aerator cost is assumed to be a fixed charge in which case it does not affect process design. This assumption is relaxed in the next example. The present-value power cost is the product of the design flow, the total dynamic head, the power price, R ($/kwh), the pumping efficiency, and the present-value factor. Here the present-value power cost is given as $1150 \ h \ R \ (1-e^{-rn})/r$. The design flow is 1 mgd; the total dynamic head is assumed to equal the aerator height, h (ft), and the pumping efficiency is assumed to be 100%. Substituting p, the surrogate decision variable, for h by using Equation 12 and summing the present-value reactor and power costs, we obtain the total objective function:

$$695 \ St + 18,500 \ \frac{R}{(k_La)^2} \left( \frac{1 - e^{-rn}}{r} \right) \ln^2 \left( \frac{1.1 \times 10^{-5} - j}{1.1 \times 10^{-5} - p} \right) \qquad (16)$$

This expression includes the decision variable t and surrogate decision variable p. The mass transfer coefficient $(k_La)$ is assumed here to be outside the designer's control and is considered to be a parameter together with S, R, r, n, and j.

## Optimization Model

The design problem is to minimize Equation 16 while satisfying Equation 15 with E equal to 0.9. Solution can be obtained using the substitution method or Lagrangian analysis. The substitution method has been used here. The resulting optimality conditions in terms of t* and p* are

$$0 = 695 \frac{S}{R} (1.1 \times 10^{-5})(t^*)^{\frac{1}{2}} - 0.0273m(t^*) + \frac{505\ m}{(k_La)^2} \left[ \frac{1 - e^{-rn}}{r} \right]$$

$$\left( \frac{1.1 \times 10^{-5} - j}{1.1 \times 10^{-5} - 0.0273\ m(t^*)^{\frac{1}{2}}} \right) \tag{17}$$

$$p^* = 0.0273\ m\ (t^*)^{\frac{1}{2}} \tag{18}$$

## Results

The effects of the alkalinity of the raw water on optimal design for Fe(II) oxidation are presented in Figure 4.5. Optimal detention time, optimal aerator head, and total present-value cost are plotted as functions of alkalinity. Results are presented for two initial pH levels corresponding to two initial $CO_2$ concentrations. These results are obtained for n = 20 yrs, r = 5%, S = \$0.20/gal, R = \$0.02 kwh, and $k_La$ = 2000 hr$^{-1}$. These and similar calculations are alternatively presented in Figure 4.6 where the optimal design is plotted as a function of the initial pH (initial $CO_2$ concentration) for an alkalinity of 500 mg/l as $CaCO_3$.

## Discussion

The effects of economic parameters on the trade-off between capital and operating costs when aeration and detention are used are similar to those presented previously for neutralization and detention. Based on the results presented in Figures 4.5 and 4.6, the following statements can be made regarding the effects of chemical parameters on the optimal design using aeration:

1. For raw waters with the same initial pH but with different alkalinities, the optimal design is almost independent of alkalinity (Figure 4.5). This is in marked contrast to the effects of alkalinity on optimal design when lime is used for pH adjustment (Figure 4.3). For aeration, it appears that the optimal $CO_2$ removal efficiency of the aerator [j - (p*)/j] is independent of the alkalinity of the water, at least in the range $10^{-3}$ eq/l < m < $10^{-2}$ eq/l. It follows that p* is proportional to m, and the optimal detention time is then independent of alkalinity. From Equation 18, t* is proportional to the ratio (p*/m). Since p* is in turn directly proportional to m, t* is then independent of m.

2. Waters with low pH (pH < 6.8) require long reaction times and tall aerators. For example, a water with a pH of 6.4 requires a detention time of 115 min and an aerator head of 54 ft (Figure 4.5). Waters

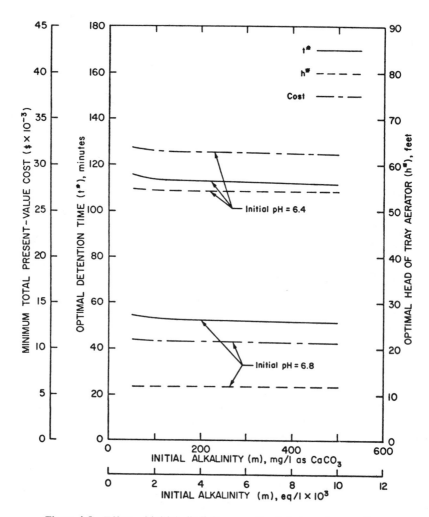

**Figure 4.5.** Effect of initial alkalinity on optimal design using aeration.
$r = 5\%$, $n = 20$ yr, $S = \$0.20/\text{gal}$, $R = \$0.02/\text{kwh}$, $k_L a = 2000$ hr,
oxidation efficiency = 90%.

with initial pH values of 6.8 or greater require facilities which are
more consistent with present practice. For pH = 6.8, $t^*$ is about
53 min and $h^*$ is 12 ft (Figure 4.5).

3. Optimal design depends upon the pH of the raw water. Detention
   time, aerator head, and total present-value costs all decrease with
   increasing pH, regardless of raw-water alkalinity (Figure 4.6).

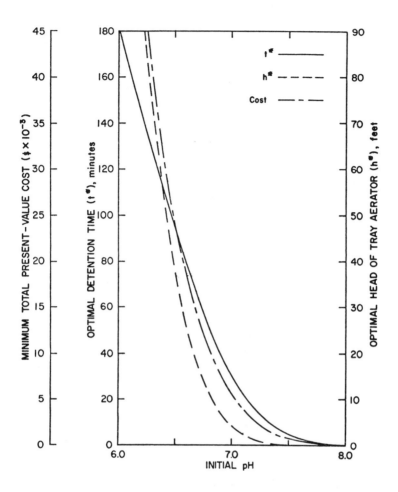

**Figure 4.6.** Effect of initial pH on optimal design using aeration.
$r = 5\%$, $n = 20$ yr, $S = \$0.20/gal$, $R = \$0.02/kwh$,   oxidation efficiency = 90%,
alkalinity = 500 mg/l as $CaCO_3$, $k_L a = 2000/hr$.

We have previously noted that optimal design depends upon economic
and chemical parameters, so that design standards are not particularly
applicable to systems with different economic or water quality conditions.
Comparison of the results of Cases 1 and 2 demonstrates that the optimal
design of one component of a treatment system depends upon other treatment units. Consider the detention basin which is common to both processes. The reactions that occur in this basin are independent of the
method of pretreatment used, but the optimal detention time to be

provided in the basin depends upon whether aeration or neutralization is used to adjust pH. Consider, for example, a water having an initial pH of 6.2 and an alkalinity of 500 mg/l as $CaCO_3$. Using lime, an optimal reaction time of 375 min is needed (Figure 4.4), while only 145 min are required if pH adjustment is accomplished by aeration (Figure 4.6).

Useful patterns emerge from a comparison of aeration and neutralization as processes for pH adjustment prior to Fe(II) oxidation. One such comparison is presented in Table 4.1. Raw water supplies are divided into four

### Table 4.1. Comparison of Neutralization and Aeration

| 1. High pH, High Alkalinity (pH = 7.4, m = 500 mg/l as $CaCO_3$) | | 2. Low pH, High Alkalinity (pH = 6.4, m = 500 mg/l as $CaCO_3$) | |
|---|---|---|---|
| **Lime** | **Air** | **Lime** | **Air** |
| $B^* = 0$ | $h^* = 0.24$ ft | $B^* = 1.1 \times 10^{-3} M$ | $h^* = 54.2$ ft |
| $t^* = 210$ min | $t^* = 7.4$ min | $t^* = 292$ min | $t^* = 111$ min |
| Cost = $30,800 | Cost = $1,100 | Cost = $65,000 | Cost = $31,300 |
| | | | |
| 3. High pH, Low Alkalinity (pH = 7.4, m = 50 mg/l as $CaCO_3$) | | 4. Low pH, Low Alkalinity (pH = 6.4, m = 50 mg/l as $CaCO_3$) | |
| **Lime** | **Air** | **Lime** | **Air** |
| $B^* = 1.2 \times 10^{-5} M$ | $h^* = 0.2$ ft | $B^* = 1.2 \times 10^{-4} M$ | $h^* = 54.6$ ft |
| $t^* = 4$ min | $t^* = 7.7$ min | $t^* = 5.7$ min | $t^* = 116$ min |
| Cost = $790 | Cost = $1,120 | Cost = $4,300 | Cost = $31,900 |

categories in Table 4.1: (1) high pH and high alkalinity, (2) low pH and high alkalinity, (3) high pH and low alkalinity, and (4) low pH and low alkalinity. Process selection is observed to depend primarily upon raw-water alkalinity. For high alkalinity waters (Cases 1 and 2 in Table 4.1, with m = 500 mg/l as $CaCO_3$), aeration plus detention is the preferred treatment process due to its lower costs. For low-alkalinity waters (Cases 3 and 4 in Table 4.1, with m = 50 mg/l as $CaCO_3$), neutralization plus detention is less expensive and so is the process of choice. Total present-value costs depend upon alkalinity and pH. Highest costs are associated with low pH, high-alkalinity waters (Case 2), while lowest costs are expended for high pH, low-alkalinity waters (Case 3).

It is plausible that some cost savings might be achieved by combining aeration and neutralization for pH adjustment prior to oxidation. This procedure is suggested by others.[5] The combined process first uses aeration for high $CO_2$ waters to take advantage of the high supersaturation of $CO_2$ as a driving force for gas transfer. Neutralization follows, with the lime dose depending upon the concentration of $CO_2$ remaining in solution after aeration. To examine this question, and also to determine more quantitative values for "high" and "low" as adjectives for initial pH and alkalinity, we examine next the case of aeration plus neutralization plus detention.

## CASE 3–AERATION PLUS NEUTRALIZATION PLUS DETENTION

### Technological Function

Again we begin by expressing oxidation efficiency as a function of decision variables and parameters. This involves replacing $k'$ in Equation 5 with an expression that describes the effects of aeration and lime addition on pH.

First we consider the effects of aeration on $CO_2$ concentration described by Equation 12. Following aeration, the water flows to chemical mixing facilities where lime (B) is added. The effect of lime is to raise the hydroxide ion concentration as described by Equation 6 in the first model. By combining Equations 6 and 12 an expression is obtained for the effect of aerator height (h) and lime dosage (B) on [OH⁻] in a water whose initial $CO_2$ concentration and alkalinity are j and m, respectively. Note that in this process, the alkalinity of the water does not change during the aeration step; it remains equal to m, the raw water alkalinity. During the subsequent neutralization step, the alkalinity is increased by the addition of lime to a value of (m + 2B). The $CO_2$ in the raw water decreases during aeration from j to p, and is further reduced in the neutralization step by lime addition.

$$[OH^-] = \frac{2 \times 10^{-8} \, (m + 2B)}{1.1 \times 10^{-5} - 2B - (1.1 \times 10^{-5} - j) \, e^{-0.249 \, k_L a \, h^{1/2}}} \tag{19}$$

Equation 19 describes the [OH⁻] in the water after aeration and lime addition; the contents of the reactor within which Fe(II) oxidation occurs will have the same [OH⁻]. Consequently, the pseudo first-order rate constant ($k'$, Equation 4) can be expressed as a function of B, h and parameters. Finally, $k'$ can be substituted into the expression for oxidation efficiency (Equation 5) to obtain a function in terms of decision variables and parameters.

$$E = \frac{6.72 \times 10^{-3} \, t \, \{(m + 2B)/[1.1 \times 10^{-5} - 2B - (1.1 \times 10^{-5} - j) e^{-0.249 \, k_L a \, h^{1/2}}]\}^2}{1 + 6.72 \times 10^{-3} \, t \, \{(m + 2B)/[1.1 \times 10^{-5} - 2B - (1.1 \times 10^{-5} - j) e^{-0.249 \, k_L a \, h^{1/2}}]\}^2}$$

For mathematical simplicity, E is left in terms of the surrogate variable p as in Case 2:

$$E = \frac{6.72 \times 10^{-3} \, t \, [(m + 2B)/(p - 2B)]^2}{1 + 6.72 \times 10^{-3} \, t \, [(m + 2B)/(p - 2B)]^2} \tag{20}$$

## Objective Function

The objective function to be minimized is the sum of the present-value costs of the reactor, the aerator, power for aeration, and lime. Previously developed present-value costs functions are assumed for all but the aerator. This facility is simply assumed to cost Q dollars per foot of height. The resulting objective function (in terms of the surrogate variable p) is

$$\underbrace{695 \, St}_{\text{(reactor)}} + \underbrace{\frac{16.1 \, Q}{(k_L a)^2} \ln^2\left[\frac{1.1 \times 10^{-5} - j}{1.1 \times 10^{-5} - p}\right]}_{\text{(aerator)}} + \underbrace{\frac{18{,}500 \, R}{(k_L a)^2}\left[\frac{1 - e^{-rn}}{r}\right] \ln^2\left[\frac{1.1 \times 10^{-5} - j}{1.1 \times 10^{-5} - p}\right]}_{\text{(power for aeration)}}$$

$$+ \underbrace{1.72 \times 10^8 \, p \left[\frac{1 - e^{-rn}}{r}\right] B}_{\text{(lime)}} \tag{21}$$

## Optimization Model

The design problem considered here is to minimize total present-value cost (Equation 2) while satisfying a performance constraint (Equation 20 with E = 0.9). Using the substitution method, Equation 20 is solved for t in terms of B, p and parameters. Substitution into Equation 21 assumes that the efficiency constraint of 90% is met and additionally results in a function of only two variables to be minimized. This minimization is accomplished by setting the partial derivatives with respect to B and p equal to zero. The following are obtained:

$$1.87 \times 10^6 \, S \frac{(p^* - 2B^*)}{(m + 2B^*)^2} + \left\{\frac{32.2 \, Q + 37{,}000 \, R(1 - e^{-rn})/r}{(k_L a)^2 \, (1.1 \times 10^{-5} - p^*)}\right\} \ln\left\{\frac{1.1 \times 10^{-5} - j}{1.1 \times 10^{-5} - p^*}\right\} = 0 \tag{22}$$

$$172 \, P\left(\frac{1 - e^{-rn}}{r}\right) - 3.72\left(\frac{(p^* - 2B^*)(m + p^*)}{(m + 2B^*)^3}\right) = 0 \tag{23}$$

Equations 22 and 23 are solved simultaneously using Newton's method. With p* and B* known, t* is obtained by rearranging Equation 20 and solving.

$$t^* = 1.34 \times 10^3 \ [(p^* - 2B^*)/(m + 2B^*)]^2 \tag{24}$$

Finally, h* is calculated using Equation 12.

## Results

The effects of initial alkalinity on optimal design for waters having an initial pH of 6.0 are presented in Figure 4.7. Optimal detention time (t*),

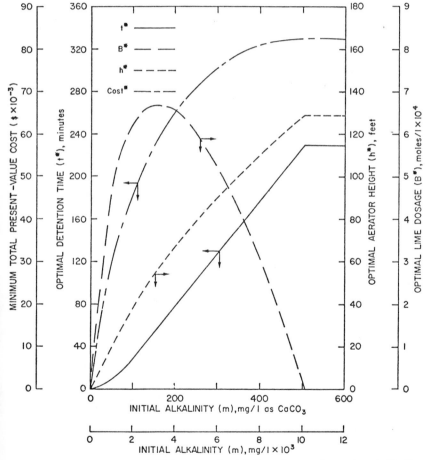

**Figure 4.7.** Effect of initial alkalinity on optimal design using aeration and neutralization. $r = 5\%$, $n = 20$ yr, $S = \$0.20/gal$, $R = \$0.02/kwh$, $P = \$0.02/lb$, $Q = \$100/ft$, oxidation efficiency = 90%, $k_La = 2000/hr$, initial pH = 6.0.

optimal lime dosage (B*), optimal aerator height (h*), and minimum total present-value cost are plotted as functions of the alkalinity of the untreated water (m). These and similar results are presented in another way in Figure 4.8 where optimal design is plotted as a function of initial pH for waters having an initial alkalinity of 2 x $10^{-3}$ eq/l (100 mg/l as $CaCO_3$).

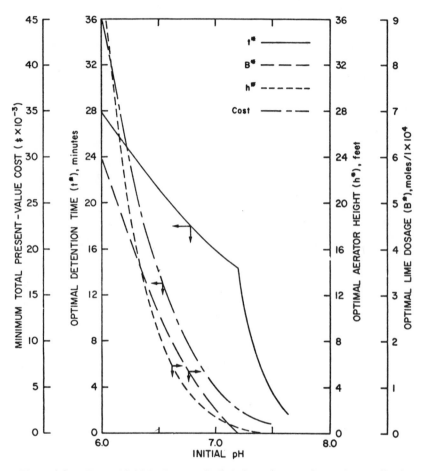

**Figure 4.8.** Effect of initial pH on optimal design using aeration and neutralization. r = 5%, n = 20 yr, S = $0.20/gal, R = $0.02/kwh, P = $0.02/lb, Q = $100/ft, oxidation efficiency = 90%, $k_La$ = 2000/hr, initial alkalinity = 2 x $10^{-3}$ eq/l (100 mg/l as $CaCO_3$)

Optimal values of lime dosage (B*) are presented as a function of initial alkalinity and pH in Figure 4.9. Optimal aerator height (h*) and detention time (t*) are presented as functions of these parameters in Figures 4.10 and

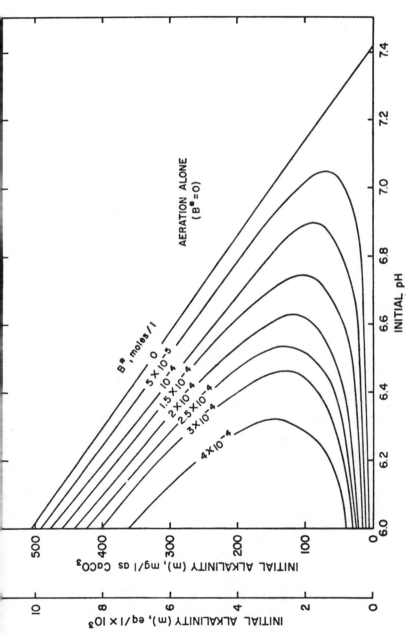

**Figure 4.9.** Isoquants of optimal lime dosage as a function of initial alkalinity and initial pH, using aeration and neutralization. $r = 5\%$, $n = 20$ yr, $S = \$0.20/\text{gal}$, $P = \$0.02/\text{lb}$, $R = \$0.02/\text{kwh}$, $Q = \$100\text{ft}$, $k_{L}a = 2000/\text{hr}$, oxidation efficiency $= 90\%$.

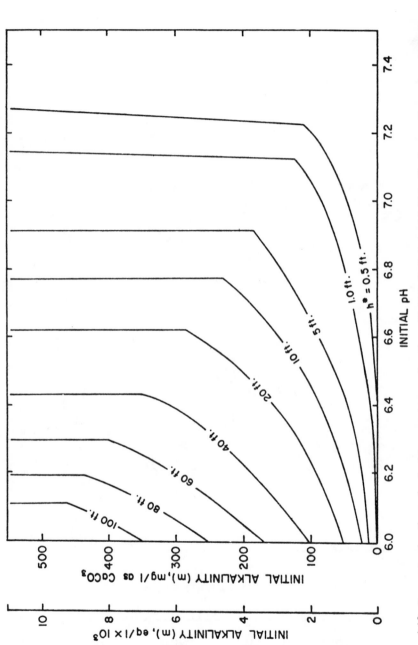

Figure 4.10. Isoquants of optimum head of tray aerator as a function of initial alkalinity and initial pH using aeration and neutralization.

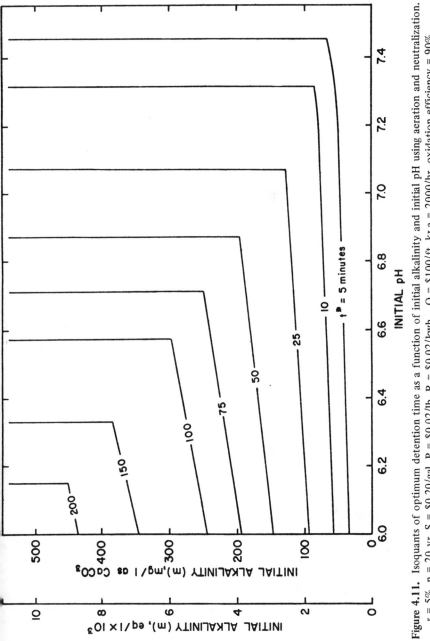

**Figure 4.11.** Isoquants of optimum detention time as a function of initial alkalinity and initial pH using aeration and neutralization. r = 5%, n = 20 yr, S = $0.20/gal, P = $0.02/lb, R = $0.02/kwh, Q = $100/ft, $k_La$ = 2000/hr, oxidation efficiency = 90%.

4.11, respectively. The results presented in Figures 4.7 to 4.11 are based on r = 5%, n = 20 yr, S = $0.02/lb, R = $0.02/kwh, Q = $100/ft, $k_La$ = 2000/hr, and E = 0.9.

**Discussion**

The results presented in Figures 4.7 and 4.8 permit statements about the effects of chemical parameters on this process which are similar to those presented earlier for Case 1, in which neutralization and detention were considered. In summary, waters with high initial alkalinities require larger detention times, aerator heights, and costs than those with low alkalinities (Figure 4.7); waters with low pH require greater detention times, lime dosages, aerator heights, and costs than those with higher pH levels (Figure 4.8); and for every alkalinity, there exists a pH below which lime addition is economical and above which aeration plus detention provides the least cost solution.

Figures 4.9 to 4.11 provide some useful insights and guidelines for the design of facilities for iron removal. First, it is possible to define a region of raw water quality (pH and initial alkalinity) within which aeration alone is the preferred method of pH adjustment (Figure 4.9). The boundary of this region (the line for B* = 0 in Figure 4.9) is not in any way related to a particular $CO_2$ concentration. For example, waters with a pH of 6.3 and an alkalinity of 400 mg/l as $CaCO_3$ or greater should also be treated with aeration alone. These waters have $CO_2$ concentrations of 3.5 mg/l or greater. Hence, "rules of thumb" that state that aeration should be used to reduce $CO_2$ down to some level (*e.g.*, 10 mg/l), after which neutralization should be provided, are not appropriate when iron removal is considered. In general, low pH and low alkalinity waters should be treated with neutralization plus aeration for pH adjustment; other waters might use only aeration. A quantitative measure of "low" and "high" as applied to pH and alkalinity is provided by Figure 4.9.

Second, optimal detention time is independent of initial alkalinity and strongly dependent upon pH when aeration alone should be used (Figure 4.10). In contrast, when aeration plus neutralization should be used in pretreatment, t* is strongly dependent upon initial alkalinity and almost independent of pH. Similar observations may be made about aerator height. When aeration alone is optimal, h* is independent of alkalinity and strongly dependent upon pH. When neutralization and aeration are optimal, h* does depend upon both initial and pH (Figure 4.10).

Finally, the treatment of high alkalinity-low pH waters requires large aerator heights, long reaction times, and hence considerable costs. Neutralization for pH adjustment is not economical. Under these circumstances, consideration should be given to some other treatment process, such as oxidation by chlorine or permanganate. The procedures presented in this chapter could be applied to these processes.

## REFERENCES

1. Stumm, W. and G. F. Lee. "Oxygenation of Ferrous Iron," *Ind. Eng. Chem.* **53**, 143 (1961).
2. Levenspiel, O. *Chemical Reaction Engineering.* (New York: John Wiley and Sons, 1962).
3. O'Melia, C. R. and D. T. Lauria. "Reactions, Reactors, and Sanitary Engineering Design," *Proc. 17th South. Water Res. and Poll. Cont. Conf.* (Chapel Hill: University of North Carolina, 1968), p. 264.
4. Stumm, W. and J. J. Morgan. *Aquatic Chemistry.* (New York: John Wiley and Sons, 1970).
5. Fair, G. M., J. C. Geyer, and D. A. Okun. *Water and Wastewater Engineering*, Vol. 2. (New York: John Wiley and Sons, 1968).

# CHAPTER 5

# MODELING AND SIMULATION OF CLARIFICATION AND THICKENING PROCESSES

Richard I. Dick

Department of Civil Engineering
University of Delaware
Newark, Delaware

Makram T. Suidan

Department of Civil Engineering
University of Illinois
Urbana, Illinois

## INTRODUCTION

Unwanted materials most frequently are separated from water and wastewater in the form of suspended solids. If they do not exist initially as suspended solids, chemical or physical processes commonly are used to convert them to that form to permit removal by solids separation techniques—most commonly, gravity sedimentation.

Because of this approach to water and wastewater treatment, treatment plants are plagued with the problems of treatment, reclamation, and disposal of sludges comprised of the suspended solids removed from water. The cost effectiveness of most sludge handling methods is a function of the concentration of suspended solids in the sludge. Thus, techniques are used to increase the suspended solids of the sludge—again, most commonly gravity sedimentation.

Thus, gravity sedimentation probably is the unit operation or process most frequently used both in water and wastewater, and in sludge treatment. Separation of solids from water (clarification) and concentration of the solids (thickening) in preparation for subsequent sludge handling schemes may be accomplished in a single settling basin, or separate clarifiers and thickeners may be used. However, in either case, each settling basin is expected to achieve some desired degree of solids capture (clarification) and solids consolidation (thickening) and, thus, both functions must be considered in design of any settling basin.

Because of this prevalence of the gravity sedimentation operation, firm understanding of the performance of settling tanks is essential to under-graduate and graduate training in water quality control. Many of the basic concepts can be illustrated readily through use of simulation techniques. However, some factors affecting performance of actual settling tanks have not yet been subjected to rigorous mathematical analysis and, as discussed later, the utility of the simulation is limited by these shortcomings. Still, the approach serves to illustrate the fundamental concepts involved and to point out limitations with the fundamental approach and the areas in which further work is needed. The nature of some of these deviations from basic theory is noted in the following sections and extension of the basic model presented here to include allowance for one or more of the additional effects noted would be an exemplary graduate classroom exercise and a means for interpreting experimental data from actual settling basins.

The purpose of this chapter is to review briefly the basic concepts in design and operation of sedimentation basins to accomplish clarification and thickening, to illustrate the demonstration of these concepts for teaching purposes using a simulation model and a digital computer, and to point out the limitations of the fundamental model and the need for additional work leading to mathematical description of actual process performance. A "black box" approach to computer programming will be used to demonstrate use of free-form input of data, although it is not inferred that the approach necessarily would be best for particular class exercises and other applications.

## THEORY OF SEDIMENTATION TANK PERFORMANCE

Major design considerations for both the clarification and thickening functions are the establishment of proper cross-sectional area and depth. Whereas identification of the function of sedimentation tanks, which controls sizing, might be obvious in some cases, as a general concept it is well to consider both the clarification and thickening requirements in design of any settling tank. The proper area of a settling tank is then the larger of the requirements for thickening and clarification (or better, selection of the area should be optimized on the basis of its effect on both clarification and thickening). Similarly, the total depth of the settling tank must be adequate for both thickening and clarification. The basic theoretical concepts governing thickening and clarification are briefly reviewed.

## Clarification

Of the two functions of settling basins, clarification has received
the most attention. Rigorous analysis of clarification is established on a
firm and old foundation laid by Hazen.[1] Gradually, Hazen's concepts have
been adopted (largely through the efforts of Camp[2-4]) until the influence
of his basic approach to analysis of clarification now is reflected in conventional design parameters. Because these basic clarification concepts are so
well-established and are described in many water quality control textbooks
(*e.g.*, Fair *et al.*[5]), they are reiterated here only briefly.

Hazen's basic approach was to relate the design of the clarification portion of sedimentation basins to the settling velocity of the particles to be
removed. This settling velocity is, in turn, a function of particle size and
specific gravity, and fluid density and viscosity. The effective weight in
water of a particle of diameter, d, and specific gravity, S, is $(S - 1)\rho g(\pi d^3/6)$,
where $\rho$ is the mass density of water and g is the gravity constant. After a
period of acceleration,[6] this effective weight is exactly equalled by resistance
of the water to sedimentation, and a terminal settling velocity is reached.
From Newton, the drag on a particle settling at velocity, v, is $C_D(\pi d^2/4)(\rho v^2/2)$
where $C_D$ is the coefficient of drag. Equating the net force on the settling
particle to zero yields

$$v = \left[ \frac{4dg(S - 1)}{3C_D} \right]^{1/2} \tag{1}$$

as a general expression for the terminal settling velocity of a particle in water.

Determination of particle-settling velocity is complicated by the fact that
the coefficient of drag reflects varying relative amounts of skin friction and
form drag and only has been evaluated experimentally (as a function of
Reynolds' number, $(\rho vd/\mu)$, where $\mu$ is the absolute fluid viscosity). The
relationship between Reynolds' number, $N_R$, and $C_D$ is displayed in many
textbooks (*e.g.*, Fair[5]), although mathematical expression of the experimentally determined relationship varies slightly (but usually unimportantly)
from author to author. For purposes of this chapter, the value of $C_D$ will
be taken as $24/N_R$ at values of $N_R < 1$ where viscous shear forces constitute
almost all of the drag (in which case, Equation 1 becomes the familiar
Stokes' Law). At high Reynolds' numbers (above $10^3$ for purposes of this
chapter), pressure drag constitutes virtually all of the drag, and $C_D$ assumes
a constant value (0.4 here) independent of v or $N_R$. The value of the coefficient of drag decreases in the turbulent range (at $N_R$ greater than about
$2.5 \times 10^5$), but sedimentation in this range is not normally encountered in
water quality control, and is not considered here. At intermediate Reynolds'
numbers (the "transitional range" where $1 < N_R < 10^3$), varying relative

amounts of friction and pressure drag exist[7] and $C_D$ has been taken as $\dfrac{24}{N_R} + \dfrac{3}{(N_R)^{1/2}} + 0.34$.[5] If a clarification problem begins with specification

of particle size and specific gravity, then the first solution step is to establish Reynolds' number associated with sedimentation of the particle and the particle settling velocity. Techniques for experimental evaluation of the settling velocity of real particles are described in standard texts (*e.g.*, Rich[6] and Fair, *et al.*[5]).

Calculation of the expected removal of a particle with known settling velocity in a clarifier is facilitated by removing from consideration the troublesome inlet, outlet, and thickening portions of the tank and by considering idealized hydraulic performance of that portion of the tank (Camp's "ideal settling tank") that remains. In such a tank, the horizontal fluid velocity ($v_h = Q/h_o w$, where Q is the flow rate through an ideal tank of depth $h_o$ and width w) is considered to be uniform at the inlet and outlet cross sections of the tank and at every point in between. The trajectory of each particle is considered to be comprised of this horizontal component and a vertical component equivalent to the particle settling velocity. Particles are considered to be distributed uniformly vertically and horizontally at the inlet, and to be removed if their trajectory takes them to the bottom of the ideal tank before they reach the outlet end. According to this conceptual model, the hardest particles to remove will be those that enter at the top of the tank. They will reach the bottom only if their settling velocity divided into the tank depth is equal to or less than the horizontal fluid velocity divided into the tank length, L. Thus, the settling velocity, $v_o$, of the slowest settling particle which is completely removed is

$$v_o = v_h \frac{h_o}{L} = \frac{Q}{wh_o} \frac{h_o}{L} = \frac{Q}{wL} = \frac{Q}{A} \qquad (2)$$

where A is the surface area of the ideal settling tank. Thus, the hydraulic surface loading, Q/A, is the basic parameter controlling removal of discrete particles in sedimentation basins—not retention time as might intuitively be anticipated.

In addition to complete removal of particles with $v \geqslant Q/A$, some slower settling particles which cross the inlet boundary of horizontal flow sedimentation basins at elevations sufficiently below the water surface will be able to reach the tank bottom below the inlet. If the distance, h, which slow settling particles must settle to reach the tank bottom is less than the product of their settling velocity, v, and the clarifier retention time, t, then they will be removed. Because t is $h_o/v_o$, the fraction, f, of particles settling at velocities less than $v_o$ which are removed is given by

$$f = \frac{vt}{v_0 t} = \frac{v}{v_0} \tag{3}$$

Equations 1, 2 and 3 provide the basis for calculating theoretical clarification performance of a settling tank. If the particle size distribution, flow rate, tank area, particle density, and fluid viscosity are known, the settling velocity of each particle can be calculated from Equation 1 and compared to $v_0$ from Equation 2. If the particle settles at a rate equal to or greater than $v_0$, it will be completely removed; the extent of removal of slower particles is given by Equation 3. Clearly, because of the trial-and-error solution of Equation 1 and the multiplicity of possible particle sizes, computerized solution of the calculations will permit far more effective use of student efforts to understand clarification.

Well-established means for modifying the idealized approach outlined above to take into account factors that cause actual performance to differ from predictions of the model, do not exist. Some causes for deviations include flocculation, hindered settling, nonuniform fluid velocity distribution, turbulence, scour, and density currents. Analysis of the effect of these factors are of interest, but are not included in this review of basic established clarification theory. However, efforts to modify the basic model to allow for one or more of these effects are extremely rewarding in graduate courses, and sufficient fundamental bases for formulating such effects are available to make the exercise meaningful. A starting point toward including the effect of turbulent diffusion is the basic convective diffusion equation familiar to those involved with analysis of the fate of wastes discharged into natural waters.[8-10] Of appreciable concern, inasmuch as most particles in water quality control plants are flocculent, is the influence of particle agglomeration on the validity of the basic model. Camp's ideal model would indicate that tank depth (or volume, or retention time) has no influence on sedimentation tank performance (Equation 2). However, "popular theory" has maintained that retention time was important because flocculation (with accompanying higher particle sedimentation rates) occurs as a function of time. Because of this view, clarifier design standards invariably contain some minimum hydraulic retention time. However, it must be recognized that particle collisions and, hence, flocculation are caused by differential particle settling velocities and by differential fluid velocities and, while long retention time benefits flocculation by the first mechanism, it might impair flocculation because of the second mechanism. This is due to the reduced fluid velocity gradients and turbulent diffusion in deep tanks. Simulation of this complex system is difficult, but Smoluchowski's work provides a start,[11] aided by that of Fair and Gemmell[12] and Argaman and Kaufman.[13] The analysis is complicated by the potential breakup of particles due to high velocity gradients,[14,15] and the dependency of particle density on size.[16] The Richardson-Zaki equation provides a basis for including effects of hindered settling,[17] but changes in the size and specific gravity of flocculent particles with changes in concentration[18] limit its applicability. Anticipated effects of possible types of "dead spots" readily can be considered by blocking off portions of the vertical or horizontal sections of the tank, and the amount of resuspension of previously deposited solids due to high bottom velocities might be predicted from basic sediment transport work.[19]

## Thickening

Historically, the thickening function has received appreciably less attention in design and operation of settling tanks than has the clarification function. However, the basic concepts associated with thickener design are well established,[20,21] and have been applied to analysis of settling tanks used in waste treatment.[22,23]

The area required for thickening in sedimentation basins assures that the time rate of application of solids per unit area (the applied solids flux) does not exceed the rate at which solids can be transmitted to the bottom of the settling tank. In a continuous thickener, solids move to the bottom because of the attraction of gravity and because of the bulk downward movement in the tank due to sludge removal. If a layer of sludge at concentration $c_i$ is considered, then the total possible downward flux, G, because of these two mechanisms, is

$$G = c_i v_i + c_i u \qquad (4)$$

where $v_i$ is the characteristic settling velocity of $c_i$, and u is the bulk downward velocity caused by sludge removal.

Considering uniform distribution of downward slurry velocities,

$$u = \frac{Q_u}{A} \qquad (5)$$

where $Q_u$ is the thickened sludge discharge rate from a settling tank of area, A. It is seen that, in Equation 4, $c_i v_i$ is the flux due to gravity alone, or the "batch flux" and $c_i u$ is the underflow flux.

If the relationship between $c_i$ and $v_i$ is determined experimentally[24] and a value of u is selected, then the magnitude of G from Equation 4 can be calculated for all values of $c_i$. Typically a result such as depicted in Figure 5.1 is obtained. It is seen that if sludge does not exist in the settling tank at a concentration less than $c_m$ and if thickened sludge is withdrawn at concentration $c_u$, then sludge at concentration $c_L$ controls the rate at which

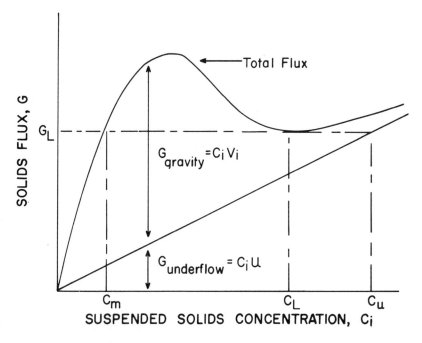

**Figure 5.1.** Total possible flux for sludge concentrations that could exist in a continuous thickener.

solids can reach the tank bottom and hence, the allowable applied flux. Further elaboration of this concept which establishes the thickening capabilities of settling basins may be found elsewhere.[24,25]

The simulation model of thickening presented here is founded on the concepts outlined in preceding paragraphs and is as presented by Dick and Young.[26] Note, however, that alternatively, the simulation model could have been based on extrapolation of tangents to the batch flux curve at various values of $c_L$.[21]

The relationship between settling velocity and suspended solids concentration is considered here to be

$$v_i = ac_i^{-n} \tag{6}$$

where a and n are experimentally determined constants, although other relationships have been proposed.[23] The case in which thickened solids are recycled from the settling tank back, ultimately to the settling tank feed (as in the activated sludge process) is considered. The recycle rate, r, is expressed as a fraction of the flow through the process, Q. Combining Equations 5 and 6 into Equation 4 and setting $Q_u$ equal to rQ (ignoring sludge wastage) yields

$$G = ac_i^{(1-n)} + \frac{rQ}{A} c_i \qquad (7)$$

Differentiation of Equation 7, with respect to concentration, gives

$$\frac{\partial G}{\partial c_i} = a(1-n)c_i^{-n} + \frac{rQ}{A} \qquad (8)$$

and, from Figure 5.1, it is seen that the concentration, $c_L$, exhibiting the limiting flux, occurs when $\partial G/\partial c = 0$ or

$$c_L = \left[ \frac{a(n-1)}{r} \frac{A}{Q} \right]^{1/n} \qquad (9)$$

Also, to confirm that the value of $c_L$ in Equation 9 is a minimum, it must be demonstrated that $\partial^2 G/\partial c_i^2 > 0$. From Equation 8,

$$\frac{\partial^2 G}{\partial c^2} = \frac{a\, n(n-1)}{c_i^{(1+n)}} \qquad (10)$$

It is seen that Equation 10 is positive when a is positive and $n > 1$, and a true minimum is realized. Experimentally determined values of a and n conform to these requirements. However, as described by Dick and Young,[26] Equation 7 does not pass through a maximum as in Figure 5.1. This is because Equation 6 is invalid at very low concentrations. As a practical matter, this defect is not considered to limit the model, but requires future improvement.

Substituting $c_L$ from Equation 9 into Equation 8 gives the limiting flux, $G_L$, in terms of the surface loading, $Q/A$, the recycle rate, r, and the basic settling properties of the sludge, a and n.

$$G_L = [a(n-1)]^{1/n} \left[ \frac{n}{n-1} \right] \left[ r\frac{Q}{A} \right]^{n-1/n} \qquad (11)$$

The underflow concentration, $c_u$, from a thickener of area, A, loaded at $G_L$ can be calculated from a mass balance to be

$$c_u = \frac{G_L A}{rQ} \qquad (12)$$

and similarly, in the special case when an activated sludge aeration tank is involved, the mixed liquor suspended solid concentration (ignoring synthesis and resulting wastage of solids, and ignoring any suspended solids in the influent) is

$$MLSS = \frac{c}{1+r} c_u \qquad (13)$$

As with the clarification function, the thickening model has some practical limitations. One source of limitations is the use of Equation 6, and another is the errors that can occur in the experimental determination of sludge-settling velocities needed to obtain a and n from that equation.[26] Other limitations originate from the convenient assumptions that solids and downward velocities are uniformly distributed throughout the cross-sectional area of settling basin. With highly compressible sludges, $v_i$ is a function of sludge depth, and modification of the basic equations could be desirable. Also, sufficient thickened sludge obviously must be maintained at point(s) of sludge withdrawal so that no "coning" occurs to allow overlying water to be removed with the sludge. Also, the equations could be modified for synthesis of biological solids and sludge wastage. However, the increased solids loading on the final tank would be partially compensated by the increase in the solids handling capacity created by the increased rate of sludge removal. Few data are available to assess the significance of these possible deviations. Data presented by Dick and Young for a pilot scale final settling tank 8 ft in diameter are shown in Figure 5.2, and would suggest that the practical limitations of available theoretical analyses of the clarification function.[26]

## DEVELOPMENT OF PROGRAMS

### Free-Form Input

To illustrate an alternative approach to simulation for classroom or other use, the programs developed here involve free-form input of data. The SCAN and TESTER (developed by Dr. L. Lopez, University of Illinois at Urbana-Champaign) subroutines used permit clarification and thickening concepts to be illustrated with minimal attention to details of computer programming. The two subroutines are presented in Appendices A and B, respectively. Also, a third subroutine, CDREAD, for reading data cards is common to the clarification and thickening programs, and is presented as Appendix C with comment cards to explain its function.

As illustrated later, careful and flexible programming allow the user of the simulations presented here to input his data in ordinary language and in any random order. Additionally, any of a wide variety of units can be used for the input, and the output automatically will be in the same units. The exact manner in which a variable is to be changed can be specified so that uniform increments of change, as in a DO loop, are not necessary. Many of the variables will be supplied by the program as default values if not specified by the user. Also, several mutually independent variations of input data may be explored in a single computer run by using the SOLVE command as explained later.

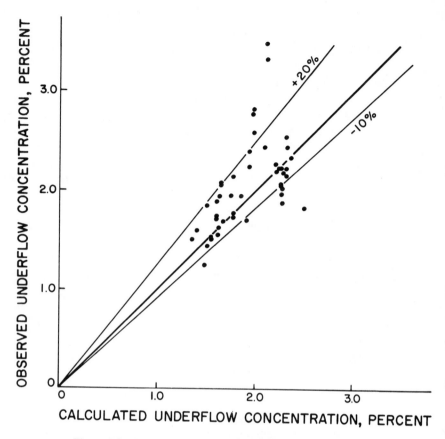

**Figure 5.2.** Comparison of continuous thickener performance with predictions of model.[24]

While the SCAN and TESTER subroutines permit very convenient illustration of sedimentation concepts without distraction by troublesome aspects of computer programming, some disadvantages are associated with the approach. The program to facilitate free-form input is complex and it is not intended that its development would be a classroom undertaking. Because analysis of characters in the input requires time, execution is more costly than when efficient conventional FORTRAN IV formats are used. Also, the subroutines are machine-dependent with the version presented here being designed to operate on an IBM/360 Model 70. Another disadvantage is that the approach deprives students of the insight associated with program development. The same basic approach to settling-tank simulation readily could be programmed in more conventional language. By substituting FORTRAN IV read statements in place of the subroutines SCAN, TESTER, CDREAD, and the statements calling these subroutines, the program could be considerably reduced in size. However, this change would require much more care when introducing data into the program.

## Clarification Function

The main program for analysis of the clarification portion of a settling tank is shown as Appendix D. As presented, the program is comprised primarily of comment cards to assist the reader. Additional clarification of the approach used in the program for calculation of the fractional removal of suspended solids of any given particle size distribution follows:

(a) The effective diameter, d, of a group of particles with extreme diameters $d_1$ and $d_2$ is

$$d = \sqrt{d_1 d_2} \tag{14}$$

(b) The effect of temperature on viscosity is taken as

$$\mu = 0.017 \, e^{-0.02524 \, T} \tag{15}$$

where T is the temperature in Centigrade. The slight effect of temperature on the mass density of water is ignored in computation of particle settling velocity.

(c) Assume that $N_R < 1$ so that $C_D$ is $24/N_R$. Then, the settling velocity of particles of effective diameter d is given by Equation 1. Note that as a substitute for this assumption and the following steps, the Newton-Raphson iterative technique[27] could be used for direct solution of particle settling velocity. A program for using this solution is contained in comment cards in the program (see Appendix D).

(d) Find $N_R$ from the values of d, $\mu$ and v from steps (a), (b), and (c).

(e) If $N_R < 1$, go to (k). Otherwise,

(f) Assume that the particle settles in the transition zone, and let $v_p$ from step (c) be the trial value, $v_t$, of the particle settling velocity.

(g) Compute $v_p$ from Equation 1 using $C_D = 24/N_R + 3/(N_R)^{1/2} + 0.34$ with the Reynolds' number computed from the value of $v_t$ in step (f).

(h) If $(v_p - v_t)/v_p < 0.1\%$, go to (i). Otherwise, let $v_p = (v_p + v_t)/2$ and go to (f).

(i) Compute $N_R$. If it is less than 1000, the assumption that the particle settled in the transition region was correct. If so, go to (k); otherwise go to (j).

(j) The particle settles in Newton's range. Let $C_D = 0.4$ in Equation 1 and compute $v_p$.

(k) If $v_p \geq v_o$, or Q/A, the overflow rate, the particles represented by this value of d are completely removed. If $v_p < Q/A$, the fraction of particles in the range which are removed is $v_p/(Q/A)$.

(l) The total removal is the sum of the fractions of the particle size distribution represented by the size fraction completely removed plus the partial removal of slower settling fractions.

Following is a summary of the form in which data cards should be inputed for analysis of the clarification function of settling basins. Additionally, illustrative clarification input appears as Appendix E. Information underlined on the data cards which follow is essential for the successful operation of the program whereas input not underlined is optional. Periods appearing between words designating units are an essential feature of the program. Data can be in integer or real number mode.

*The Water Temperature*—The temperature of the water should appear on a separate card. The form of this card is

THE *TEMP*ERATURE OF THE WATER IS *X Y*

where X is the value of the temperature and Y represents the units of the temperature which could be in

Degrees, Centigrade

or    Degrees, Fahrenheit

If the units are absent, the temperature is assumed to be in Centigrade. If no information on temperature is presented, then a default value of 20°C is used.

*The Rate of Flow*—The form of this card is

THE *RATE* OF FLOW INTO THE SETTLING TANK IS *X Y*

where X is the flow rate and Y represents any of the following units of the flow

Million Gallons Per Day

or    Cubic Feet Per Second

or    Cubic Meters Per Day

If the units are absent, mgd would be assumed.

*The Specific Gravity of the Particles*—The form of this card is

THE *SPEC*IFIC GRAVITY OF THE PARTICLES IS *X*

where X is the value of the specific gravity. If this card is missing, then the specific gravity of the particles takes a default value of 2.65 (a high value for all but grit chambers).

*The Area of the Tank*—The form of this card is

THE *AREA* OF THE SETTLING TANK IS *X Y*

where X is the area and Y represents the units of area—either

Square Feet

or    Square Meters

If the units are not designated, square feet would be assumed.

*The Feed Concentration*—This is the only basic data card that has a re-striction as to position. When used, it should appear before the size dis-tribution data cards. If this card is missing, then the concentration of par-ticles in each range on the size distribution cards should be expressed in mg/l. The form of the card is

THE <u>CONC</u>ENTRATION OF THE SUSPENDED PARTICLES
IN THE FEED IS <u>X Y</u>

where X is the value of the concentration and Y is either

Milligrams Per Liter
or Pounds Per Cubic Foot
or Percent

If the units are absent, mg/l would be used.

*The Size Distribution of the Particles*—This part of the data is made up of more than one card. The first card is a heading, and has the following form:

THE <u>SIZE</u> DISTRIBUTION OF THE PARTICLES IS AS FOLLOWS

It is followed by a separate card for each particle size range. The form of these cards is

<u>X Y</u> OF THE PARTICLES HAVE (HAS A) DIAMETER FROM <u>R</u> TO <u>S T</u>

where X is the amount (by weight) of the particles falling in that size range in units, Y, of either

Percent
or Milligrams Per Liter

Percent here indicates the amount of solids in a particular size range as a fraction of the total amount of solids present. The percent form can be used only if the feed concentration card is used. R and S express the two boundary particle sizes of the range. T, the units of the particle diameters, could be in either one of four forms:

Centimeters
or Microns
or Inches
or Millimeters

If the units are not indicated, centimeters would be used.

If all particles are of uniform size, one particle size range card should be used, and on it R and S both have the same value.

*The Commands SOLVE and EXIT*–SOLVE causes the program to execute the data that it has read. This work should appear on a separate card placed after all the other data cards comprising one set of data. More than one data set could appear in the same program but each data set must terminate with the word SOLVE. EXIT causes execution to terminate and should appear on the last data card.

### Thickening Function

Equations 6, 11, 12, and 13 form the basis of the main thickening program (Appendix F). Like the clarification program, it makes use of the SCAN, TESTER, and CDREAD subroutines (Appendices A, B, and C). Basically, the following steps are involved in computing final settling tank underflow concentration and the resulting mixed liquor solids concentration.

> (a) From a log plot of settling velocity as a function of concentration, find the value of -n (the slope). Then, using any point on the curve, solve for the constant "a" using Equation 6. This approach is used to automatically make the units of "a" compatible with those of c. Alternatively, a subroutine could be used to obtain the values of "a" and n by fitting of log $c_i$ vs log $v_i$ data.
> (b) Find $G_L$ from Equation 11 from given values of Q/A and r.
> (c) Find $c_u$ from Equation 12.
> (d) Find MLSS from Equation 13.

A description of the form in which data cards are to be prepared for the thickening program follows, and illustrative input is included as Appendix G. As with the clarification program, material shown here in lower case is irrelevant to successful operation of the program, periods appearing between words designating units are essential, and data can be in integer or real number mode.

*The Overflow Rate*–The form of this card is

THE *OVER*FLOW FLOW RATE IS <u>X Y</u>

where X is Q/A and Y represents any of the following units of X:

Gallons Per Day Per Square Foot (or, Gallons Per Square Foot Per Day)
or Cubic Feet Per Day Per Square Foot (or Cubic Feet Per Square Foot Per Day)

*The Value of n*–A separate data card is used to input the slope of the log $c_i$ vs log $v_i$ plot in the following form:

THE <u>SLOPE</u> OF THE SETTLING DATA WHEN PLOTTED
ON A LOG LOG SCALE IS <u>X</u>

where X is the magnitude of the slope.

*A Point on the Concentration vs Velocity Curve*—This data card contains the information used in evaluating "a" in Equation 6. Its form is

THE *SET*TLING VELOCITY IS X Y WHEN
THE CONCENTRATION IS Z W

where X is the settling velocity and Y represents one of the following forms of units:

> Feet Per Minutes
> or Feet Per Hour
> or Feet Per Day

If Y is not specified, feet per day is assumed.

Z is the sludge concentration and W represents one of the following units of concentration:

> Milligrams Per Liter
> or Percent
> or Pounds Per Cubic Foot

If W is not specified, pounds per cubic foot is assumed. The last half of the information on this card can be omitted, and concentration will be taken as one pound per cubic foot.

*The Recycle Ratio*—The form of this card is
THE *RE*CYCLE RATIO IS X

where X is the value of the recycle ratio as a percentage or a ratio.

*SOLVE and EXIT*—Same as clarification program.

## ILLUSTRATIVE OUTPUT

Illustrative output from clarification program input such as illustrated in Appendix E are shown as Figures 5.3 and 5.4. In Figure 5.3, influent and effluent particle size distributions for solids of various specific gravities are shown for a heavily loaded settling tank. Figure 5.4 shows anticipated effluent settling velocity distributions for a suspension of heavy particles in settling tanks with various overflow rates. The figures illustrate that essentially no removal of particles in or approaching the colloidal range can be expected and demonstrate the purpose of coagulation. In a similar fashion, the effect of temperature, influent particle size distribution, for example, could be demonstrated.

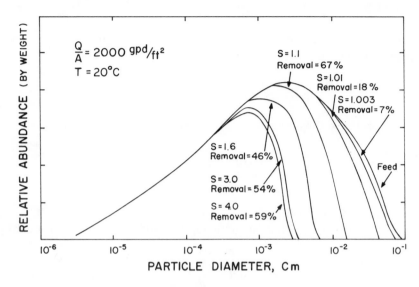

**Figure 5.3.** Illustrative output from clarification model showing removal of particles from suspension with broad particle size distribution in heavily loaded tank.

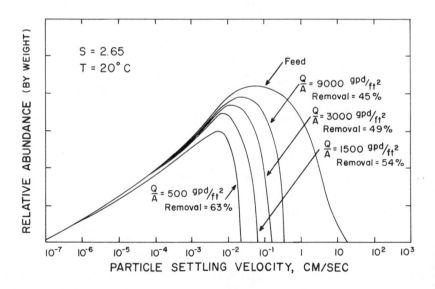

**Figure 5.4.** Illustrative output from clarification model showing effect of overflow rate on removal of particles from a suspension of heavy particles with a broad distribution of settling velocities.

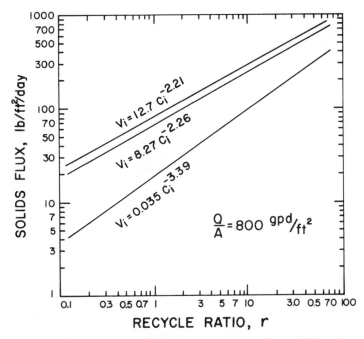

**Figure 5.5.** Illustrative thickening output showing the thickening capacity of a final settling tank for three different activated sludges.

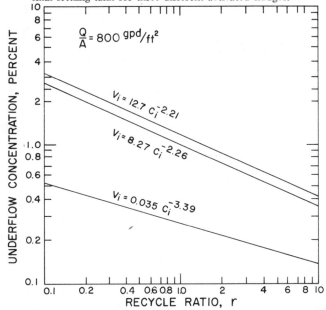

**Figure 5.6.** Illustrative thickening output showing the anticipated thickened sludge concentrations for three different activated sludges in similarly loaded tanks.

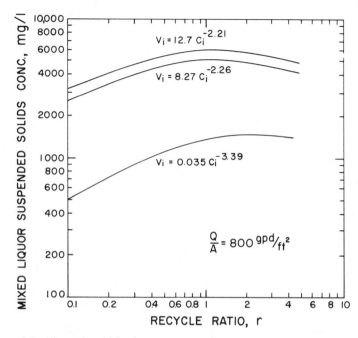

**Figure 5.7.** Illustrative thickening output showing the mixed liquor suspended solids concentration for three different sludges as a function of recycle rate.

Figures 5.5, 5.6, and 5.7 show typical output from thickening program input such as illustrated in Appendix G. The three curves on each figure reflect the sedimentation characteristics of three different activated sludges.[28] The settling velocity equations are in the form of Equation 6, and the units of $v_i$ and $c_i$ are ft/day and mg/l, respectively. Integration of data of this type with a simulation of the activated sludge process is a rewarding classroom undertaking and provides means for optimal combination of the two components of the process.

## REFERENCES

1.  Hazen, A. "On Sedimentation," *Transactions, ASCE* **53**, 45 (1904).
2.  Camp, T. R. "A Study of the Rational Design of Settling Basins," *Sewage Works J.* 8(9), 742 (1936).
3.  Camp, T. R. "Sedimentation and Design of Settling Tanks," *Transactions, ASCE* **111**, 895 (1946).
4.  Camp, T. R. "Studies of Sedimentation Design," *Sewage Ind. Wastes* 25(1), 1 (1953).
5.  Fair, G. M., J. C. Geyer, and D. A. Okun. *Water and Wastewater Engineering, Volume 2, Water Purification and Wastewater Treatment and Disposal* (New York: John Wiley and Sons, 1968).

6. Rich, L. G. *Unit Operations of Sanitary Engienering.* (New York: John Wiley and Sons, 1961), p. 308.
7. Shapiro, A. H. *Shape and Flow.* (New York: Doubleday, 1961), p. 186.
8. Dobbins, W. E. "Effects of Turbulence on Sedimentation," *Transactions, ASCE* **109**, 629 (1944).
9. Takamatsu, T. and M. Naito. "Effects of Flow Conditions on the Efficiency of a Sedimentation Vessel," *Water Res.* **1**(6), 433 (1967).
10. El-Baroudi, H. M. "Characterization of Settling Tanks by Eddy Diffusion," *J. San. Eng. Div., ASCE* **95**(SA3), 527 (1969).
11. Smoluchowski, M. "Versuch einer Mathematischen Theorie der Koagulationskinetic Kolloider Losungen," *Z. Physik. Chem.* **92**, 129 (April 1917).
12. Fair, G. M. and R. S. Gemmell. "A Mathematical Model of Coagulation," *J. Colloid Interface Sci.* **19**, 360 (1964).
13. Argaman, Y. and W. J. Kaufman. "Turbulence and Flocculation," *J. San. Eng. Div., ASCE* **96**(SA2), 223 (1970).
14. Bradley, R. A. and R. B. Krone. "Shearing Effects of Settling of Activated Sludge," *J. San. Eng. Div., ASCE* **97**(SA1), 59 (1971).
15. Parker, D. S., W. J. Kaufman, and D. Jenkins. "Floc Breakup in Turbulent Flocculation Processes," *J. San. Eng. Div., ASCE* **98**(SA1), 79 (1972).
16. Lagvankar, A. L. and R. S. Gemmell. "A Size-Density Relationship for Flocs," *J. Amer. Water Works Assoc.* **60**, 1040 (1968).
17. Richardson, J. F. and Zaki, W. N. "Sedimentation and Fluidization, Part I," *Trans. Inst. Chem. Eng.* **32**, 35 (1954).
18. Javaheri, A. R. and Dick, R. I. "Aggregate Size Variations During Thickening of Activated Sludge," *J. Water Poll. Control Fed.* **41**(5), 197 (1969).
19. Graf, W. H. *Hydraulics of Sediment Transport.* (New York: McGraw-Hill Book Co., 1971), p. 513.
20. Kynch, G. J. "A Theory of Sedimentation," *Trans. Faraday Soc.* **48**, 166 (1952).
21. Yoshioka, N., Y. Hotta, S. Tanaka, S. Naitu, and S. Tsugami. "Continuous Thickening of Homogenous Flocculated Slurries," *Chem. Eng.* (Tokyo) **21**(2), 66 (1957).
22. Vesilind, P. A. "Design of Prototype Thickeners from Batch Settling Tests," *Water Sewage Works* **115**(7), 302 (1968).
23. Dick, R. I. and B. B. Ewing. "Discussion Closure to Evaluation of Activated Sludge Thickening Theories," *J. San. Eng. Div., ASCE* **95** (SA2), 333 (1969).
24. Dick, R. I. "Gravity Thickening of Waste Sludges," *Proc. Filtration Soc., Filtration and Separation* **9**(2), 177 (1972).
25. Dick, R. I. "Role of Activated Sludge Final Settling Tanks," *J. San. Eng. Div., ASCE*, **96**(SA2), 423 (1970).
26. Dick, R. I. and K. W. Young. "Analysis of Thickening Performance of Final Settling Tanks," *Proc. 27th Ind. Waste Conf.*, Purdue University, 1972, Engineering Extension Series, No. 141, 33-53 (1974).
27. Stoecker, W. F. *Design of Thermal Systems.* (New York: McGraw-Hill Book Co., 1971).
28. Dick, R. I. and B. B. Ewing. "Evaluation of Activated Sludge Thickening Theories," *J. San. Eng. Div., ASCE*, **93**(SA4), 9 (1967).

## NOMENCLATURE

| Symbol | Definition and Comments[a] |
|---|---|
| a | Constant in Equation 6 (COEF), $L^3/F$ |
| A | Surface Area of Settling Tank (AREA), $L^2$ ($cm^2$) |
| $c_e$ | Effluent Concentration of Suspended Solids (TESS), $F/L^3$ (mg/l) |
| $c_f$ | Feed Concentration of Suspended Solids (TCONSS), $F/L^3$ (mg/l) |
| $c_i$ | Concentration of Suspended Solids, $F/L^3$ |
| $c_L$ | Limiting Concentration of Suspended Solids, $F/L^3$ |
| $c_D$ | Coefficient of Drag (CD), Dimensionless |
| d | Diameter of a Particle (DIAM), L (cm) |
| f | Fractional Removal of Particles Settling Slower than $v_0$, Dimensionless |
| g | Gravitational Constant (G), $L/T^2$ ($cm/sec^2$) |
| G | Solids Flux, $F/L^2 T$ |
| $G_L$ | Limiting Solids Flux (GL), $F/L^2 T$ |
| L | Length of Settling Tank, L |
| n | Constant of Equation 6 (SLOPE), Dimensionless |
| $N_R$ | Reynolds' Number (RN), Dimensionless |
| Q | Flow Rate Into Settling Tank (FLOW), $L^3 T$ ($cm^3$/sec) |
| Q/A | Overflow Rate (OVFLVE), L/T (cm/sec) |
| $Q_u$ | Underflow Rate, $L^3 T$ |
| S | Specific Gravity of Particles (Given in Default Value of 2.65) (SPGROP), Dimensionless |
| t | Hydraulic Retention Time, T |
| T | Temperature of the Water (Given a Default Value of 20°C) (TEMP) (°C) |
| u | Downward Velocity Due to Sludge Removal, L/T |
| v | Velocity of Particle (VP), L/T (cm/sec) |
| $v_i$ | Settling Velocity of Sludge at Concentration $c_i$, L/T |
| $v_n$ | Horizontal Velocity in Settling Tank |
| $V_o$ | Settling Velocity of Slowest Settling Particle Completely Removed (OVFLVE), L/T |
| W | Width of Settling Tank, L |
| $\mu$ | Absolute Viscosity of Water (VISCOS), $Ft/L^2$ |
| $\rho$ | Mass Density of Water (SPGROW), $T^2 F/L^4$ |

[a]Units in parentheses indicate those actually used in execution of the program. However, as described in the text, a variety of units can be accommodated in the input.

# APPENDIX A

## Scan Subroutine

```
C   ......................        SINGLE USER  ........................
C
C
C
C
C
C                               CLASS
C
C                       D                   A               B
C                       I                   L               L
C                       G                   P           S   A
C                       I                   H           E   N
C                       T     +    -    E   A    '   .  P   K    $
C
C                 CLASS  1    2    3    4    5    6   7   8    9   10
C           STATE
C  START       0         1   11   11    3    3    7   2   6    0   20
C  INTEGER     1         1   19   19    5    5   19   2  19   19   19
C  REAL        2         2   19   19    9    5   19   5  19   19   19
C  LABEL       3         3   19   19    3    3   19   4  19   19   19
C  NAME        4         4   19   19    4    4   19   4  19   19   19
C  ANYTEXT     5         5   19   19    5    5   19   5  19   19   19
C  SEPARATOR   6        19   19   19   19   19   19  19  19   19   19
C  STRING      7         7    7    7    7    7   19   7   7    7   19
C  GETSTR      8         1   11   11    3    3   19   2   6    8   19
C  EXP1        9        10    9    9   19   19   19   5  19   19   19
C  EXP2       10        10   19   19   19   19   19   5  19   19   19
C  + -        11         1   19   19   19   19   19   2  19   19   19
C
C
C
C
C
C          SEPARATOR CODE              VAL...SEPARATOR
C
C
C     1...CENTS            10...*            19...GREATER THAN
C     2...PERIOD           11...)            20...QUESTION MARK
C     3...-                12...SEMICOLON    21...COLON
C     4...(                13...NOT SIGN     22...SHARP
C     5...+                14...LESS THAN    23...EACH
C     6...OR SIGN          15.../            24...'
C     7...&                16...,            25...=
C     8...EXCLAMATION      17...PERCENT      26...QUOTE
C     9...$                18...UNDERLINE
C
C
C
C
C
C
      SUBROUTINE WDINIT(INUF,MARK,IPRINT)
      LOGICAL*1 BUFR,ANS,BLANK ,TER
      LOGICAL*4 XBUFR,SWIT
      INTEGER*2 ICLASS,ISTATE,NNO, ITAB
      INTEGER*4 OFF,ON
      DIMENSION BUFR(84),BLANK(4),  ICLASS(256),ISTATE(120),
     1  NNO(40),TER(4),ITAB(256)
      EQUIVALENCE (FLT,INT)
      EQUIVALENCE (XBUFR,L),(XBLANK,BLANK(1)),(TER(1),XT)
C
C
C
```

```
      DATA ISTATE /  1,11,11,3,3,7,2,6,0,20,    1,2*19,2*5,19,2,3*19,
     1                2,2*19, 9,5,19,5, 3*19,    3,2*19, 2*3,19,4, 3*19,
     2                4,2*19, 4,4,19,4, 3*19,    5,2*19, 2*5,19,5, 3*19,
     3                10*19,                     5*7,19, 3*7,19,
     4                1,11,11,3,3, 19,2,6,8,19,
     A                10,9,9,3*19, 5,3*19,
     5                10,5*19,5,3*19,
     7                1, 5*19, 2, 3*19 /
C
C
C



      DATA ICLASS /  64*0,     9, 9*0,  8, 7, 8, 8, 2,  8,  8,  9*0,  8,
C
     1   10,  8,  8, 8,  8, 3,  8, 9*0,  8,  8,  8,  8,  8,  10*0,  8,
C
C          #  @  '  =  "
     2    8,  8,  6,  8,  6, 0, 5, 5, 5, 5, 5, 5, 5, 5, 5,
C
C
     3 7*0, 5, 5, 5, 5, 5, 5, 5, 5, '5,  8*0, 5, 5, 5, 5, 5, 5, 5, 5,
C
C                A  B  C  D  E  F  G  H  I         J  K  L  M  N  O  P
     4 22*0,  0, 5, 5, 5, 5, 4, 5, 5, 5, 5, 7*0,  5, 5, 5, 5, 5, 5, 5,
C
C        Q  R           S  T  U  V  W  X  Y  Z           0  1  2  3  4  5
     5 5, 5,  8*0,  5, 5, 5, 5, 5, 5, 5, 5,  6*0,  1, 1, 1, 1, 1, 1,
C
C        6  7  8  9
     6  1, 1, 1, 1,  6*0 /
C
C
C
C
C



      DATA ITAB / 74*ZFFFF , 1,2,14,  4,5,6, 7,      9*ZFFFF,
C
     1    8, 9, 10,11,12,13, 3, 15,     9*ZFFFF,     16,17,18,19,20,
C
     2   10*ZFFFF,   21,22,23,24,25,26,   128*ZFFFF  /
C
C
C
C
      DATA  BUFR /80*' '/
      DATA XT /'     ' / ,DOL/'$$   ' /
      DATA XBLANK/'     '/,T1/'C   '/
      DATA SWIT,ON/.FALSE.,'ON  '/
      DATA MAXCL,IWORD,MQUOTE      / 10,19,0 /
      COMMON/SCANER/ANS(80),MODE,VAL,JWORD
      DATA INUNIT,IOUT/5,6/
      DIMENSION ILIST(10)
      DIMENSION LIST(10)
      INTEGER FILPT
      DATA FILPT/0/
      LOGICAL LABSW
      DATA LABSW / .FALSE./
      SWIT = .TRUE.
      RETURN
C
C     GET LIST OF FILES FOR INPUT STREAM
C
```

```
        ENTRY SETFIL(ILIST,NFIL)
160 II=NFIL
        IF(FILPT+NFIL.LE.10)GO TO 180
        II=10-FILPT
150 WRITE(IOUT,165) II
165 FORMAT('0 TOO MANY FILES IN INPUT LIST. ONLY',I5,'   CAN BE USED')
180 J=FILPT+II+1
        DO 170 K=1,II
170 LIST(J-K)=ILIST(K)
        FILPT=FILPT+II
        INUNIT=LIST(FILPT)
        RETURN
C
C       SET INPUT AND OUTPUT UNITS
C
        ENTRY SETIN(ILIST)
        NFIL=1
        GO TO 160
        ENTRY SETOUT (IO)
        IOUT = IO
        RETURN
C
C       GET A LABEL FOR THE NEXT LINE TO BE ECHO PRINTED
C
        ENTRY ACCLAB(LABEL)
        ILABEL=LABEL
        LABSW=.TRUE.
        RETURN




C
C
C
C
C       READ ANOTHER RECORD
C
        ENTRY RDLINE
        IF(SWIT) GO TO 1
        INUF = 10
        MARK = 72
        IPRINT = ON
    1 READ (INUNIT,2, END=4) (BUFR(I), I=1,80)
    2 FORMAT(80A1)
C
C       ECHO PRINT THE INPUT RECORD
C
        IF(IPRINT.NE.ON) GO TO 203
        IF(LABSW) GO TO 204
        WRITE(IOUT,205)(BUFR(I),I=1,80)
205 FORMAT(15X,80A1)
        GO TO 206
204 WRITE(IOUT,207) ILABEL,(BUFR(I),I=1,80)
207 FORMAT(4X,I5,6X,80A1)
206 LABSW=.FALSE.
C
C       SKIP IF COMMENT CARD
C
203 TER(1)= BUFR(1)
        TER(2)=BUFR(2)
        IF(XT.EQ.T1) GO TO 1
        LAST = MARK +1
        XT = DOL
        BUFR(LAST) = TER(1)
        GO TO 6
C
C
C       END OF FILE EXIT        SWITCH FILES IF USER REQUESTED IT
C
```

```
    195 WRITE (IOUT, 5)
      5 FORMAT ('0ALL DATA PROCESSED')
        CALL EXIT
      4 FILPT=FILPT-1
        IF(FILPT.EQ.0)GO TO 195
        INUNIT=LIST(FILPT)
        GO TO 1
C
C
C
C           INITIALIZE VARIABLES
C
        ENTRY RESET
      6 JSRT =1
        KWD = 0
        JLST=JSRT
        IST=0
        IBLANK = 0
        RETURN
C
C
C
C
C           SCAN FOR NEXT ITEM
C
C
        ENTRY SCAN
        KWD = KWD +1
        NNO(KWD)  =   JSRT
        ISA = IST
        IBLANK = 0
C
C
C           COUNT THE BLANKS
C
      7 XBUFR = BUFR(JLST)
        ICL = ICLASS(L+1)
        IST= ISTATE(MAXCL*IST + ICL)
        IF(IST .NE. 0 .AND. IST .NE. 8) GO TO 10
        IBLANK = IBLANK +1
        IF(IBLANK .GE. INUF ) GO TO 50
        JLST = JLST +1
        JSRT = JLST
        GO TO 7


C
C           LOOP TO FIND STATE
C
      9 XBUFR = BUFR(JLST)
        ICL = ICLASS(L+1)
        IST = ISTATE(MAXCL*IST + ICL)
     10 IF(IST-IWORD) 11,14,50
     11 JLST = JLST +1
        ISA = IST
        GO TO 9
C
C
C           END QUOTE ENCOUNTERED WHILE UNDER GETSTR
C
     13 MQUOTE =0
        MODE=8
        JWORD = 1
        JLST = JLST +1
        ANS(1)=BUFR(JSRT)
        GO TO 68
C
C           FIND THE LENGTH OF THE ITEM AND FOLLOW ' STATE GO TO ' .
C
```

```
   14 JWORD = JLST - JSRT
      NNO(KWD) = JSRT
      IF(ISA.EQ.0) GO TO 50
      JLAS = JLST-1
      GO TO (15,30,60,45,60,52,51,13,30,30,52  ),ISA
C
C                    BUILD THE VALUE OF AN INTEGER
C
   15 MODE =1
      IVAL = 0
      LOW=JSRT
      ISIG =1
      XBUFR=BUFR(JSRT)
      ICL = ICLASS(L+1)
      IF(ICL-2) 22,21,20
   20 ISIG=-1
   21 LOW=JSRT+1
   22 DO 25 I=LOW,JLAS
      XBUFR = BUFR(I)
   25 IVAL = IVAL*10 +L-240
      INT= IVAL * ISIG
      VAL = FLT
      GO TO 65
C
C                    BUILD THE VALUE OF A  REAL
C
   30 MODE = 2
      SIGN = 1.0
      VAL = 0.0
      POW = 10.0
      LOW=JSRT
      DEC = 1.0
      FAC = 1.0
C
C                 CHECK FOR + - SIGNS
      XBUFR=BUFR(JSRT)
      ICL = ICLASS(L+1)
      IF(ICL.EQ.3) SIGN = -1.0
      IF(ICL.EQ.2.OR.ICL.EQ.3)  LOW= JSRT+1
      JNO = LOW
      DO 33 I=LOW,JLAS
      DEC = DEC * FAC
      XBUFR = BUFR(I)
      ICL = ICLASS(L+1)
      JNO = JNO +1
      XL = L -240
C
C                 CHECK FOR . OR E
      IF(ICL-4) 31,35,32
   31 VAL=VAL*POW + XL/DEC
      GO TO 33
C                       . ENCOUNTERED
   32 POW = DEC
      FAC = 10.0
   33 CONTINUE
      VAL = VAL *SIGN
      GO TO 65


C
C
C                   E ENCOUNTERED    -    BUILD EXPONENT
C
   35 XBUFR = BUFR(JNO)
      ICL = ICLASS(L+1)
      ISIG = 1
      IEX = 0
C
C                    CHECK FOR + - SIGNS
      IF(ICL-2) 38,37,36
```

```
     36  ISIG=-1
     37  JNO = JNO +1
     38  DO 40 I=JNO,JLAS
         XBUFR = BUFR(I)
     40  IEX=IEX*10 +L -240
         VAL = VAL * SIGN *10.0**(IEX*ISIG)
         GO TO 65
C
C
C        PICK OUT FIRST LETTERS TO REPRESENT A NAME
C
     45  ANS(1)= BUFR(JSRT)
         JSRT = JSRT +1
         ID = 1
         DO 47 I=JSRT,JLAS
         XBUFR =BUFR(I)
         ICL = ICLASS(L+1)
         IF(ICL-7) 47,46,47
     46  ID = ID +1
         ANS(ID) = BUFR(I+1)
     47  CONTINUE
         JWORD = ID
         MODE  = 3
         GO TO 68
C
C
C             END OF RECORD     END OF LINE
C
C  CHECK FOR END OF RECORD DURING STRING OR GETSTRING MODES
     50  IF(ISA.EQ.8) GO TO 13
         IF(ISA.EQ.7) GO TO 51
         MODE = 9
         JLST = JSRT
         JWORD = 0
         GO TO 68
C
C
C             CHOP  QUOTATION  MARKS OFF THE STRING
C
     51  JWORD = JWORD -1
         JSRT=JSRT+1
         JLST = JLST+1
         GO TO 60
C
C
C         SEPARATOR ENCOUNTERED
C
     52  MODE = 6
         XBUFR= BUFR(JSRT)
         INT = ITAB(L+1)
         VAL = FLT
         ANS(1)=BUFR(JSRT)
         GO TO 68
C
C
C             SET MODE  AND STUFF CHARACTERS INTO THE ANSWER VECTOR
C
     60  MODE = ISA
C
     65  DO 66 I = 1,JWORD
     66  ANS(I)= BUFR(I-1+JSRT)
C
     68  DO 70 I=1,4
     70  ANS(JWORD +I)= BLANK(1)
     75  IST = 0
         IF(MQUOTE.EQ.2) IST=8
         JSRT = JLST
         RETURN
```

```
C
C
C              CHANGE THE TABLE   'ITAB'  .
C
      ENTRY TABLE(JTAB)
      DIMENSION JTAB(1)
      DO 76 I=1,256
   76 ITAB(I) = JTAB(I)
      RETURN
C
C
C         GET THE CONTENTS OF THE STRING JUST FOUND
C
      ENTRY GETSTR
      JSRT=NNO(KWD)+1
      MQUOTE = 2
      JLST=JSRT
      IST = 0
      RETURN
C
C
C
C         SKIP THE REMAINING CONTENTS OF THIS STRING
C
      ENTRY SKPSTR
      DO 80 I=JSRT,MARK
      XBUFR=BUFR(I)
      ID = I
      IF(ICLASS(L+1) .EQ. 6 ) GO TO 82
   80 CONTINUE
   82 JSRT = ID +1
      JLST = JSRT
      MQUOTE=0
      IST = 0
      RETURN
C
C
C
C         BACK UP    I    ITEMS
C
      ENTRY BACKSP(I)
      IF(KWD.LT.I) GO TO 6
      KWD = KWD-I
      JSRT = NNO(KWD+1)
      JLST = JSRT
      RETURN
C
C
      END
```

## APPENDIX B

### Tester Subroutine

```
SUBROUTINE TESTER
LOGICAL ISCAN,KMODE
DATA ISCAN/.TRUE./
LOGICAL LMODE
LOGICAL*1 ITYPE
LOGICAL *1 IBUFR,IMODE
DIMENSION IBUFR(80),IMODE(1)
COMMON/SCANER/ENTITY(20),MODE,IVALUE,NCHAR,NWD
EQUIVALENCE(IBUFR(1),ENTITY(1)),(MMODE,KMODE),(NMODE,LMODE)
ENTRY READ
CALL RDLINE
ISCAN=.TRUE.
RETURN
ENTRY INTEGR(*)
MMODE=1
GO TO 50
ENTRY REAL(*)
MMODE=2
GO TO 50
ENTRY LABEL(*)
MMODE=3
GO TO 50
ENTRY NAME(*)
MMODE=4
GO TO 50
ENTRY ANYTXT(*)
MMODE=5
GO TO 50
ENTRY SEP(*)
MMODE=6
GO TO 50
ENTRY STRING(*)
MMODE=7
GO TO 50
ENTRY ENDSTR(*)
MMODE=8
GO TO 50
ENTRY END(*)
MMODE=9
GO TO 50
ENTRY TRUE(*)
ISCAN=.TRUE.
RETURN 1
ENTRY BLANK(*)
ISCAN=.FALSE.
RETURN 1
50 IF(ISCAN)CALL SCAN(ENTITY(1),MODE,IVALUE,NCHAR)
NWD=(NCHAR+3)/4
IF(MODE-MMODE)22,23,22
23 ISCAN=.TRUE.
RETURN 1
22 ISCAN=.FALSE.
IF(MODE.EQ.9)ISCAN=.TRUE.
RETURN
ENTRY MATCH(IMODE,ICHAR,*)
IF(ISCAN)CALL SCAN(ENTITY,MODE,IVALUE,NCHAR)
NWD=(NCHAR+3)/4
IF(NCHAR.GE. ICHAR)GO TO 20
GO TO 22
20 DO 21 I=1,ICHAR
KMODE=IBUFR(I)
LMODE=IMODE(I)
```

```
      IF(MMODE.NE.NMODE)GO TO 22
   21 CONTINUE
      ISCAN=.TRUE.
      RETURN1
      END
```

## APPENDIX C

## CDREAD Subroutine

```
C
C
C     THE SUBROUTINE 'CDREAD' IS CALLED UPON TO READ A NUMBER AND ITS ASSOCIATED
C     UNITS FROM A DATA CARD .  THE NUMBER COULD BE EITHER INTEGER OR REAL .
C     THE UNITS COULD BE A SINGLE WORD  EX. POUNDS. ,OR UP TC FOUR WORDS
C     EX. MILLION.GALLONS.PER.DAY , THE POINTS APPEARING AFTER THE SINGLE WORD
C     OR BETWEEN THE TWO , THREE ,OR FOUR WORDS ARE ESSENTIAL .  IN CASE THE
C     UNITS ARE MISSING THE SUBROUTINE RETURNS WITH BLANKS IN PLACE OF UNITS .
C     IN CASE THE NUMBER IS MISSING THE SUBROUTINE PRINTS AN ERROR MESSAGE AND
C     STOPS THE PROGRAM .  THE UNITS COULD EITHER PRECEED OR FOLLOW THE NUMBER ,
C     THE NUMBER AND ITS UNITS COULD BE ON THE SAME CARD OR ON SEPARATE CARDS ,
C     BUT NOTHING ELSE SHOULD APPEAR IN BETWEEN
C
C
C
      SUBROUTINE CDREAD(X,BLANK)
      COMMON /SCANER/ ENTITY(20),MODE,IVALUE,NCHAR,NWD
      EQUIVALENCE (VALUE,IVALUE)
      DATA BITE/'     '/
      BLANK = BITE
      I = 0
    6 CALL END(&1)
   10 CALL REAL(&2)
      CALL INTEGR(&3)
    8 CALL LABEL(&4)
      WRITE (6,5)
    5 FORMAT(10X,'PROGRAM CANNOT READ ONE OF INPUT CARDS')
      STOP
    4 BLANK = ENTITY(1)
      IF(I.EQ.0)GO TO 6
      GO TO 15
    3 X = IVALUE
      GO TO 7
    2 X =   VALUE
    7 I = 1
      CALL END(&15)
      GO TO 8
    1 CALL READ
      CALL END(&9)
      GO TO 10
    9 WRITE(6,11)
   11 FORMAT(10X,'MISSING DATA AFTER ENTRY TITLE')
      STOP
   15 RETURN
      END
```

# APPENDIX D

## Main Clarification Program

```
C
C
C      THE MAIN PROGRAM FOR CLARIFICATION
C
C
       DIMENSION UNITS(2,16),A(50,8),B(2,16),C(5,3)
C
C
C      THE FOLLOWING THREE CARDS ARE USED TO ASSOCIATE SCANER WITH THE PROGRAM
C
C
       COMMON/SCANER/ENTITY(20),MODE,IVALUE,NCHAR,NWD
       EQUIVALENCE(VALUE,IVALUE)
       CALL WDINIT(80,80,'OFF ')
C
C
C      THE ARRAY 'C' IS USED IN THE PRINT OUT TO SPECIFY WHETHER THE SETTLING
C      PARTICLE WAS IN STOKES REGION , THE TRANSITION REGION OR IN NEWTON'S
C      REGION.
C
C
       DATA C/'STOK','ES R','EGIO','N   ','    ', 'TRAN','SITI','ON R',
      1'EGIO','N   ','NEWT','ONS ','REGI','ON  ','    '/
C
C
C      THE ARRAY 'B' IS USED IN THE PRINT OUT OF THE UNITS OF THE VARIABLES
C
C
       DATA B/'M.G.','D.  ','C.F.','S.  ','C.M.','P.D.','SQ.F','T.  ',
      1'SQ.M','TRS.','M.G.','P.L.','P.P.','C.F.','PER ','CENT','C.M.',
      2'    ','METE','RS  ','INCH','ES  ','M.M.','    ','DEG.',' C. ',
      3'DEG.',' F. ','PER ','CENT','M.G.','P.L.'/
C
C
C      THE SPECIFIC GRAVITY OF THE PARTICLE 'SPGROP' IS GIVEN A DEFAULT VALUE
C      OF 2.65
C      THE TEMPERATURE OF THE WATER 'TEMP' IS GIVEN A DEFAULT VALUE OF 20.
C      DEGREES CENTIGRADE.
C      THE GRAVITATIONAL CONSTANT 'G' IS GIVEN A VALUE OF 980. CM/SEC**2
C      THE SPECIFIC GRAVITY OF THE WATER IS ASSIGNED A VALUE OF 1.
C
C
       DATA SPGROP/2.65/,TEMP/20./, BLANK1/'    '/,BLANK2/'    '/,G/980./
      1,SPGROW/1./,I4/0 /,I1/13/
C
C
C      THE ARRAY 'UNITS' IS USED TO ASSOCIATE THE DIFFERENT UNITS WITH THEIR
C      CORRESPONDING CONVERSION FACTORS . THE UNIT IN WHICH THE VALUE IS READ IS
C      SEARCHED FOR IN THE UPPER ROW OF THE ARRAY . THE CORRESPONDING UNIT
C      BELOW IT WILL BE USED TO CONVERT THE VALUE TO UNITS WHICH THE PROGRAM
C      IS CAPABLE OF WORKING WITH.   AN INDICATOR OF THE ORIGINAL UNITS IS
C      STORED IN THE SET I1 , I2 , I3 , I4 , I5 , AND I6 , AND IN THE PRINT OUT
C      THE UNITS ARE CONVERTED BACK TO THE ORIGINAL FORM
C
C
       DATA UNITS/'MGPD',43817.5536,'CFPS',28320.,'CMPD',1000000.,'SF  ',
      2929.0304,'SM  ',10000.,'MGPL',1.,'PPCF',16050.,'PC  ',10000.,
      3'CM  ',1.,'M   ',0.0001,'I   ',2.54,'MM  ',0.1,'DC  ',1.,'DF  ',
      40.,'PC  ',1.,'MGPL',1./
C
C
C      THE COMMAND CALL READ CAUSES A DATA CARD TO BE READ .
C
C
     1 CALL READ
```

```
C
C
C        THE WORD 'THE' IS THE FIRST WORD SEARCHED FOR IN A CARD , BUT WHETHER I
C        IS FOUND OR NOT HAS THE SAME EFFECT OF TRANSFERRING CONTROL TO THE NEXT
C        STATEMENT
C
C
      2 CALL MATCH('THE',3,&3)



C
C
C        THE PROGRAM CHECKS FOR THE WORDS 'TEMPERATURE' , 'RATE' , 'AREA' ,
C        'SPECIFIC' , 'CONCENTRATION' , 'SIZE' , AND 'SOLVE' . IF ANY OF THE AB
C        WORDS MATCHES WITH THE FIRST OR SECOND WORD OF A BASIC DATA CARD , THE
C        CONTROL IS TRANSFERRED TO THAT PORTION OF THE PROGRAM SPECIALIZED IN
C        READING THAT CARD . OTHERWISE , CONTROL IS TRANSFERRED TO THE NEXT
C        STATEMENT . IF NONE OF THE ABOVE WORDS MATCHES , THE PROGRAM STOPS WI
C        AN ERROR MESSAGE .
C
C
      3 CALL MATCH('TEMP',4,&50)
        CALL MATCH('RATE',4,&60)
        CALL MATCH('AREA',4,&80)
        CALL MATCH('SPEC',4,&90)
        CALL MATCH('CONC',4,&100)
        CALL MATCH('SIZE',4,&120)
        CALL MATCH('SOLV',4,&155)
        CALL MATCH('EXIT',4,&999)
        WRITE(6,4)
      4 FORMAT(10X,'PROGRAM DOES NOT RECOGNIZE EITHER THE FIRST OR THE SEC
       1OND WORD ON A DATA CARD')
C
C
C        THIS IS THE SECTION OF THE PROGRAM THAT READS THE TEMPERATURE AND CON
C        IT TO DEGREES CENTIGRADE IN CASE IT IS GIVEN IN DIFFERENT UNITS .
C
C
     50 CALL MATCH('OF',2,&51)
     51 CALL MATCH('THE',3,&52)
     52 CALL MATCH('WAT',3,&53)
     53 CALL MATCH('IS',2,&54)
     54 CALL CDREAD(TEMP,BLANK)
        I1 = 13
        IF(BLANK.EQ.BLANK1)GO TO 1
        IF(BLANK.EQ.UNITS(1,13))GO TO 1
        IF(BLANK.EQ.UNITS(1,14))GO TO 55
        WRITE(6,56)
     56 FORMAT(10X,'THE UNITS OF TEMPERATURE ARE NOT CATERED FOR IN THIS P
       1ROGRAM')
        GO TO 999
     55 TEMP = (TEMP-32.)*5./9.
        I1 = 14
        GO TO 1
C
C
C        THIS IS THE SECTION OF THE PROGRAM THAT READS THE FLOW INTO THE TANK
C        CONVERTS IT TO MILLI LITERS PER SECOND .
C
C
     60 CALL MATCH('OF',2,&61)
     61 CALL MATCH('FLOW',4,&62)
     62 CALL MATCH('INTO',4,&63)
     63 CALL MATCH('THE',3,&64)
     64 CALL MATCH('SET',3,&65)
     65 CALL MATCH('TANK',4,&66)
     66 CALL MATCH('IS',2,&73)
     73 CALL CDREAD(FLOW,BLANK)
        I2 = 1
        IF(BLANK.EQ.BLANK1) GO TO 67
        DO 68 K=1,3
        IF(BLANK.EQ.UNITS(1,K)) GO TO 69
```

```
   68 CONTINUE
      WRITE(6,70)
   70 FORMAT(10X,'THE UNITS OF FLOW ARE NOT CATERED FOR IN THIS PROGRAM'
     1)
      GO TO 999
   69 I2 = K
      FLOW = FLOW*UNITS(2,K)
      GO TO 1
   67 FLOW = FLOW*UNITS(2,1)
      WRITE(6,71)
   71 FORMAT(10X,'FLOW WAS GIVEN NO UNITS AND IT WAS ASSUMED TO BE IN MI
     1LLION GALLONS PER DAY')
      GO TO 1
C
C
C     THIS IS THE SECTION OF THE PROGRAM WHICH READS THE AREA OF THE SETTLING
C     TANK AND CONVERTS IT TO SQUARE CENTIMETERS .
C
C
   80 CALL MATCH('OF',2,&81)
   81 CALL MATCH('THE',3,&82)
   82 CALL MATCH('SET',3,&83)
   83 CALL MATCH('TANK',4,&84)
   84 CALL MATCH('IS',2,&85)
   85 CALL CDREAD(AREA,BLANK)
      I3 = 4
      IF(BLANK.EQ.BLANK1)GO TO 86
      DO 87 K =4,5
      IF(BLANK.EQ.UNITS(1,K))GO TO 88
   87 CONTINUE
      WRITE(6,89)
   89 FORMAT(10X,'THE UNITS OF AREA ARE NOT CATERED FOR IN THIS PROGRAM'
     1)
      GO TO 999
   88 AREA = AREA*UNITS(2,K)
      I3 = K
      GO TO 1
   86 AREA = AREA*UNITS(2,4)
      WRITE(6,72)
   72 FORMAT(10X,'THE AREA OF THE SETTLING TANK WAS GIVEN NO UNITS AND I
     1T WAS ASSUMED TO BE IN SQUARE FEET')
      GO TO 1
C
C
C     THIS IS THE SECTION OF THE PROGRAM WHICH READS THE SPECIFIC GRAVITY OF THE
C     PARTICLES .
C
C
   90 CALL MATCH('GRAV',4,&91)
   91 CALL MATCH('OF',2,&92)
   92 CALL MATCH('THE',3,&93)
   93 CALL MATCH('PART',4,&94)
   94 CALL MATCH('IS',2,&95)
   95 CALL CDREAD(SPGROP,BLANK)
      GO TO 1
C
C
C     THIS IS THE SECTION OF THE PROGRAM WHICH READS THE CONCENTRATION OF THE
C     SUSPENDED PARTICLES INFLUENT TO THE TANK , IT CONVERTS THE CONCENTRATION
C     TO MG/L IN CASE IT IS GIVEN IN DIFFERENT UNITS .
C
C
  100 CALL MATCH('OF',2,&101)
  101 CALL MATCH('THE',3,&102)
  102 CALL MATCH('SUS',3,&103)
  103 CALL MATCH('PART',4,&104)
  104 CALL MATCH('IN',2,&105)
  105 CALL MATCH('THE',3,&106)
  106 CALL MATCH('FEED',4,&107)
  107 CALL MATCH('IS',2,&108)
```

```
      108 CALL CDREAD(TCONSS,BLANK)
          I4 = 6
          IF(BLANK.EQ.BLANK1)GO TO 109
          DO 110 K=6,8
          IF(BLANK.EQ.UNITS(1,K))GO TO 111
      110 CONTINUE
          WRITE(6,112)
      112 FORMAT(10X,'THE UNITS OF CONCENTRATION ARE NOT CATERED FOR IN THIS
          1 PROGRAM')
          GO TO 999
      111 TCONSS = TCONSS*UNITS(2,K)
          I4 = K
          GO TO 1
      109 TCONSS = TCONSS*UNITS(2,6)
          WRITE(6,113)
      113 FORMAT(10X,'THE CONCENTRATION WAS GIVEN NO UNITS AND IT WAS ASSUME
          1D TO BE IN MILLIGRAMS PER LITER')
          GO TO 1
    C
    C
    C     THIS IS THE SECTION OF THE PROGRAM WHICH READS THE SUSPENDED PARTICLES
    C     SIZE DISTRIBUTION . IF THE FRACTION OF PARTICLES IS GIVEN IN % IT CONVE
    C     IT TO MILLIGRAMS PER LITER , OTHERWISE THE PORTION SHOULD BE GIVEN IN
    C     MILLI GRAMS PER LITER .
    C     THE SIZE OF THE PARTICLES IS CONVERTED TO CENTIMETERS IF IT IS GIVEN IN
    C     DIFFERENT UNITS .
    C
    C

      120 CALL MATCH('DIST',4,&121)
      121 CALL MATCH('OF',2,&122)
      122 CALL MATCH('THE',3,&123)
      123 CALL MATCH('PART',4,&124)
      124 CALL MATCH('IS',2,&125)
      125 CALL MATCH('AS',2,&126)
      126 CALL MATCH('FOLL',4,&127)
      127 N = 0
    C
    C
    C     THE ARRAY 'A' IS A MATRIX OF 8 COLUMNS ,
    C     - THE FIRST COLUMN GIVES THE PORTION OF THE SUSPENDED SOLIDS WITHEN A
    C     CERTAIN SIZE RANGE IN MG/L
    C     - THE SECOND & THIRD COLUMNS GIVE THE UPPER AND LOWER BOUNDS OF PARTIC
    C     SIZES WHICH CONTAIN THE AMOUNT IN COLUMN ONE
    C     - THE FOURTH COLUMN CONTAINS THE GEOMETRIC MEAN OF THE TWO BOUNDARY
    C     PARTICLE SIZES
    C     - THE FIFTH COLUMN IS RESERVED FOR THE SETTLING VELOCITY OF THE PARTIC
    C     OF THAT SIZE RANGE .
    C     - THE SIXTH COLUMN IS RESERVED FOR THE % REMOVAL OF THAT PARTICLE SIZE
    C     RANGE .
    C     - THE SEVENTH COLUMN IS RESERVED FOR THE EFFLUENT CONCENTRATION OF THE
    C     PARTICLES OF THAT PARTICLE SIZE RANGE .
    C     - THE EIGHTH COLUMN CONTAINS AN INDICATOROF WHETHER THE PARTICLES OF T
    C     SIZE RANGE SETTLE IN THE STOKES , TRANSITION , OR NEWTON'S RANGE .
    C
    C
      143 CALL READ
          CALL REAL(&128)
          CALL INTEGR(&129)
          GO TO 2
      128 N = N+1
          A(N,1) = VALUE
          GO TO 130
      129 N = N+1
          A(N,1) = IVALUE
      130 BITE = BLANK2
          CALL LABEL(&131)
      131 BITE = ENTITY(1)
          DO 132 K = 15,16
          IF(BITE.EQ.UNITS(1,K))GO TO 133
```

```
  132 CONTINUE
      WRITE(6,134)N
  134 FORMAT(10X,'THE UNITS EXPRESSING FRACTION OF PARTICLES FALLING WIT
     1HIN A CERTAIN RANGE ARE MISSING FROM THE',I4,'CARD OF THAT SET')
      GO TO 999
  133 I5 = K
      IF(K.EQ.16)GO TO 135
      A(N,1) = TCONSS*A(N,1)/100.
  135 L = 2
      CALL MATCH('OF',2,&136)
  136 CALL MATCH('THE',3,&137)
  137 CALL MATCH('PART',4,&138)
  138 CALL MATCH('HA',2,&139)
  139 CALL MATCH('A',1,&140)
  140 CALL MATCH('DIAM',4,&141)
  141 CALL MATCH('FROM',4,&142)
  142 CALL REAL(&144)
      CALL INTEGR(&145)
      WRITE(6,146)
  146 FORMAT(10X,'SIZE OF PARTICLES IS IMPROPERLY DEFINED')
      GO TO 999
  144 A(N,L) = VALUE
      GO TO 147
  145 A(N,L) = IVALUE
  147 IF(L.EQ.3) GO TO 150
      L = 3
      CALL MATCH('TO',2,&142)
      GO TO 142
  150 BITE = BLANK2
      CALL LABEL(&151)
  151 BITE = ENTITY(1)
      IF(BITE.EQ.BLANK1) GO TO 158
      DO 152 K=9,12
      IF(BITE.EQ.UNITS(1,K))GO TO 153
  152 CONTINUE
      WRITE(6,154)  N
  154 FORMAT(10X,'THE UNITS OF THE SIZES OF PARTICLES ARE NOT DEFINEF ON
     1 THE APPROPRIATE CARD',I4)
      GO TO 999


  153 I6 = K
      DO 157 I=2,3
      A(N,I) = A(N,I)*UNITS(2,K)
  157 CONTINUE
      GO TO 143
  158 I6 = 9
      GO TO 143
C
C
C     THE CONCENTRATION OF THE INFLUENT SUSPENDED SOLIDS IS RECALCULATED FROM
C     THE PARTICLE SIZE DISTRIBUTION DATA
C
C
  155 TCONSS = 0.
      DO 156 LL = 1,N
      TCONSS = TCONSS + A(LL,1)
  156 CONTINUE
C
C
C     VISCOSITY IS EVALUATED FROM A FITTED FUNCTION OF VISCOSITY VERSUS
C     TEMPERATURE .
C
C
      VISCOS = 1.7*EXP(-TEMP*0.02524)*0.01
C
C
C     'OVFLVE' REPRESENTS THE HYDRAULIC LOADING RATE OF THE SETTLING TANK IN
C     CM/SEC
C
C
```

```
      OVFLVE = FLOW/AREA
      DO 200 I = 1,N
      A(I,4) = SQRT(A(I,2)*A(I,3))
      DIAM = A(I,4)
C
C
C     THE NEXT 20 CARDS (UP TO 998) ARE AN ITERATIVE PROCEDURE FOR SOLVING
C     THE SETTLING VELOCITY OF THE PARTICLES USING THE EQUATION
C     CD = 24./RN + 3./SQRT(RN) + 0.34   WHERE RN IS REYNOLD'S NUMBER
C     AND THE GENERAL EQUATION   V = SQRT(4.*G*(SPGROP-SPGROW)*DIAM)
C                                       (    3.*CD*SPGROW          )
C     THE TWO ABOVE EQUATIONS ARE SOLVED BY THE NEWTON   RAPHSON'S ITERATIVE
C     TECHNIQUE . IN  ASE THIS METHOD OF SOLUTION IS DESIRED THE 'C' SHOULD
C     REMOVED FROM THE NEXT CARD UP TO CARD 998
C
C
C     VP = 0.5
C  11 V  = VP
C     FX = VP-(((4./3.)*DIAM*G*(SPGROP-SPGROW))/((24.*VISCOS/(SPGROW*VP*
C    1DIAM))+(3./SQRT(SPGROW*VP*DIAM/VISCOS))+0.34)**0.5)
C     FIX = 1.-0.5*(((4./3.)*DIAM*G*(SPGROP-SPGROW)/(24.*VISCOS/(SPGROW*
C    1VP*DIAM)+3./SQRT(SPGROW*VP*DIAM/VISCOS)+0.34)**(-0.5))*(4.*DIAM*G
C    2*(SPGROP-SPGROW)/3.)*(-24.*VISCOS/(SPGROW*DIAM*(VP**2))-3.*SQRT(
C    3VISCOS/(SPGROW*DIAM))/(0.5*(VP**1.5)))/((24.*VISCOS/(SPGROW*VP*
C    4DIAM)+3./SQRT(SPGROW*VP*DIAM/VISCOS)+0.34)**2)
      MACK = 1
C     VP = VP-FX/FIX
C     IF(ABS((VP- V)/VP).GT.0.001)GO TO 11
C     RN = SPGROW*VP*DIAM/VISCOS
C     IF(RN.LT.1.) GO TO 12
C     IF(RN.GT.250000.)GO TO 13
C     II = 2
C     GO TO 199
C  12 II = 1
C     GO TO 199
C  13 II = 3
C 998 GO TO 199
C
C
C     THE SECOND MODE OF SOLUTION
C
C     THE PARTICLE IS TESTED TO CHECK WHETHER ITS SETTLING CHARACTERISTICS
C     FALL IN STOKES RANGE . IF IT DOES ANOTHER PARTICLE SIZE IS TESTED ,
C     DOES NOT TRANSFER IS SET TO TEST THE TRANSITION ZONE . IN THE TRANSI
C     ZONE , THE SAME TEST IS PERFORMED . IF THE TEST FAILS AGAIN . THEN T
C     PARTICLE SETTLES IN NEWTONS ZONE
C
C

  998 VP = G*(SPGROP - SPGROW)*( DIAM**2)/(18.*VISCOS)
      RN = SPGROW*VP*DIAM/VISCOS
      IF(RN.GT.1)GO TO 161
      II = 1
      GO TO 199
  161 V1 = VP
      RN = SPGROW*VP*DIAM/VISCOS
      CD = 24./RN + 3./SQRT(RN) + 0.34
      VP = SQRT(4.*G*(SPGROP - SPGROW)*DIAM/(3.*CD*SPGROW))
      COMP = ABS((VP - V1)/VP)
      IF(COMP.LT.0.001) GO TO 170
      VP = (VP+V1)/2.
      GO TO 161
  170 IF(RN.GT.250000.)GO TO 180
      II = 2
      GO TO 199
  180 VP = SQRT(4.*G*(SPGROP-SPGROW)*DIAM/(3.*0.4*SPGROW))
      II = 3
  199 A(I,5) = VP
C
C
```

```
C         THE FRACTION REMOVED OF A PARTICLE SIZE RANGE IS EQUAL TO THE RATIO OF THE
C         SETTLING VELOCITY OF THE PARTICLE TO THE OVERFLOW RATE .
C
C
          IF(VP.LT.OVFLVE)GO TO 181
          A(I,6) = 100.
          GO TO 182
  181 A(I,6) = VP*100./OVFLVE
  182 A(I,7) = A(I,1)*(100.-A(I,6))/100.
          A(I,8) = II
  200 CONTINUE
          TESS = 0.
          DO 201 IL = 1,N
          TESS = TESS+A(IL,7)
  201 CONTINUE
C
C
C         THE SUSPENDED SOLIDS FRACTIONIS CCNVERTED BACK TO ITS ORIGINAL UNITS .
C
C
          IF(I5.EQ.16)GO TO 210
          DO 211 LM = 1,N
          A(LM,1) = A(LM,1)*100./TCONSS
  211 CONTINUE
C
C
C         THE PARTICLE SIZE IS CONVERTED BACK TO ITS ORIGINAL UNITS .
C
C
  210 IF(I6.EQ.9)GO TO 212
          DO 213 LM = 1,N
          DO 213 LN = 2,4
          A(LM,LN) = A(LM,LN)/UNITS(2,I6)
  213 CONTINUE
C
C
C         THE EFFLUENT CONCENTRATION IS CCNVERTED TO THE UNITS OF THE INFLUENT
C         CONCENTRATION .
C
C
  212 IF(I4.EQ.0) GO TO 216
          IF(I4.EQ.6) GO TO 216
          DO 217 LM = 1,N
          A(LM,7) = A(LM,7)/UNITS(2,I4)
  217 CONTINUE
          TESS = TESS/UNITS(2,I4)
  216 WRITE(6,236)
  236 FORMAT(1H1,48X,'THE RESULTING DATA ARE',/,49X,'--------------------
     1---',//)
          WRITE(6,237)(B(I,I6),I=1,2),(B(I,I6),I=1,2),(B(I,I6),I=1,2),
     1(B(I,I4),I=1,2),(B(I,I5),I=1,2)
  237 FORMAT(1X,'FRACTION OF',4X,'BOUNDARY SIZES',4X,'GECMETRIC MEAN',3X
     1,'SETTLING VELOCITY',3X,'% REMOVAL',3X,'EFFLUENT CONC',3X,'REGION
     2OF',/,2X,'INFLUENT',7X,4A4,6X,2A4,10X,'CM/SEC',22X,2A4,7X,'SETTLIN
     3G',/,2X,2A4,/ )
          DO 240 LL = 1,N
          WRITE(6,238)(A(LL,I),I=1,7),(C(K,A(LL,8)),K = 1,5)


  238 FORMAT(1X,F8.5,4X,F9.8,2X,F9.8,4X,F9.8,6X,F13.10,4X,F11.7,3X,F9.4,
     13X,5A4,/)
  240 CONTINUE
C
C
C         THE TEMPERATURE IS CONVERTED BACK TO THE ORIGINAL UNITS. AND IS PRINTED
C         OUT
C
C
          IF(I1.EQ.14) TEMP = TEMP*9./5.+32.
          WRITE(6,230) TEMP,(B(K,I1),K=1,2)
```

```
  230 FORMAT(///,10X,'THE TEMPERATURE OF THE WATER IS',F5.1,2X,2A4,//)
      IF(I1.EQ.14) TEMP = (TEMP-32.)*5./9.
C
C
C     THE FLOW INTO THE SETTLING TANK IS CONVERTED TO THE ORIGINAL UNITS AND
C     IT IS PRINTED OUT .
C
C
      FLOW = FLOW/UNITS(2,I2)
      WRITE(6,231) FLOW,(B(K,I2),K = 1,2)
  231 FORMAT(10X,'THE RATE OF FLOW INTO THE SETTLING TANK IS',F8.2,2X,
     12A4,//)
      FLOW = FLOW * UNITS(2,I2)
C
C
C     THE AREA OF THE SETTLING TANK IS CONVERTED BACK TO THE ORIGINAL UNITS AN
C     IS PRINTED OUT
C
C
      AREA = AREA/UNITS(2,I3)
      WRITE(6,232) AREA,(B(K,I3),K=1,2)
  232 FORMAT(10X,'THE AREA OF THE SETTLING TANK IS',F10.2,2X,2A4,//)
      AREA = AREA*UNITS(2,I3)
C
C
C     THE SPECIFIC GRAVITY OF THE PARTICLES IS PRINTED OUT .
C
C
      WRITE(6,233) SPGROP
  233 FORMAT(10X,'THE SPECIFIC GRAVITY OF THE PARTICLES IS',F8.5,//)
C
C
C     THE INFLUENT SUSPENDED SOLIDS CONCENTRATION IS CONVERTED BACK TO THE
C     ORIGINAL UNITS AND IS PRINTED OUT .
C
C
      IF(I4.EQ.0) I4 = I5
      TCONSS = TCONSS/UNITS(2,I4)
      WRITE(6,234) TCONSS,(B(K,I4),K=1,2)
  234 FORMAT(10X,'THE CONCENTRATION OF THE SUSPENDED PARTICLES IN THE FE
     1ED IS',F9.2,2X,2A4,//)
      TCONSS = TCONSS*UNITS(2,I4)
C
C
C     THE EFFLUENT SUSPENDED SOLIDS CONCENTRATION IS CONVERTED TO THE UNITS O
C     THE INFLUENT SUSPENDED SOLIDS CONCENTRATION AND IT IS PRINTED OUT .
C
C
      TESS = TESS/UNITS(2,I4)
      WRITE(6,235) TESS,(B(K,I4),K=1,2)
  235 FORMAT(10X,'THE CONCENTRATION OF THE SUSPENDED PARTICLES IN THE EF
     1FLUENT IS',F8.2,2X,2A4,//)
C
C
C     THE OVERFLOW RATE FROM THE SETTLING TANK IS PRINTED OUT
C
C
      WRITE(6,239) OVFLVE
  239 FORMAT(10X,'THE OVERFLOW RATE FROM THE SETTLING TANK IS',F8.4,2X,
     1'CM/SEC',//)
C
C
C     THE VISCOSITY OF THE WATER IS PRINTED OUT .
C
C
      WRITE(6,251) VISCOS
  251 FORMAT(10X,'THE VISCOSITY OF THE WATER IS',2X, F8.6)
      IF(I5.EQ.16) GO TO 255
      DO 250 LM = 1,N
```

```
      A(LM,1)  =  A(LM,1)*TCONSS/100.
250 CONTINUE
255 IF(I6.EO.9)GO TO 256
      DO 257 LM = 1,N
      DO 257 LN = 2,4
      A(LM,LN)  =  A(LM,LN)*UNITS(2,I6)
257 CONTINUE
256 GO TO 1
999 CONTINUE
      STOP
      END
```

## APPENDIX E

## Illustrative Clarification Input

```
THE TEMPERATURE OF THE WATER IS 20 DEGREES.CENTIGRADE
THE RATE OF FLOW INTO THE SETTLING TANK IS 0.5  M.G.P.D
THE AREA OF THE SETTLING TANK IS 1000 SQUARE.FEET
THE SPECIFIC GRAVITY OF THE PARTICLES IS 2.65
THE CONCENTRATION OF THE SUSPENDED PARTICLES IN THE FEED IS 150 M.G.P.L
THE SIZE DISTRIBUTION OF THE PARTICLES IS AS FOLLOWS
0.5 P.C  OF THE PARTICLES HAVE A DIAMETER FROM    0.1     TO   0.08     C.M
 1  P.C  OF THE PARTICLES HAVE A DIAMETER FROM    0.08    TO   0.06     C.M
 3  P.C  OF THE PARTICLES HAVE A DIAMETER FROM    0.06    TO   0.05     C.M
5.5 P.C  OF THE PARTICLES HAVE A DIAMETER FROM    0.05    TO   0.04     C.M
 8. P.C  OF THE PARTICLES HAVE A DIAMETER FROM    0.04    TO   0.03     C.M
 10 P.C  OF THE PARTICLES HAVE A DIAMETER FROM    0.03    TO   0.01     C.M
 15 P.C  OF THE PARTICLES HAVE A DIAMETER FROM    0.01    TO   0.005    C.M
17. P.C  OF THE PARTICLES HAVE A DIAMETER FROM    0.005   TO   0.001    C.M
 15 P.C  OF THE PARTICLES HAVE A DIAMETER FROM    0.001   TO   0.0005   C.M
 10 P.C  OF THE PARTICLES HAVE A DIAMETER FROM    0.0005  TO   0.0001   C.M
 8  P.C  OF THE PARTICLES HAVE A DIAMETER FROM    0.0001  TO   0.00005  C.M
 5  P.C  OF THE PARTICLES HAVE A DIAMETER FROM    0.00005 TO   0.00001  C.M
 2  P.C  OF THE PARTICLES HAVE A DIAMETER FROM    0.00001 TO   0.000005 C.M
    SOLVE
THE RATE OF FLOW INTO THE SETTLING TANK IS  1 MILLION.GALLONS.PER.DAY
    SOLVE
THE RATE OF FLOW INTO THE SETTLING TANK IS 1.5  M.G.P.D
    SOLVE
THE RATE OF FLOW INTO THE SETTLING TANK IS 5.0  M.G.P.D
    SOLVE
 THE TEMPERATURE OF THE WATER IS 1 DEGREES.CENTIGRADE
    SOLVE
THE TEMPERATURE OF THE WATER IS 30 DEGREES.CENTIGRADE
    SOLVE
 THE SPECIFIC GRAVITY OF THE PARTICLES IS 1.001
    SOLVE
 THE SPECIFIC GRAVITY OF THE PARTICLES IS 1.04
    SOLVE
THE SPECIFIC GRAVITY OF THE PARTICLES IS 2.4
    SOLVE
    EXIT
```

## APPENDIX F

## Main Thickening Program

```
C
C
C        THE MAIN PROGRAM FOR THICKENING
C
C
         DIMENSION UNITS(2,6),UNITS1(3,4)
C
C
C        THE FOLLOWING THREE CARDS ARE USED TO ASSOCIATE SCANER WITH THE PROGRAM
C
C
         COMMON/SCANER/ENTITY(20),MODE,IVALUE,NCHAR,NWD
         EQUIVALENCE(VALUE,IVALUE)
         CALL WDINIT(80,80,'OFF ')
C
C
C        THE ARRAY 'UNITS' IS USED TO ASSOCIATE THE DIFFERENT UNITS WITH THEIR
C        CORRESPONDING CONVERSION FACTORS . THE UNIT IN WHICH THE VALUE IS READ IS
C        SEARCHED FOR IN THE UPPER ROW OF THE ARRAY .  THE CORRESPONDING UNIT
C        BELOW IT WILL BE USED TO CONVERT THE VALUE TO UNITS WHICH THE PROGRAM
C        IS CAPABLE OF WORKING WITH.
C
C
C        THE ARRAY 'UNITS1' IS USED TO ASSOCIATE THE DIFFERENT UNITS WITH THEIR
C        CORRESPONDING CONVERSION FACTORS . THIS ARRAY SERVES FOR THE UNDERFLOW
C        DATA ONLY .
C
C
         DATA UNITS/'MGPL',0.0000624,'PPCF',1.,'PC  ',0.624, 'FPM ',1440.,
        1'FPH ',24.,'FPD ',1./
         DATA UNITS1/'GPDP','SF  ',0.1336,'CFPD','PSF ',1., 'GPSF','PD  ',
        10.1336,'CFPS','FPD ',1./
C
C
C        THE SOLIDS CONCENTRATION AT WHICH THE SETTLING VELOCITY IS INPUTED IS
C        ASSUMED TO BE 1 POUND PER CUBIC FOOT UNLESS THE USER INPUTS A DIFFERENT
C        CONCENTRATION .
C
C
         DATA BLANK1/'    '/, CONC1/1./
C
C
C        THE COMMAND CALL READ CAUSES A DATA CARD TO BE READ .
C
     1 CALL READ
C
C
C        THE WORD 'THE' IS THE FIRST WORD SEARCHED FOR IN A CARD , BUT WHETHER IT
C        IS FOUND OR NOT HAS THE SAME EFFECT OF TRANSFERRING CONTROL TO THE NEXT
C        STATEMENT
C
C
     2 CALL MATCH('THE',3,&3)
C
C
C        THE PROGRAM CHECKS FOR THE WORDS 'SLOPE','SETTLING','OVERFLOW',
C        'RECYCLE','SOLVE',  AND 'EXIT'
C        IF ANY OF THE ABOVE WORDS MATCHES WITH THE FIRST OR SECOND WORD OF A
C        DATA CARD . THEN CONTROL IS TRANSFERRED TO THAT SECTION OF THE PROGRAM
C        SPECIALIZED IN READING THAT CARD . OTHERWISE , CONTROL IS TRANSFERRED TO
C        THE NEXT STATEMENT . IF NONE OF THE ABOVE WORDS MATCH  , THE PROGRAM
C        STOPS WITH AN ERROR MESSAGE .
C
C
```

```
C
    3 CALL MATCH('SLOP',4,&20)
      CALL MATCH('SET',3,&40)
      CALL MATCH('OVER',4,&70)
      CALL MATCH('RECY',4,&90)
      CALL MATCH('SOLV',4,&100)
      CALL MATCH('EXIT',4,&999)
      WRITE(6,4)
    4 FORMAT(10X,'THE FIRST OR SECOND WORD ON A DATA CARD IS NOT CATERED
     1 FOR IN THIS PROGRAM')


C
C
C       THIS IS THE SECTION OF THE PROGRAM THAT READS THE SLOPE OF THE SETTLIN
C       CURVE WHEN PLOTTED ON A LOG LOG SCALE .
C
C
   20 CALL MATCH('OF',2,&21)
   21 CALL MATCH('THE',3,&22)
   22 CALL MATCH('SET',3,&23)
   23 CALL MATCH('CURV',4,&24)
   24 CALL MATCH('WHEN',4,&25)
   25 CALL MATCH('PLOT',4,&26)
   26 CALL MATCH('DRAW',4,&27)
   27 CALL MATCH('ON',2,&28)
   28 CALL MATCH('A',1,&29)
   29 CALL MATCH('LOG',3,&30)
   30 CALL MATCH('LOG',3,&31)
   31 CALL MATCH('CURV',4,&32)
   32 CALL MATCH('SCAL',4,&33)
   33 CALL MATCH('IS',2,&34)
   34 CALL REAL(&35)
      CALL INTEGR(&36)
      WRITE(6,37)
   37 FORMAT(10X,'THE SLOPE OF THE SETTLING CURVE IS IMPROPERLY DEFINED'
     1)
      GO TO 999
   35 SLOPE =  VALUE
      GO TO 38
   36 SLOPE = IVALUE
   38 IF(SLOPE.LT.0.) SLOPE = - SLOPE
      GO TO 1
C
C
C       THIS IS THE SECTION OF THE PROGRAM WHICH READS THE SETTLING VELOCITY
C       THE PARTICLE 'VEL1' , AND CONVERTS IT TO FEET PER DAY . IT ALSO READS
C       CONCENTRATION 'CONC1' , AT WHICH THE SETTLING VELOCITY IS MEASURED AN
C       CONVERTS IT TO POUNDS PER CUBIC FOOT .
C
C
   40 CALL MATCH('VEL',3,&41)
   41 CALL MATCH('IS',2,&42)
   42 CALL CDREAD(VEL1,BLANK)
      IF(BLANK.EQ.BLANK1)GO TO 50
      DO 43 K = 4,6
      IF(BLANK.EQ.UNITS(1,K)) GO TO 45
   43 CONTINUE
      WRITE(6,44)
   44 FORMAT(10X,'THE UNITS OF VELOCITY ARE NOT CATERED FOR IN THIS PROG
     1RAM')
      GO TO 999
   45 VEL1 = VEL1*UNITS(2,K)
   50 CALL END(&60)
      CALL MATCH('WHEN',4,&51)
   51 CALL MATCH('THE',3,&52)
   52 CALL MATCH('CONC',4,&53)
   53 CALL MATCH('IS',2,&54)
   54 CALL CDREAD(CONC1,BLANK)
      IF(BLANK.EQ.BLANK1) GO TO 60
      DO 55 K = 1,3
      IF(BLANK.EQ.UNITS(1,K)) GO TO 57
```

```
   55 CONTINUE
      WRITE(6,56)
   56 FORMAT(10X,'THE UNITS OF CONCENTRATION ARE NOT CATERED FOR IN THIS
     1 PROGRAM')
      GO TO 999
   57 CONC1 = CONC1*UNITS(2,K)
C
C
C
C     THE COEFFICIENT 'COEF' OF THE EQUATION EXPRESSING THE SETTLING VELOCITY
C     AS A FUNCTION OF CONCENTRATION IS COMPUTED FROM THE VALUES OF 'CONC1' ,
C     'VEL1' AND 'SLOPE'
C
C
   60 COEF = VEL1/(CONC1**(-SLOPE))
      WRITE(6,82) CCEF,SLOPE
   82 FORMAT(1H1,//,45X,'THE EXPRESSION FOR THE SETTLING VELOCITY IS',/,
     145X,'VELOCITY =',F8.3,' * CONCENTRATION **-',F5.3,/,45X,
     2'------------------------------------------------',//)
      WRITE(6,83) (UNITS1(I,J),I=1,2)
   83 FORMAT(24X,'RECYCLE FRACTION',5X,'SOLID FLUX',5X,'UNDERFLOW CONC.'
     1,5X,'MLSS',10X,'OVERFLOW RATE',//,44X,'LB/DAY-FT**2' ,8X,'MG/L',11X
     2,'MG/L', 12X,2A4,//)


      GO TO 1
C
C
C     THIS IS THE SECTION OF THE PROGRAM THAT READS THE OVERFLOW RATE 'OVLR'
C     CARD AND CONVERTS THE OVERFLOW RATE TO CUBIC FEET PER DAY PER SQUARE FOOT
C
C
   70 CALL MATCH('FLOW',4,&71)
      J = 2
   71 CALL MATCH('RATE',4,&72)
   72 CALL MATCH('IS',2,&73)
   73 CALL REAL(&74)
      CALL INTEGR(&75)
      WRITE(6,76)
   76 FORMAT(10X,'THE VALUE OF FLOW IS NOT PROPERLY DEFINED')
      GO TO 999
   74 OVFLR = VALUE
      GO TO 77
   75 CVFLR = IVALUE
   77 OVFLR1 = OVFLR
      CALL LABEL(&79)
   79 CD = BLANK1
      CD = ENTITY(1)
      IF(CD.EQ.BLANK1) GO TO 1
      DO 78 K = 1,4
      IF(CD.EQ.UNITS1(1,K)) GO TO 80
   78 CONTINUE
      WRITE(6,81)
   81 FORMAT(10X,'THE UNITS FOR THE OVERFLOW RATE ARE NOT CATERED FOR IN
     1 THIS PROGRAM')
      GO TO 999
   80 OVFLR = OVFLR*UNITS1(3,K)
      J = K
      GO TO 1
C
C
C
C     THIS THE SECTION OF THE PROGRAM THAT READS THE RECYCLE RATIO AND CONVERTS
C     IT TO A FRACTION
C
C
   90 CALL MATCH('RAT',3,&91)
   91 CALL MATCH('IS',2,&92)
   92 CALL CDREAD(RCYCLE,BLANK)
      IF(RCYCLE.GT.10) RCYCLE = RCYCLE/100.
      GO TO 1
C
C
```

```
C          THE SOLID FLUX LOADING RATE IN POUNDS PER DAY PER CUBIC FOOT IS COMPUTED
C          FROM THE EXPRESSION GIVEN BY DICK AND YOUNG IN MAY 1972
C
C
  100 GL = ((COEF*(SLOPE-1.))**(1./SLOPE))* (SLOPE/(SLOPE-1.))*
     1((RCYCLE*OVFLR)**((SLOPE-1.)/SLOPE))
C
C
C          THE UNDERFLOW CONCENTRATION 'CU' IS COMPUTED FROM MASS BALANCE
C          CONSIDERATIONS AND IT IS CONVERTED TO MG/L
C
C
      CU = GL*16000./(RCYCLE*OVFLR)
C
C
C          THE MIXED LIQUER SUSPENDED SOLIDS CONCENTRATION 'XMLSS' IS COMPUTED BY
C          ASSUMING ALL THE SOLIDS PRESENT IN THE INFLUENT FLOW TO BE REMOVED
C          IN THE UNDERFLOW .
C
C
      XMLSS = CU*RCYCLE/(1. + RCYCLE)
      WRITE(6,101) RCYCLE,GL,CU,XMLSS,OVFLR1
  101 FORMAT(27X,F8.5,10X,F10.5,6X,F10.2,4X,F10.2, 8X,F10.2,/)
      GO TO 1
  999 CONTINUE
      STOP
      END
```

## APPENDIX F

### Illustrative Thickening Input

```
THE OVERFLOW RATE IS 800  GALLONS.PER.DAY.PER.SQUARE.FOOT
THE SLOPE OF THE SETTLING CURVE WHEN PLOTTED ON A LOG LOG SCALE IS  -2.26
THE SETTLING VELOCITY IS 1   FOOT.PER.HOUR WHEN THE CONCENTRATION IS 1.   P.C
 THE RECYCLE RATE IS 0.1
 SOLVE
 THE RECYCLE RATE IS 0.2
SOLVE
 THE RECYCLE RATE IS 1.0
 SOLVE
THE SLOPE OF THE SETTLING CURVE WHEN PLOTTED ON A LOG LOG SCALE IS  -2.21
THE SETTLING VELOCITY IS 1.5 FEET.PER.HOUR WHEN THE CONCENTRATION IS 1.   P.C
 THE RECYCLE RATE IS 0.1
SOLVE
 THE RECYCLE RATE IS 0.2
SOLVE
 EXIT
```

# CHAPTER 6

## DEEP GRANULAR FILTERS, MODELING AND SIMULATION

John L. Cleasy

Department of Civil Engineering
Iowa State University
Ames, Iowa

## INTRODUCTION

The objective of this chapter is to present typical problems related to deep granular filter design, operation, and performance, which are appropriate for student problems. Not all the problems are complex nor do they all need computer solution. The one problem most suited to computer solution, *i.e.,* performance prediction, is probably the least convenient for real world design use due to the need for substantial time to develop the empirical data. The other simpler problems are presented because their mastery is also important to the student, and they can be applied in real world problems today without a time requirement for data collection that is unrealistic.

## FIXED BED PROBLEMS IN DEEP GRANULAR FILTERS

### Head Loss Through a Clean-Graded Filter Bed

The models for this problem are well-established and ready for practiced usage. Familiar equations such as the Kozeny equation can be used.[1,2] The biggest weakness is estimation of shape factor or sphericity of the media. For practical use, this must be estimated by visual observation of the media. More precise methods are available but are not economically practical for the design engineer. However, the results using an estimated sphericity are usually sufficiently close for the intended uses.

193

Uses of this model include calculation of initial head loss through filter beds, ion exchange resin beds, supporting gravel layers, and flow through porous media in ground water development. An example problem is presented in Appendix A.

## FLUIDIZED BED PROBLEMS

### Head Loss Through the Fluidized Bed

The head loss through a fluidized bed is simply and precisely the weight of the filtering media in water.

$$\Delta P = D(1 - \epsilon)(G_{s,m} - G_{s,w}) \tag{1}$$

where

$\Delta P$ = pressure drop across the fluidized bed in ft of water
D = unexpanded bed depth
$\epsilon$ = porosity of unexpanded bed
$G_{s,m}$ and $G_{s,w}$ = specific gravity of the filter medium and of water, respectively

This model can be applied in designing the backwashing facilities for deep granular filters; for example, this is one component of the head required for backwash pumps or backwash tanks.

### Expansion Flow Rate Relationships

The approach suggested here is the empirical approach published by Amirtharajah and Cleasby.[2] It was shown to be effective for sand. The extension of the method to garnet sand and crushed anthracite coal is near completion. The method consists of the following steps.

The minimum fluidization velocity, $v_f$, is calculated from the empirical nonhomogeneous equation,

$$V_f(\text{gpm/ft}^2) = \frac{0.0038(d_{60\%})^{1.82}[\omega_s(\omega_m - \omega_s)]^{0.94}}{\mu^{0.88}} \tag{2}$$

where

$d_{60\%}$ = 60% finer size of the sand, mm
$\omega_m$ = specific weight of the sand, lb/ft$^3$
$\omega_s$ = specific weight of the water, lb/ft$^3$
$\mu$ = water viscosity, centipoise (cP)

The dimensionless Reynolds number $Re_f = \rho_\varrho v_f d_{60\%}/\mu$ corresponding to a minimum fluidization velocity and the 60% finer sand size, is then calculated. If the $Re_f$ is greater than 10, a multiplying correction factor

$K_R$, must be applied to the above value of $v_f$, given by

$$K_R = 1.775 \, Re_f^{-0.272} \tag{3}$$

The unhindered settling velocity, $v_s$, of the hypothetical average particle is then calculated from the following expression:

$$v_s = 8.45v_f \tag{4}$$

The Reynolds number for this particle, based on the unhindered settling velocity conditions, $Re_o = \rho_\varrho v_s d_{60\%}/\mu$, is then determined for use in calculating the expansion coefficient, $n_e$, which is defined by $n_e = 4.45 \, Re_o^{-0.1}$.

The value of $n_e$ is then used in Equation 5 to find the constant $K_e$ for the particular system, using the values of velocity and porosity at minimum fluidization, $v_f$ (Equation 2) and $\epsilon$ (porosity of the unexpanded bed):

$$v = K_e(\bar{\epsilon})^{n_e} \tag{5}$$

where:

$v$ = superficial flow velocity of the water above the sand

$\bar{\epsilon}$ = porosity of the expanded bed

The equation for $n_e$ is valid for $1 < Re_o < 500$ and when the ratio of particle diameter to the diameter of the fluidization column is negligible. These conditions generally apply in a backwashed sand filter. Equation 5 is a modification of an expression developed for fluidization,[4] using the value of $K_e$ determined as indicated above. Equation 5 is then used again to calculate the expanded porosity at any desired face velocity, $v$.

The expanded depth is calculated from

$$\frac{D_e}{D} = \frac{(1 - \epsilon)}{(1 - \bar{\epsilon})} \tag{6}$$

where:

$D$ = depth of unexpanded bed

$D_e$ = depth of expanded bed

The expansion of graded filter sands at different temperatures has been predicted to an excellent degree by the above calculation procedure. However, its application to graded coal beds yielded poor expansion prediction. An example of the use of this model is presented in Appendix A.

The application of the method to other materials, such as crushed garnet and crushed anthracite coal, appears imminent and will involve modification of the relationship between $n_e$ and $Re_o$. The equation for $n_e$ presented above was developed for spherical particles. Recent data collected at Iowa

State University indicates that for nonspherical particles, this $n_e$ relationship with $Re_o$ is substantially different. However, the basic relationship between v and $\bar{\epsilon}$ is valid for all fluidized beds as demonstrated by the linearity of log v versus log $\bar{\epsilon}$ plots for all such systems.

### Uniformity of Backwash Distribution

The approach to this design problem is to make the head loss through underdrain orifices substantial compared with head losses and velocity head recovery in the underdrain distribution manifold. The losses in the underdrain manifold consist of minor losses of entrance and friction losses. The friction losses in a perforated pipe manifold can be shown to be one-third the loss expected, if the full flow travelled the full length of the manifold.[1] The friction and minor losses are partly offset by the recovery of velocity head along the manifold.

The design of an underdrain system for a filter, checking the uniformity of distribution, and sketching the position of the hydraulic grade line (HGL) and total energy line (TEL) provides an interesting challenge to the students. Such a problem is presented in Figure 6.1.

The position of the HGL and TEL can be calculated assuming uniform backwash water distribution. Then the precision of that assumption can be checked by noting the difference in the orifice head loss available at any

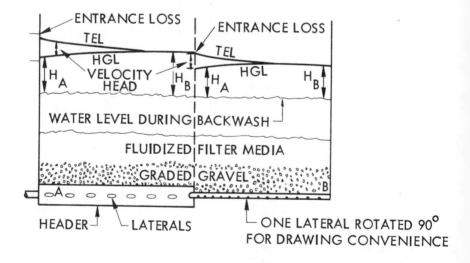

**Figure 6.1.** Cross section of deep filter during backwashing.

desired orifices. Since orifice discharge is proportional to the square root of the head loss across the orifice, the ratio of discharge between orifice A and B ($Q_A$ and $Q_B$) in Figure 6.1 would be

$$\frac{Q_A}{A_B} = \left(\frac{H_A}{H_B}\right)^{1/2} \tag{7}$$

## Prediction of Intermixing Between Two Filter Media

The ability to predict the extent of intermixing between two filter media of different specific gravity from measurable properties of the media assumes increasing importance as filters are built with new sizes or unusual combinations of media. New sizes of media are being used in domestic and industrial wastewater filters.

The mathematical model for such prediction is not yet adequately developed, so no example is presented in this chapter. The most promising approach at this time is based on the work of Pruden[5] and LeClair,[6] who noted that intermixing of two media at a particular flow rate was observed if the bulk density of each of the media calculated independently was nearly equal. They also noted that some combinations of media actually reverse (invert) their layered positions in the bed over a range of upflow rates. Since the above approach uses bulk density of the individual media as its base, the ability to predict expansion versus upflow rate for all common filter media is an essential first step to using this method.

## Filtration Rate and Water Level Changes

For several years the use of influent flow splitting and variable declining rate filters has been advocated. The advantages of these systems compared with the conventional constant rate filters with rate controllers has also been discussed.[7]

The transient response of a group of filters designed for the influent flow splitting system of rate control when the delivery of water to the filter is changed is an interesting problem for students. For example, when one filter is removed from service for backwashing, or when the total plant input is increased, the filter must respond to that change and seek a new equilibrium water level. The approach to the problem is based on two assumptions: (1) that flow through the filter is laminar before and after the change, *i.e.*, head loss is directly proportional to flow rate (this assumption has been shown to be valid[8]) and (2) the rate increases resulting from the change do not cause substantial dislodgment of deposited solids with resulting changes in permeability of the dirty bed at the time of the change. (This assumption may not be valid in all cases.)

In the influent flow splitting system, all operating filters receive an equal portion of the flow. The effluent pipe is uncontrolled. Thus, when an increased flow is delivered to a filter box, the water level in the filter box rises slowly to develop the necessary head to filter at the new filtration rate. The same type of response in the reverse world may be expected if the flow to a filter were decreased. An example problem is presented in Appendix A.

## PERFORMANCE MODELS FOR DEEP GRANULAR FILTERS

### Model of K. J. Ives[9]

Several workers have attempted to model filtration mathematically in deep granular filters. The most common approach has been to develop a model relating filtrate quality and head loss with time, incorporating measurable macroscopic variables of filtration such as filtration rate, grain size, and water viscosity. The relationships have been developed empirically by filtering various suspensions and observing filtrate quality and head loss for different filter depths. These models do not attempt to consider the dominant mechanisms responsible for observed removals. Other workers have attempted to develop models of expected filter behavior by considering the basic mechanisms responsible for removal and then synthesizing models which describe the appropriate mechanisms.

All macroscopic physical models of filtration are based on the fact that the rate of removal per unit depth of filter is proportional to the local concentration of suspended solids, as originally noted by Iwasaki:[10]

$$- \frac{\partial C_{ss}}{\partial L} = \lambda C_{ss} \tag{8}$$

where $C_{ss}$ is the concentration of suspended solids at any time and depth in the filter, L is the length (depth) of the filter, and $\lambda$ is the filter coefficient, which varies with time and depth in the filter.

The variation of the filter coefficient, $\lambda$, with the volume of specific deposit accumulated in a filter layer, has been the subject of considerable research, with different workers proposing different relationships. Ives has developed a general relationship between $\lambda$ and specific deposit that can accommodate the relationships developed by other workers.[9] It is based upon the hypothesis that the filter coefficient, $\lambda$, is a function of the changing specific filter surface (*i.e.*, surface area per unit filter volume) available for deposition and the increasing interstitial velocity. The specific surface in the early part of a filter run is represented by a porous bed of individual spheres. As the spheres are coated with deposit, the specific surface in the filter increases and $\lambda$ also increases. As the run progresses and the deposit accumulates, the flow paths tend to straighten and the specific surface is represented by an assembly of individual cylindrical capillaries. The specific surface decreases as the capillaries become smaller and $\lambda$ decreases. The interstitial velocity is modified by the average amount of deposit present in any small depth element. As the deposits accumulate, the interstitial velocity increases and $\lambda$ decreases.

The details of the model will not be presented here since they are readily available elsewhere.[9,11] The model does lend itself to incremental solution using a digital computer. However, the problem of relating the volume of the specific deposit in the filter pores to measurable concentrations of the influent and effluent is unresolved, limiting the usefulness of the model.

### Model of O'Melia and Yao

O'Melia[12] and Yao, *et al.*[13] have focused principally on primary conceptual filtration models incorporating basic mechanisms of removal. They consider filtration to be dependent upon particle transport and particle attachment. Attachment depends upon destabilization of the particles, as for destabilization in coagulation. Particle transport is required to move the particle from the bulk of the fluid in the interstices to the immediate vicinity of the solid-liquid interface such as a grain surface or a previously deposited particle. Particle transport in filtration is analogous to particle transport in flocculation processes.

The models are based upon work done in air filtration, which has shown that particles of less than about 1 $\mu$ in size are transported by Brownian diffusion. The larger particles must be transported by other mechanisms; settling and interception are considered dominant for the removal of larger particles in water filtration. Separate equations for each transport mechanism (diffusion, interception, sedimentation) have been developed. These conceptual models are useful in indicating the effect of filtration variables on filter efficiency. However, they are not ready for practical use in predicting filtrate quality and head loss development.

### Model of Hsiung and Cleasby

Hsiung and Cleasby[14] proposed a model for prediction of granular filter performance by using pilot plant operating data and the analogy of the filtration process to the statistical chi-square distribution. A brief summary of the model is presented here. (The reader is referred to the original paper for the complete development.) This model is based on the fact that in filtration suspended solids are transported through the granular filter by a fluid which changes direction in a random manner seeking a relatively unobstructed pathway. The particles are brought within the range of the van der Waal's forces of the granular filter medium or previously deposited particles and attach themselves to surfaces exhibiting such forces. The point at which a suspended particle becomes attached in the filter bed is determined in a random manner.

A typical plot for the chi-square distribution is shown in Figure 6.2.

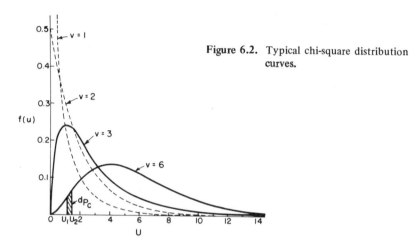

**Figure 6.2.** Typical chi-square distribution curves.

Two plots of experimental filtration data are shown in Figures 6.3 and 6.4.[15,16] The similarity between these figures and Figure 6.2 was noted and is the basis of this empirical approach. Each curve in Figures 6.3 and 6.4 represents a typical distribution for a specific time, t. The ordinates of such a curve represent

$$\frac{\Delta C}{C_0 \Delta L} \tag{9}$$

where $C_0$ equals the initial solids concentration in the water and C equals the concentration at any time, and L represents the depth of the filter media measured from the surface.

The area under each curve between $L_1$ and $L_2$ is equal to

$$\frac{\Delta C}{C_0 \Delta L} \Delta L = \frac{\Delta C}{C_0} \tag{10}$$

If small increments of C and L are considered, the area under the distribution curve between L and L + dL is equal to

$$\frac{dC}{C_0} \tag{11}$$

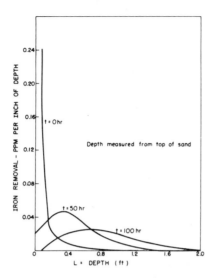

**Figure 6.3.** Iron removal per unit depth versus depth at various filtration time, Eliassen.

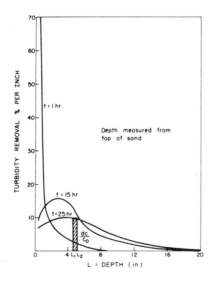

**Figure 6.4.** Turbidity removal per inch versus depth at various filtration time, Ling.

Analysis of Figure 6.2 shows that for a given v the area under each curve between $U_1$ and $U_2$ is equal to f(U). $\Delta U$ or between U and U + dU is equal to f(U)dU. By definition, the cumulative probability $P_c$ is equal to $\int$ f(U)dU for a continuous random variable U. Therefore, the area under each curve between U and U + dU is equal to $dP_c$.

The similarity between filtration data collected from pilot filters and the chi-square distribution was used to establish the relationship between U, v, and $P_c$ of the chi-square distribution and their filtration counterparts, L, t, and $C/C_0$, respectively. The exact mathematical relationship between the filtration data and the chi-square probability distribution parameters was not developed since such an approach was not considered practical. Instead, a graphical method was used, whereby the variate U of the chi-square distribution was used as a parameter for predicting filter performance. The authors concluded that a graphical analysis based on the chi-square distribution was adequate to provide sufficient and useful information to predict filter performance and for a rational filter design.

The use of the chi-square filter performance prediction technique was illustrated with filtration data collected in the laboratory using three thin-layer filters. The suspended solid used in these tests resulted from the addition of ferrous sulfate ($FeSO_4$) to aerated tap water (a hard well-water containing about 500 mg/l total dissolved solids, 320 mg/l alkalinity, and 450 mg/l of total hardness). Water temperature was maintained at $77°F$ throughout the filter runs. The tests were conducted using several sand sizes, several sand depths, different flow rates, and different influent iron concentrations ($C_0$). The resulting effluent water quality (C) and head loss at different filter depths was measured at different filtration times.

The data collected were used to develop empirical filter performance prediction equations for the removal of iron from suspension. These equations are presented and used below to illustrate application of this performance model. The equations can be used to predict the performance of alternative filter designs, including various combination of media size and depth, filtration rate, and influent concentration. The equations can also be used to determine which combination of design variables will provide equivalent performance. It should be emphasized that these performance prediction equations are only applicable to the water from which they were developed, in this case, $FeSO_4$ added to Iowa State University tap water at $77°F$. However, similar expressions can be developed for other waters using the model. The use of the equations for class problems will permit students to observe the effect of changing the variables on filter performance.

The program presented in Appendix B, based on the Hsiung-Cleasby model, can be used to select filter designs that provide equivalent performance (as illustrated in the sample output). This is one of the potential uses of the model. Another use would be to predict effluent quality and head loss development with time for any combination of the other variables. This problem is not illustrated in the program output.

## EQUIVALENT PERFORMANCE PROGRAM

Two filters may be said to provide equivalent performance when they produce the same quantity and quality of filtered water from the same water source during the same time period, *e.g.*, one day. A computer program to select filter designs providing equivalent performance is presented in Appendix B along with sample output. The computer program involves several steps:

1.  The values of the variate U in the chi-square distribution are fed into the program as built-in input data, as shown in Table 6.1, in which the variate U is related to the filtration data as $C/C_0 = P_c$ (cumulative probability) and length of run, T, in hours equals v, degree of freedom, *i.e.*, 1 hr = 1 degree of freedom. For a given influent water quality and any desired effluent water quality, the $C/C_0$ ratio, the ratio of effluent suspended solids concentration to the influent suspended solids concentration is calculated by the computer and used as an index in picking up the corresponding value of U in the chi-square distribution, which has been built into the program.

Table 6.1. Values of U, Which are Derived from Variate of Chi-Square Distribution Based on 1 hr = 1 Degree of Freedom

| $C/C_0$ T HR | 0.01 | 0.02 | 0.03 | 0.04 | 0.05 | 0.06 | 0.07 | 0.08 | 0.09 | 0.10 |
|---|---|---|---|---|---|---|---|---|---|---|
| 10 | 2.56 | 3.06 | 3.42 | 3.71 | 3.99 | 4.20 | 4.40 | 4.58 | 4.74 | 4.90 |
| 12 | 3.59 | 4.20 | 4.60 | 4.95 | 5.21 | 5.50 | 5.70 | 5.90 | 6.10 | 6.30 |
| 14 | 4.65 | 5.40 | 5.84 | 6.20 | 6.75 | 6.80 | 7.10 | 7.35 | 7.60 | 7.80 |
| 16 | 5.84 | 6.60 | 7.15 | 7.54 | 7.90 | 8.20 | 8.50 | 8.80 | 9.00 | 9.30 |
| 18 | 7.10 | 8.00 | 8.60 | 9.00 | 9.40 | 9.80 | 10.10 | 10.40 | 10.60 | 10.90 |
| 20 | 8.30 | 9.30 | 10.00 | 10.50 | 11.00 | 11.30 | 11.70 | 12.00 | 12.20 | 12.50 |
| 22 | 9.50 | 10.60 | 11.30 | 11.90 | 12.30 | 12.80 | 13.10 | 13.50 | 13.90 | 14.10 |
| 24 | 10.80 | 11.90 | 12.60 | 13.20 | 13.70 | 14.10 | 14.50 | 14.90 | 15.10 | 15.50 |
| 26 | 12.10 | 13.30 | 14.20 | 14.90 | 15.30 | 15.80 | 16.20 | 16.60 | 17.00 | 17.20 |
| 28 | 13.50 | 14.80 | 15.60 | 16.20 | 16.90 | 17.30 | 17.80 | 18.10 | 18.40 | 18.90 |
| 30 | 14.90 | 16.20 | 17.10 | 17.90 | 18.40 | 19.00 | 19.40 | 19.90 | 20.20 | 20.60 |
| 32 | 16.40 | 17.90 | 18.90 | 19.60 | 20.20 | 20.80 | 21.20 | 21.60 | 22.00 | 22.30 |
| 34 | 17.90 | 19.30 | 20.20 | 21.00 | 21.70 | 22.20 | 22.80 | 23.10 | 23.60 | 24.00 |
| 36 | 19.20 | 20.90 | 22.00 | 22.80 | 23.30 | 24.00 | 24.40 | 25.00 | 25.20 | 25.80 |
| 38 | 20.60 | 22.20 | 23.20 | 24.00 | 24.90 | 25.30 | 26.00 | 26.40 | 26.90 | 27.20 |
| 40 | 22.10 | 23.90 | 25.00 | 26.00 | 26.60 | 27.20 | 27.90 | 28.40 | 28.90 | 29.20 |
| 42 | 23.80 | 25.60 | 26.80 | 27.80 | 28.30 | 29.00 | 29.70 | 30.00 | 30.60 | 31.00 |
| 44 | 25.00 | 27.00 | 28.20 | 29.00 | 29.80 | 30.40 | 31.00 | 31.50 | 32.00 | 32.40 |
| 46 | 26.60 | 28.40 | 29.90 | 30.80 | 31.40 | 32.00 | 32.80 | 33.20 | 33.90 | 34.10 |
| 48 | 28.20 | 30.00 | 31.50 | 32.40 | 33.00 | 34.00 | 34.50 | 35.00 | 35.60 | 36.00 |
| 50 | 29.80 | 31.70 | 33.00 | 33.90 | 34.80 | 35.30 | 36.00 | 36.40 | 37.00 | 37.80 |
| 52 | 31.20 | 33.40 | 34.80 | 35.80 | 36.50 | 37.00 | 38.00 | 38.40 | 39.00 | 39.40 |

2. Two filter performance curves for removing iron from a particular water at $77°F$ [14] are programmed to predict the filtration rate, Q, and the terminal head loss, $H_t$, at a given influent quality, $C_0$, and terminal effluent quality, C, the given sand depth, L, and the given length of run, T. The two performance curves (I, II) can be represented by two mathematical models; one is used to predict filtration rate and the second is used to predict terminal head loss.

The first performance curve is expressed mathematically as

$$\log \frac{U}{L} = -0.208 + 1.950 \log \left(\frac{G}{L^{1.2}}\right) - 0.645 \left[\log \left(\frac{G}{L^{1.2}}\right)\right]^2 \tag{12}$$

where:

U = the variate from Table 6.1 corresponding to any desired ratio of $C/C_0$ (dimensionless) and run time T (hr)

L = the filter depth (in.)

G = a grouped term defined as $Q^{0.29}d^{0.62}T$ for this particular influent suspension of iron in water at temperature of $77°F$

Q = filtration rate (gpm/sq ft)

d = media size (mm)

In the computer program, the value of $G/L^{1.2}$ is calculated from Equation 12, in which U/L is a known value. $G/L^{1.2}$ is set equal to $(Q^{0.29}d^{0.62}T)/L^{1.2}$. Solving this relation for Q, one obtains the following expression which can be used to calculate the filtration rate:

$$Q = \left[\left(\frac{G}{L^{1.2}}\right)^{3.448} L^{4.1376}\right] \Big/ \left[d^{2.1377} T^{3.448}\right] \tag{13}$$

The second performance curve is expressed mathematically as

$$\log \frac{R}{L^{1.6}} = -3.25 + 1.013 \log \left(\frac{G}{L^{1.2}}\right) - 0.036 \left[\log \left(\frac{G}{L^{1.2}}\right)\right]^2 \tag{14}$$

This relation can be used to solve for $R/L^{1.6}$ from the value of $G/L^{1.2}$ found above where:

R is a group term defined as $d^{2.5}(H_t - H_0)/Q^{1.2}C_0^{1.4}$,

$(H_t - H_0)$ is the increase in head loss (ft of water), and

$C_0$ is the influent iron concentration (mg/l).

Therefore, the head loss relationship for the filters can be related as

$$H_t - H_0 = \left[\frac{R}{L^{1.6}} Q^{1.2}C_0^{1.4}L^{1.6}\right] \Big/ \left[d^{2.5}\right] \tag{15}$$

where:

$H_t$ = terminal head loss (ft of water)
$H_o$ = head loss through the clean filter at the beginning of filtration
(ft of water)

The initial head loss, $H_o$, which is the head loss at the beginning of filtration through the clean filter, is proportional to the rate of flow and filter depth. One empirical formula for estimating the head loss for the flow of water through clean sand is:[17]

$$H_o = [27 \ Lq(73 - f)]/[10^5 d^{1.89}(t_w + 20.6)] \qquad (16)$$

where:

$H_o$ = head loss (ft)
L  = depth of sand (in.)
q  = filtration rate (MGAD)
d  = 50% sand size (mm)
$t_w$ = water temperature ($^\circ$F)
f  = porosity of sand (%)

The total terminal head loss can be predicted using Equations 15 and 16. Other relations previously cited such as the Kozeny equation could also be used for $H_o$.

The computer program (Appendix B) was designed to predict the filtration rate and terminal head loss for any combination of $C/C_o$ ratio, length or run, sand size, and sand depth. For example, assume that the following conditions are to be investigated:

C  = 0.3 mg/l or iron
$C_o$ = 5.0 mg/l or iron
d  = 0.8 mm
L  = 10 to 36 in. in 1-in. increments
T  = 10 to 50 hr in 4-hr increments
Temperature of water = 77$^\circ$F
Porosity of sand = 42%
Service flow = 3 MGD (2080 gpm)
Safety factor = 2

In using the computer program in Appendix B, the data for this example would be input in the following way:

**8 0   C O L U M N   D A T A   S H E E T**

| PROGRAM | | | JOB NO. | BY |
|---|---|---|---|---|

```
              10          20          30          40          50

  Card Type 1 ← Table of U Values — 8 Column Fields

     2, 5,6,    3,.0,6,     3,.2 4,     3,.7,2,


  Card Type 2 ← Problem Type 1 or 2, First No. — Right Situated
              ' Number of Problems — Second No. — 5 Column Fields

     1
  Card Type 3 ← Example Problem Type 1, Explained Below

   0,.3,0,  5,.0,0, 4,2,.0,0  7,7,.0,0  0,.8,0,  0,.8,0  1,0,  5 0, ,2,0,8,0,.0,0,
```

The following data explains the format of the input card:

| Column | Kind of data | Example |
|---|---|---|
| 1-6 | Desired effluent concentration (mg/l) | 0.30 mg/l |
| 7-12 | Influent concentration (mg/l) | 5.00 mg/l |
| 13-18 | Sand bed porosity (%) | 42.00% |
| 19-24 | Water temperature ($^{\circ}$F) | 77.00$^{\circ}$F |
| 25-36 | Range of sand size investigated (mm) | 0.80 mm-0.80 mm |
| 37-44 | Range of run length investigated, with increment every 4 hours (hr) | 10-50 hr |
| 45-52 | Service flow (gpm) | 2080 gpm (3 mgd) |

The computer output for this particular job for predicting filter performance is also presented in Appendix B.

## REFERENCES

1. Fair, G. M., J. C. Geyer, and D. A. Okun. *Water and Wastewater Engineering, Vol. 2: Water Purification and Wastewater Treatment and Disposal.* (New York: John Wiley & Sons, 1968).
2. Camp, T. R. "Theory of Water Filtration," *J. San. Eng. Div., ASCE* **90** (SA4), 1 (1964).
3. Amirtharajah, A. and J. L. Cleasby. "Predicting Expansion of Filters During Backwash," *J. Amer. Water Works Assoc.* **64**, 52 (1972).
4. Richardson, J. F. and W. N. Zaki. "Sedimentation and Fluidization," *Trans. Inst. Chem. Eng. (Brit.)* **32**, 35 (1954).
5. Pruden, B. B. "Particle Size Segregation in Particulately Fluidized Beds," Unpublished MASc Thesis, Library, University of British Columbia, Canada (1964).

6.  LeClair, B. P. "Two Component Fluidization," Unpublished MASc Thesis, Library, University of British Columbia, Canada (1964).
7.  Cleasby, J. L. "Filter Rate Control Without Rate Controllers," *J. Amer. Water Works Assoc.* **61**, 181 (1969).
8.  Cleasby, J. L. and E. R. Baumann. "Selection of Sand Filtration Rates," *J. Amer. Water Works Assoc.* **54**, 579 (1962).
9.  Ives, K. J. "Theory of Filtration," Special Subject No. 7, *Proc. Internat. Water Supply Cong. Exhibition*, Vienna (1969) (London, England: International Water Supply Assoc., 1969).
10. Iwasaki, T. "Some Notes on Sand Filtration," *J. Amer. Water Works Assoc.* **29**, 1591 (1937).
11. Weber, Walter J., Jr. (Ed.) *Physicochemical Processes for Water Quality Control.* (New York: Wiley-Interscience, 1972).
12. O'Melia, C. R. and W. Stumm. "Theory of Water Filtration," *J. Amer. Water Works Assoc.* **59**, 1393 (1967).
13. Yao, K., M. T. Habibian, and C. R. O'Melia. "Water and Waste Water Filtration: Concepts and Applications," Publication 250, Environmental Science and Engineering Department, University of North Carolina, Chapel Hill (1970).
14. Hsiung, K. and J. L. Cleasby. "Prediction of Filter Performance," *J. San. Eng. Div., ASCE* **94**(SA6), 1043 (1968).
15. Eliassen, R. "Clogging of Rapid Sand Filters," *J. Amer. Water Works Assoc.* **33**, 926 (1941).
16. Ling, J. T. T. "A Study of Filtration Through Uniform Sand Filters," *Proc. ASCE* **81**(751), 1 (1955).
17. Hulbert, R. and D. Feben. "Hydraulics of Rapid Filter Sand," *J. Amer. Water Works Assoc.* **25**, 19 (1933).

## APPENDIX A

### Example Problems

The examples presented below are essentially the same as some problems presented in the text, *Physiochemical Processes for Water Quality Control,* Water J. Weber, Jr., Editor (New York: John Wiley & Sons, Inc., 1972), Chapter 4.

*Problem 1*

Assume a sand to be used in a filter has a 60% finer size of 0.75 mm. Calculate the backwash rate required to achieve 20% expansion of the bed with water at 20°C using the method of Amirtharajah. Initial sand porosity is 0.40.

Solution

a. Calculate minimum fluidization velocity using Equation 2

$$v_f = \frac{0.00381(d_{60\%})^{1.82}[\omega_s(\omega_m - \omega_s)]^{0.94}}{\mu^{0.88}}$$

$$= \frac{0.00381(0.75)^{1.82}[62.4(2.65 \times 62.4 - 62.4)]^{0.94}}{1.009^{0.88}}$$

$$= 8.68 \text{ gpm/ft}^2 = 0.0193 \text{ fps}$$

b. The Reynolds number corresponding to minimum fluidization velocity is then

$$Re_f = \frac{(0.75 \times 3.28 \times 10^{-3}) \text{ft} \times 0.0193 \text{ fps} \times 1.94 \text{ lb sec}^2 \text{ft}^{-4}}{2.1 \times 10^{-5} \text{ lb sec ft}^{-2}}$$

$$= 4.4$$

Because $Re_f < 10$, no correction for $v_f$ is required.

c. From Equation 4

$$v_s = 8.45v_f = 8.45 \times 0.0193 \text{ fps} = 0.163 \text{ fps}$$

d. The Reynolds number corresponding to the unhindered settling velocity is

$$Re_0 = \frac{\rho_1 v_f d 60\%}{\mu} = 8.45 Re_f$$

$$= 37.2$$

therefore, the expansion coefficient, n, is

$$n_e = 4.45 Re_0{}^{-0.1} = 4.45(37.2)^{-0.1} = 3.1$$

e. Using Equation 5 at minimum fluidization,

$$v = K_e(\epsilon)^{n_e}$$

$$8.68 \text{ gpm/ft}^2 = K_e(0.40)^{3.1}$$

therefore, the correlation constant, $K_e$, is

$$K_e = 149 \text{ gpm/ft}^2$$

f. Calculate the desired porosity at 20% expansion with Equation 6

$$\frac{D_e}{D} = \frac{1 - \epsilon}{1 - \bar{\epsilon}} \qquad 1.2 = \frac{1 - 0.4}{1 - \bar{\epsilon}}$$

$$\bar{\epsilon} = 0.50$$

g. Reapply Equation 5 to find desired wash rate, v.

$$v = 149(0.50)^{3.1} = 17.4 \text{ gpm/ft}^2$$

*Problem 2*

Calculate the pressure required in the underdrain system of a dual media filter when backwashing at a rate of 30 ipm with water at 20°C. The filter is composed of 18 in. of crushed anthracite coal over 12 in. of sand. The coal has a specific gravity of 1.67 and porosity of 0.50. The sand has a specific gravity of 2.65 and porosity of 0.40. The dual media is supported on 8 in. of graded gravel and a Leopold type underdrain system as with 5/32-in. orifices, 15/ft$^2$ on the surface and 5/8-in. control orifices, 2/ft$^2$ in the middle web. The wash trough edges are 5 ft above the bottom of the Leopold block.

Solution

a. Calculate the head loss through the fluidized sand and coal using Equation 1:

$$\Delta P = D(1 - \epsilon)(G_{s,m} - G_{s,w})$$

Coal $\Delta P = 1.5(1 - 0.50)(1.67 - 1) = 0.50$ ft

Sand $\Delta P = 1(1 - 0.40)(2.65 - 1) = 0.99$ ft

b. Calculate the loss through the supporting gravel. Assume the gravel remains unfluidized as desired, and that it has a porosity of 0.40 and a shape factor of 7.5. Using the Kozeny equation for each layer[2]

$$\frac{h_f}{L} = J \, \frac{\nu}{g} \, \frac{(1 - \epsilon)^2}{\epsilon^3} \, v \, \frac{\sigma_s}{d_p}^2$$

$\nu = 1.088 \times 10^{-5}\,\text{ft}^2/\text{sec}$, $J = 6$, $g = 32.2$ ft/sec$^2$, $v = 30$ ipm $= 0.0417$ fps, $\sigma_s = 7.5$, $\epsilon = 0.4$.

**Table 6.2**

| Layer Position | Layer Thickness (in.) | Size Range (in.) | Mean Size $d_p$ (ft) | $(d_p)^2$ | $h_f/L$ | $h_f$ |
|---|---|---|---|---|---|---|
| Top | 2 | 1/8-10 mesh* | 0.00845 | $0.71 \times 10^{-4}$ | 0.38 | 0.063 |
| 2nd | 2 | 1/4-1/8 | 0.0157 | $2.46 \times 10^{-4}$ | 0.11 | 0.018 |
| 3rd | 2 | 1/2-1/4 | 0.0312 | $9.71 \times 10^{-4}$ | 0.028 | 0.005 |
| 4th | 2 | 3/4-1/2 | 0.052 | $27.0 \times 10^{-4}$ | 0.010 | 0.002 |
| | | | | Total loss | | 0.088 |
| | | | | | | ft water |

*10 mesh U.S. Std sieve = 2 mm = 0.0787 in.

c. Loss through the underdrain orifices. Calculate the loss with the typical orifice discharge equation, $Q = C^\circ A\sqrt{2gh_f}$, where A = area of orifice (ft$^2$), $h_f$ = pressure drop across orifice (ft), Q = flow rate (cfs), $C^\circ$ = discharge coefficient, about 0.6 for a submerged sharp crested orifice.

Flow per 5/32 in. orifice = 0.0417 cfs/ft$^2$ x 1/15 = 0.00278 cfs/orifice

Solving, $h_f = 19.4$ ft

Flow per 5/8 in. orifice = 0.0416/2 = 0.021 cfs/orifice

Solving, $h_f = 4.2$ ft

d. Total pressure needed in bottom block compartment = 5 + 0.50 + 0.99 + 0.09 + 19.4 + 4.2 = 30.2 ft water.

*Problem 3*

A small plant with 3 filters arranged for the "influent flow splitting" method of rate control is in operation at a filtration rate of 2 gpm/ft$^2$.

At a particular instant, the head loss in each filter is as follows:

| Filter no. | 1 | 2 | 3 |
|---|---|---|---|
| Head loss (ft) | 2 | 5 | 8 |

Filter 3 is removed from service for backwashing and the other two filters pick up the load equally. Assuming that laminar flow conditions prevail and that no sediment is dislodged by the rate increase (*i.e.*, coefficient of permeability remains unchanged), how long would it take for filters 1 and 2 to reach a filtration rate of 2.9 $gpm/ft^2$ since the new equilibrium rate will be 3 $gpm/ft^2$.

Solution

Change in depth in the filter box ($\Delta h_f$) must equal

$$\text{(input rate - output rate)}\Delta t$$

Mathematically,

$$dh_f = (Q_{in} - Q_{out})dt$$
$$Q_{in} = \text{constant} = 3 \text{ gpm/ft}^2 = 0.4 \text{ fpm}$$
$$Q_{out} = \text{initial rate x } \frac{\text{prevailing head loss}}{\text{initial head loss}}$$

For Filter No. 1

$$Q_{out} = \left(\frac{2 \text{ gpm/ft}^2}{7.48 \text{ gal/ft}^3}\right)\left(\frac{h_f}{2}\right) = 0.268 \frac{h_f}{2} \text{ min}^{-1}$$

$$dh_f = (0.4 - 0.268 \frac{h_f}{2}) dt \Rightarrow {_2}\!\int^{2.9} \frac{dh_f}{(0.4 - 0.134 h_f)} = {_0}\!\int^t dt$$

Integrating,

$$-\frac{1}{0.134} \ln(0.4 - 0.134 h_f) \Big]_2^{2.9} = t \Big]_0^t$$

Inserting limits and solving,

$$t = 17.7 \text{ min}$$

For Filter 2, in like manner, with $h_f$ increasing from 5 to 7.25 ft to achieve the rate-change, t = 43.4 min. Thus in a 10-15 min backwash period, neither filter would reach its new equilibrium head loss.

## APPENDIX B

## FORTRAN Computer Program for Predicting Performance of Granular Filters Using the Hsiung Cleasby Model[14] for a Particular Iron Bearing Water

```
//STEP1  EXEC  FORTGCG
//FORT.SYSIN  DD  *
C PROGRAM FOR PREDICTING SAND FILTER DESIGNS PROVIDING EQUIVALENT PERFOR
C MANCE,
C THE TOTAL COST IS COMPARED AMONG SETS OF EQUIVALENT PERFORMANCE, WHICH
C IS DEFINED AS SETS OF SAND FILTER DESIGN PRODUCING EQUAL AMOUNT OF
C FILTERED WATER DURING SAME PERIOD OF TIME AND SAME EFFLUENT QUALITY
C FROM THE SAME WATER SOURCE
      DIMENSION BW(5),GU(5),H(4),AH(4),E(21),AF(21),F(5),AF(5),W(5),AW(5
     1),AO(7),OP(7),U(10,36)
C READ IN U VALUES
      N1=22
      DO 3 N=1,N1
    3 READ (5,101) (U(M,N),M=1,10)
  101 FORMAT (10F8.2)
      WRITE(6,207)
  207 FORMAT('1',////,12X,'TABLE    VALUES OF U, WHICH ARE DERIVED FROM V
     1ARIATE OF CHI-',/,20X,'SQUARE DISTRIBUTION BASED ON 1 HR=1 DEGREE
     2OF FREEDOM',//,11X,'C/CO 0.01   0.02   0.03   0.04   0.05   0.06   0.07
     3 0.08   0.09   0.10',/,12X,'T',/,12X,'HR')
      DO 606 I=1,N1
      NN=10+(I-1)*2
      WRITE(6,607) NN,(U(J,I),J=1,10)
  607 FORMAT(12X,I2,10F6.2)
  606 CONTINUE
C IN THE CASE OF PREDICTING FILTRATION RATE AND HEAD LOSS,
C BY KNOWING THE SERVICE FLOW, C/CO, RUN LENGTH, AND SAND SIZE AND
C DEPTH, THE NO. OF 1 IS WRITTEN ON THE NUMBER FIVE COLUMN  OF THE FIRST
C INPUT DATA CARD; IF THE CASE OF PREDICTING EFFLUENT QUALITY
C OF A FILTER BY KNOWING THE SERVICE FLOW, INFLUENT
C CONCENTRATION, FILTRATION RATE, HEAD LOSS LIMIT, AND SAND SIZE AND SAND
C DEPTH, THE NO. OF 2 IS WRITTEN ON THE NUMBER FIVE COLUMN
C READ IN EFFLUENT CONC, INFLUENT CONC, POROSITY OF SAND, TEMP OF WATER,
C RANGE OF SAND SIZE, RANGE OF RUN LENGTH, AND SERVICE FLOW
      READ (5,401) KKKK,KI
  401 FORMAT (2I5)
      DO 701 LI=1,KI
      IF(KKKK-1) 402,402,808
  402 READ (5,201) CONF,CONI,PORE,TEMP,DIAM,DIAM1,ITIME1,ITIME2,SRELOW
      CCOD=CONF/CONI
  201 FORMAT (6F6.2,2I4,F8.2)
      WRITE(6,806)
  806 FORMAT( /,14X,'6. READ IN THE EFFLUENT CONC., INFLUENT CONC., PORO
     1SITY OF',/,17X,'SAND, WATER TEMP., RANGE OF SAND SIZE, RANGE OF RU
     2N LENGTH,',/,17X,'AND SERVICE FLOW',/,17X,'C MG/L CO MG/L P %  TEM
     3P F   SIZE   MM   LENGTH HR  FLOW GPM',/)
      WRITE(6,807) CONF,CONI,PORE,TEMP,DIAM,DIAM1,ITIME1,ITIME2,SRELOW
  807 FORMAT(18X,F4.2,3F7.2,F6.2,F5.2,2I5,F9.2)
  200 DO 707 ITIME=ITIME1,ITIME2,4
  832 WRITE(6,300)
  300 FORMAT('1',////,12X,'TABLE   SAND FILTER DESIGNS PROVIDING EQUIVAL
     1ENT PERFORMANCE AT',/,20X,'WATER TEMPERATURE 77F, WITH A SAFETY FA
     2CTOR OF 2')
      WRITE(6,305) CONI,CONF,CCOD
  305 FORMAT(20X,'CO=',F4.1,'MG/L OF IRON,  C=',F3.1,'MG/L OF IRON,  C/C
     10=',F5.2,/)
      WRITE(6,306)
```

```
  306 FORMAT(18X,'LENGTH   SAND   SAND    FILTR HEAD THRU INIT HEAD TOTAL H
     1EAD',/,13X,'ALT. OF RUN SIZE  DEPTH  RATE    SAND      LOSS
     2 LOSS',/,20X,' T      D      L      Q      HT-HO      HO        HT'
     3,/,20X,'HR      MM     IN    GPM/SF   FT                FT    FT',//)
        DO 909 IDEP=10,36
        TIME=FLOAT(ITIME)
        DEPTH=FLOAT(IDEP)
  C TO FIND THE U VALUE, WITH SAFETY FACTOR OF 2
        CCO=CONF/CONI*100.0/2.0
        CCCO=CCO/100.0*2.0
        ICCO=IFIX(CCO)
        N=ITIME/2-4
        M=ICCO
        UVAL=U(M,N)+(U(M+1,N)-U(M,N))*(CCO-FLOAT(ICCO))
  C TO FIND THE FILTRATION RATE REQUIRED
        A=UVAL/DEPTH
        B=(1.950-SQRT(1.95**2-4.*0.645*(0.208+ALOG10(A))))/(2.0*0.645)




        C=-3.25+1.013*B-0.036*B**2
        D=10.**B
        ED=10.0**C
        Q=D**3.448*DEPTH**4.1376/(DIAM**2.1377*TIME**3.448)
  C TO FIND THE HEAD LOSS ACROSS THE FILTER
        HD=ED*Q**1.2*CONI**1.4*DEPTH**1.6/DIAM**2.5
        HO=27.*DEPTH*Q*62.7264*(73.-PORE)/(10.**5*DIAM**1.89*(TEMP+20.6))
        HT=HD+HO
        IF(HT-40.0) 111,111,707
  111 NUMB=IDEP-9
  909 WRITE(6,301) NUMB,ITIME,DIAM,DEPTH,Q,HD,HO,HT
  301 FORMAT(14X,I2,4X,I2,4X,F4.2,2X,F4.1,3X,F5.2,3X,F5.2,5X,F5.2,5X,F5.
     12)
  707 CONTINUE
        GO TO 701
  808 CONTINUE
  C READ IN INFLUENT CONCENTRATION, FILTRATION RATE, HEAD LOSS, SAND SIZE
  C AND SAND DEPTH, SERVICE FLOW, POROSITY OF SAND, TEMPERATURE OF WATER,
  C AND ORDER NO. OF JOB
        READ(5,405)CONI,Q,HT,DIAM,DEPTH,SRFLOW,PORE,TEMP,NUMB
  405 FORMAT(5F5.1,F10.1,2F5.1,I5)
        HO=27.*DEPTH*Q*62.7264*(73.-PORE)/(10.**5*DIAM**1.89*(TEMP+20.6))
        HD=HT-HO
        TIME=(HD*DIAM**1.9)*(10.**4)/(Q**1.5*CONI**1.4*DEPTH**0.5*4.16)
        G=Q**0.29*DIAM**0.62*TIME
        GG=G/DEPTH**1.2
        GGG=-0.208+1.950*ALOG10(GG)-0.645*(ALOG10(GG))**2
        UVAL=10.**GGG*DEPTH
        ITIME3=IFIX(TIME)
        N=ITIME3/2-4
        IF(UVAL-U(1,N))861,860,860
  860 DO 406 M=1,10
        IF(UVAL-U(M,N))407,407,406
  406 CONTINUE
  407 ICCO1=M
        ICCO=ICCO1/100
        CCO=FLOAT(ICCO)
        GO TO 862
  861 CCO=0.0005
  862 CONF=CONI*CCO
        WRITE(6,408)
  408 FORMAT('1',//////,12X,'LENGTH OF RUN  U VALUE  EFFLUENT QUALITY',/,
     112X,'PREDICTED, HR          PREDICTED, MG/L',//)
        WRITE(6,409) TIME,UVAL,CONF
  409 FORMAT(16X,F6.2,6X,F6.2,5X,F7.4)
  701 CONTINUE
  900 STOP
        END
```

```
//GO.FT06F001   DD   SYSOUT=5
//GO.SYSIN    DD   *
    2.56      3.06      3.42      3.71      3.99      4.20      4.40      4.58      4.74      4.90
    3.59      4.20      4.60      4.95      5.21      5.50      5.70      5.90      6.10      6.30
    4.65      5.40      5.84      6.20      6.75      6.80      7.10      7.35      7.60      7.80
    5.84      6.60      7.15      7.54      7.90      8.20      8.50      8.80      9.00      9.30
    7.10      8.00      8.60      9.00      9.40      9.80     10.10     10.40     10.60     10.90
    8.30      9.30     10.00     10.50     11.00     11.30     11.70     12.00     12.20     12.50
    9.50     10.60     11.30     11.90     12.30     12.80     13.10     13.50     13.90     14.10
   10.80     11.90     12.60     13.20     13.70     14.10     14.50     14.90     15.10     15.50
   12.10     13.30     14.20     14.90     15.30     15.80     16.20     16.60     17.00     17.20
   13.50     14.80     15.60     16.20     16.90     17.30     17.80     18.10     18.40     18.90
   14.90     16.20     17.10     17.90     18.40     19.00     19.40     19.90     20.20     20.60
   16.40     17.90     18.90     19.60     20.20     20.80     21.20     21.60     22.00     22.30
   17.90     19.30     20.20     21.00     21.70     22.20     22.80     23.10     23.60     24.00
   19.20     20.90     22.00     22.80     23.30     24.00     24.40     25.00     25.20     25.80
   20.60     22.20     23.20     24.00     24.90     25.30     26.00     26.40     26.90     27.20
   22.10     23.90     25.00     26.00     26.60     27.20     27.90     28.40     28.90     29.20
   23.80     25.60     26.80     27.80     28.30     29.00     29.70     30.00     30.60     31.00
   25.00     27.00     28.20     29.00     29.80     30.40     31.00     31.50     32.00     32.40
   26.60     28.40     29.90     30.80     31.40     32.00     32.80     33.20     33.90     34.10
   28.20     30.00     31.50     32.40     33.00     34.00     34.50     35.00     35.60     36.00
   29.80     31.70     33.00     33.90     34.80     35.30     36.00     36.40     37.00     37.80
   31.20     33.40     34.80     35.80     36.50     37.00     38.00     38.40     39.00     39.40
    1       1
   0.30     5.00     42.00     77.00     0.80     0.80     10     50   2080.00
```

TABLE    SAND FILTER DESIGNS PROVIDING EQUIVALENT PERFORMANCE AT
         WATER TEMPERATURE 77F. WITH A SAFETY FACTOR OF 2
         CO= 5.0MG/L OF IRON,   C=0.3MG/L OF IRON,   C/CO= 0.06

| ALT. | LENGTH OF RUN . T HR | SAND SIZE D MM | SAND DEPTH L IN | FILTR RATE Q GPM/SF | HEAD THRU SAND HT—HO FT | INIT HEAD LOSS HO FT | TOTAL HEAD LOSS HT FT |
|---|---|---|---|---|---|---|---|
| 1 | 10 | 0.80 | 10.0 | 2.88 | 0.98 | 0.24 | 1.22 |
| 2 | 10 | 0.80 | 11.0 | 3.66 | 1.46 | 0.33 | 1.79 |
| 3 | 10 | 0.80 | 12.0 | 4.56 | 2.09 | 0.45 | 2.54 |
| 4 | 10 | 0.80 | 13.0 | 5.59 | 2.93 | 0.60 | 3.52 |
| 5 | 10 | 0.80 | 14.0 | 6.76 | 4.00 | 0.78 | 4.78 |
| 6 | 10 | 0.80 | 15.0 | 8.08 | 5.36 | 0.99 | 6.35 |
| 7 | 10 | 0.80 | 16.0 | 9.55 | 7.04 | 1.25 | 8.30 |
| 8 | 10 | 0.80 | 17.0 | 11.18 | 9.12 | 1.56 | 10.68 |
| 9 | 10 | 0.80 | 18.0 | 12.98 | 11.64 | 1.92 | 13.56 |
| 10 | 10 | 0.80 | 19.0 | 14.95 | 14.68 | 2.33 | 17.01 |
| 11 | 10 | 0.80 | 20.0 | 17.11 | 18.31 | 2.81 | 21.12 |
| 12 | 10 | 0.80 | 21.0 | 19.45 | 22.60 | 3.35 | 25.95 |
| 13 | 10 | 0.80 | 22.0 | 22.00 | 27.63 | 3.97 | 31.60 |
| 14 | 10 | 0.80 | 23.0 | 24.75 | 33.51 | 4.67 | 38.18 |

TABLE    SAND FILTER DESIGNS PROVIDING EQUIVALENT PERFORMANCE AT
WATER TEMPERATURE 77F. WITH A SAFETY FACTOR OF 2
CO= 5.0MG/L OF IRON,   C=0.3MG/L OF IRON,   C/CO= 0.06

| ALT. | LENGTH OF RUN T HR | SAND SIZE D MM | SAND DEPTH L IN | FILTR RATE Q GPM/SF | HEAD THRU SAND HT-HO FT | INIT HEAD LOSS HO FT | TOTAL HEAD LOSS HT FT |
|---|---|---|---|---|---|---|---|
| 1 | 14 | 0.80 | 10.0 | 2.23 | 0.94 | 0.18 | 1.13 |
| 2 | 14 | 0.80 | 11.0 | 2.80 | 1.38 | 0.25 | 1.63 |
| 3 | 14 | 0.80 | 12.0 | 3.46 | 1.95 | 0.34 | 2.29 |
| 4 | 14 | 0.80 | 13.0 | 4.20 | 2.69 | 0.45 | 3.14 |
| 5 | 14 | 0.80 | 14.0 | 5.04 | 3.63 | 0.58 | 4.21 |
| 6 | 14 | 0.80 | 15.0 | 5.97 | 4.81 | 0.74 | 5.55 |
| 7 | 14 | 0.80 | 16.0 | 7.01 | 6.27 | 0.92 | 7.19 |
| 8 | 14 | 0.80 | 17.0 | 8.16 | 8.04 | 1.14 | 9.18 |
| 9 | 14 | 0.80 | 18.0 | 9.42 | 10.19 | 1.39 | 11.58 |
| 10 | 14 | 0.80 | 19.0 | 10.80 | 12.75 | 1.68 | 14.43 |
| 11 | 14 | 0.80 | 20.0 | 12.30 | 15.78 | 2.02 | 17.80 |
| 12 | 14 | 0.80 | 21.0 | 13.92 | 19.35 | 2.40 | 21.75 |
| 13 | 14 | 0.80 | 22.0 | 15.68 | 23.51 | 2.83 | 26.34 |
| 14 | 14 | 0.80 | 23.0 | 17.57 | 28.34 | 3.31 | 31.66 |
| 15 | 14 | 0.80 | 24.0 | 19.59 | 33.91 | 3.86 | 37.77 |

TABLE    SAND FILTER DESIGNS PROVIDING EQUIVALENT PERFORMANCE AT
WATER TEMPERATURE 77F. WITH A SAFETY FACTOR OF 2
CO= 5.0MG/L OF IRON,   C=0.3MG/L OF IRON,   C/CO= 0.06

| ALT. | LENGTH OF RUN T HR | SAND SIZE D MM | SAND DEPTH L IN | FILTR RATE Q GPM/SF | HEAD THRU SAND HT-HO FT | INIT HEAD LOSS HO FT | TOTAL HEAD LOSS HT FT |
|---|---|---|---|---|---|---|---|
| 1 | 18 | 0.80 | 10.0 | 1.88 | 0.95 | 0.15 | 1.10 |
| 2 | 18 | 0.80 | 11.0 | 2.34 | 1.36 | 0.21 | 1.57 |
| 3 | 18 | 0.80 | 12.0 | 2.87 | 1.90 | 0.28 | 2.19 |
| 4 | 18 | 0.80 | 13.0 | 3.46 | 2.60 | 0.37 | 2.97 |
| 5 | 18 | 0.80 | 14.0 | 4.12 | 3.47 | 0.47 | 3.94 |
| 6 | 18 | 0.80 | 15.0 | 4.85 | 4.56 | 0.60 | 5.15 |
| 7 | 18 | 0.80 | 16.0 | 5.67 | 5.88 | 0.74 | 6.63 |
| 8 | 18 | 0.80 | 17.0 | 6.56 | 7.49 | 0.91 | 8.41 |
| 9 | 18 | 0.80 | 18.0 | 7.53 | 9.42 | 1.11 | 10.53 |
| 10 | 18 | 0.80 | 19.0 | 8.60 | 11.71 | 1.34 | 13.05 |
| 11 | 18 | 0.80 | 20.0 | 9.75 | 14.41 | 1.60 | 16.01 |
| 12 | 18 | 0.80 | 21.0 | 10.99 | 17.57 | 1.89 | 19.46 |
| 13 | 18 | 0.80 | 22.0 | 12.33 | 21.24 | 2.23 | 23.47 |
| 14 | 18 | 0.80 | 23.0 | 13.77 | 25.47 | 2.60 | 28.07 |
| 15 | 18 | 0.80 | 24.0 | 15.32 | 30.34 | 3.01 | 33.35 |
| 16 | 18 | 0.80 | 25.0 | 16.96 | 35.89 | 3.48 | 39.36 |

TABLE    SAND FILTER DESIGNS PROVIDING EQUIVALENT PERFORMANCE AT
WATER TEMPERATURE 77F. WITH A SAFETY FACTOR OF 2
CO= 5.0MG/L OF IRON,  C=0.3MG/L OF IRON,  C/CO= 0.06

| ALT. | LENGTH OF RUN T HR | SAND SIZE D MM | SAND DEPTH L IN | FILTR RATE Q GPM/SF | HEAD THRU SAND HT−HO FT | INIT HEAD LOSS HO FT | TOTAL HEAD LOSS HT FT |
|---|---|---|---|---|---|---|---|
| 1 | 22 | 0.80 | 10.0 | 1.59 | 0.90 | 0.13 | 1.03 |
| 2 | 22 | 0.80 | 11.0 | 1.96 | 1.27 | 0.18 | 1.45 |
| 3 | 22 | 0.80 | 12.0 | 2.38 | 1.76 | 0.23 | 2.00 |
| 4 | 22 | 0.80 | 13.0 | 2.85 | 2.38 | 0.30 | 2.69 |
| 5 | 22 | 0.80 | 14.0 | 3.37 | 3.16 | 0.39 | 3.54 |
| 6 | 22 | 0.80 | 15.0 | 3.96 | 4.11 | 0.49 | 4.60 |
| 7 | 22 | 0.80 | 16.0 | 4.60 | 5.27 | 0.60 | 5.88 |
| 8 | 22 | 0.80 | 17.0 | 5.30 | 6.67 | 0.74 | 7.41 |
| 9 | 22 | 0.80 | 18.0 | 6.06 | 8.34 | 0.89 | 9.24 |
| 10 | 22 | 0.80 | 19.0 | 6.89 | 10.32 | 1.07 | 11.39 |
| 11 | 22 | 0.80 | 20.0 | 7.79 | 12.64 | 1.28 | 13.91 |
| 12 | 22 | 0.80 | 21.0 | 8.76 | 15.33 | 1.51 | 16.84 |
| 13 | 22 | 0.80 | 22.0 | 9.79 | 18.46 | 1.77 | 20.22 |
| 14 | 22 | 0.80 | 23.0 | 10.91 | 22.05 | 2.06 | 24.11 |
| 15 | 22 | 0.80 | 24.0 | 12.10 | 26.16 | 2.38 | 28.54 |
| 16 | 22 | 0.80 | 25.0 | 13.37 | 30.83 | 2.74 | 33.57 |
| 17 | 22 | 0.80 | 26.0 | 14.71 | 36.13 | 3.14 | 39.27 |

TABLE    SAND FILTER DESIGNS PROVIDING EQUIVALENT PERFORMANCE AT
WATER TEMPERATURE 77F. WITH A SAFETY FACTOR OF 2
CO= 5.0MG/L OF IRON,  C=0.3MG/L OF IRON,  C/CO= 0.06

| ALT. | LENGTH OF RUN T HR | SAND SIZE D MM | SAND DEPTH L IN | FILTR RATE Q GPM/SF | HEAD THRU SAND HT−HO FT | INIT HEAD LOSS HO FT | TOTAL HEAD LOSS HT FT |
|---|---|---|---|---|---|---|---|
| 1 | 26 | 0.80 | 10.0 | 1.40 | 0.88 | 0.12 | 1.00 |
| 2 | 26 | 0.80 | 11.0 | 1.72 | 1.24 | 0.16 | 1.40 |
| 3 | 26 | 0.80 | 12.0 | 2.07 | 1.70 | 0.20 | 1.91 |
| 4 | 26 | 0.80 | 13.0 | 2.47 | 2.28 | 0.26 | 2.54 |
| 5 | 26 | 0.80 | 14.0 | 2.91 | 3.00 | 0.33 | 3.33 |
| 6 | 26 | 0.80 | 15.0 | 3.40 | 3.88 | 0.42 | 4.29 |
| 7 | 26 | 0.80 | 16.0 | 3.93 | 4.94 | 0.52 | 5.46 |
| 8 | 26 | 0.80 | 17.0 | 4.51 | 6.22 | 0.63 | 6.85 |
| 9 | 26 | 0.80 | 18.0 | 5.14 | 7.73 | 0.76 | 8.49 |
| 10 | 26 | 0.80 | 19.0 | 5.83 | 9.51 | 0.91 | 10.42 |
| 11 | 26 | 0.80 | 20.0 | 6.57 | 11.60 | 1.08 | 12.67 |
| 12 | 26 | 0.80 | 21.0 | 7.36 | 14.01 | 1.27 | 15.28 |
| 13 | 26 | 0.80 | 22.0 | 8.21 | 16.80 | 1.48 | 18.28 |
| 14 | 26 | 0.80 | 23.0 | 9.12 | 20.00 | 1.72 | 21.72 |
| 15 | 26 | 0.80 | 24.0 | 10.09 | 23.64 | 1.99 | 25.63 |
| 16 | 26 | 0.80 | 25.0 | 11.12 | 27.77 | 2.28 | 30.06 |
| 17 | 26 | 0.80 | 26.0 | 12.22 | 32.45 | 2.61 | 35.05 |

TABLE    SAND FILTER DESIGNS PROVIDING EQUIVALENT PERFORMANCE AT
WATER TEMPERATURE 77F. WITH A SAFETY FACTOR OF 2
$CO= 5.0MG/L$ OF IRON,   $C=0.3MG/L$ OF IRON,   $C/CO= 0.06$

| ALT. | LENGTH OF RUN T HR | SAND SIZE D MM | SAND DEPTH L IN | FILTR RATE Q GPM/SF | HEAD THRU SAND HT-HO FT | INIT HEAD LOSS HO FT | TOTAL HEAD LOSS HT FT |
|---|---|---|---|---|---|---|---|
| 1 | 30 | 0.80 | 10.0 | 1.26 | 0.87 | 0.10 | 0.97 |
| 2 | 30 | 0.80 | 11.0 | 1.53 | 1.21 | 0.14 | 1.34 |
| 3 | 30 | 0.80 | 12.0 | 1.84 | 1.64 | 0.18 | 1.82 |
| 4 | 30 | 0.80 | 13.0 | 2.18 | 2.18 | 0.23 | 2.41 |
| 5 | 30 | 0.80 | 14.0 | 2.55 | 2.84 | 0.29 | 3.13 |
| 6 | 30 | 0.80 | 15.0 | 2.96 | 3.65 | 0.36 | 4.02 |
| 7 | 30 | 0.80 | 16.0 | 3.42 | 4.63 | 0.45 | 5.08 |
| 8 | 30 | 0.80 | 17.0 | 3.91 | 5.79 | 0.54 | 6.34 |
| 9 | 30 | 0.80 | 18.0 | 4.44 | 7.17 | 0.66 | 7.82 |
| 10 | 30 | 0.80 | 19.0 | 5.01 | 8.78 | 0.78 | 9.56 |
| 11 | 30 | 0.80 | 20.0 | 5.63 | 10.66 | 0.92 | 11.59 |
| 12 | 30 | 0.80 | 21.0 | 6.30 | 12.84 | 1.08 | 13.92 |
| 13 | 30 | 0.80 | 22.0 | 7.01 | 15.33 | 1.26 | 16.60 |
| 14 | 30 | 0.80 | 23.0 | 7.77 | 18.19 | 1.47 | 19.65 |
| 15 | 30 | 0.80 | 24.0 | 8.58 | 21.44 | 1.69 | 23.13 |
| 16 | 30 | 0.80 | 25.0 | 9.44 | 25.11 | 1.94 | 27.05 |
| 17 | 30 | 0.80 | 26.0 | 10.35 | 29.25 | 2.21 | 31.46 |
| 18 | 30 | 0.80 | 27.0 | 11.31 | 33.90 | 2.50 | 36.41 |

TABLE    SAND FILTER DESIGNS PROVIDING EQUIVALENT PERFORMANCE AT
WATER TEMPERATURE 77F. WITH A SAFETY FACTOR OF 2
$CO= 5.0MG/L$ OF IRON,   $C=0.3MG/L$ OF IRON,   $C/CO= 0.06$

| ALT. | LENGTH OF RUN T HR | SAND SIZE D MM | SAND DEPTH L IN | FILTR RATE Q GPM/SF | HEAD THRU SAND HT-HO FT | INIT HEAD LOSS HO FT | TOTAL HEAD LOSS HT FT |
|---|---|---|---|---|---|---|---|
| 1 | 34 | 0.80 | 10.0 | 1.17 | 0.88 | 0.10 | 0.98 |
| 2 | 34 | 0.80 | 11.0 | 1.41 | 1.21 | 0.13 | 1.34 |
| 3 | 34 | 0.80 | 12.0 | 1.68 | 1.63 | 0.17 | 1.80 |
| 4 | 34 | 0.80 | 13.0 | 1.98 | 2.15 | 0.21 | 2.36 |
| 5 | 34 | 0.80 | 14.0 | 2.32 | 2.79 | 0.27 | 3.05 |
| 6 | 34 | 0.80 | 15.0 | 2.68 | 3.56 | 0.33 | 3.89 |
| 7 | 34 | 0.80 | 16.0 | 3.07 | 4.49 | 0.40 | 4.89 |
| 8 | 34 | 0.80 | 17.0 | 3.50 | 5.58 | 0.49 | 6.07 |
| 9 | 34 | 0.80 | 18.0 | 3.97 | 6.88 | 0.59 | 7.47 |
| 10 | 34 | 0.80 | 19.0 | 4.47 | 8.39 | 0.70 | 9.09 |
| 11 | 34 | 0.80 | 20.0 | 5.01 | 10.15 | 0.82 | 10.97 |
| 12 | 34 | 0.80 | 21.0 | 5.59 | 12.17 | 0.96 | 13.13 |
| 13 | 34 | 0.80 | 22.0 | 6.20 | 14.49 | 1.12 | 15.61 |
| 14 | 34 | 0.80 | 23.0 | 6.86 | 17.13 | 1.29 | 18.43 |
| 15 | 34 | 0.80 | 24.0 | 7.56 | 20.13 | 1.49 | 21.62 |
| 16 | 34 | 0.80 | 25.0 | 8.30 | 23.52 | 1.70 | 25.22 |
| 17 | 34 | 0.80 | 26.0 | 9.08 | 27.32 | 1.94 | 29.26 |
| 18 | 34 | 0.80 | 27.0 | 9.91 | 31.58 | 2.19 | 33.78 |
| 19 | 34 | 0.80 | 28.0 | 10.78 | 36.34 | 2.48 | 38.81 |

TABLE    SAND FILTER DESIGNS PROVIDING EQUIVALENT PERFORMANCE AT
         WATER TEMPERATURE 77F. WITH A SAFETY FACTOR OF 2
         CO= 5.0MG/L OF IRON,   C=0.3MG/L OF IRON,   C/CO= 0.06

| ALT. | LENGTH OF RUN T HR | SAND SIZE D MM | SAND DEPTH L IN | FILTR RATE Q GPM/SF | HEAD THRU SAND HT-HO FT | INIT HEAD LOSS HO FT | TOTAL HEAD LOSS HT FT |
|---|---|---|---|---|---|---|---|
| 1 | 38 | 0.80 | 10.0 | 1.09 | 0.88 | 0.09 | 0.97 |
| 2 | 38 | 0.80 | 11.0 | 1.30 | 1.20 | 0.12 | 1.32 |
| 3 | 38 | 0.80 | 12.0 | 1.54 | 1.60 | 0.15 | 1.75 |
| 4 | 38 | 0.80 | 13.0 | 1.81 | 2.09 | 0.19 | 2.28 |
| 5 | 38 | 0.80 | 14.0 | 2.10 | 2.69 | 0.24 | 2.94 |
| 6 | 38 | 0.80 | 15.0 | 2.42 | 3.42 | 0.30 | 3.72 |
| 7 | 38 | 0.80 | 16.0 | 2.77 | 4.29 | 0.36 | 4.65 |
| 8 | 38 | 0.80 | 17.0 | 3.15 | 5.32 | 0.44 | 5.76 |
| 9 | 38 | 0.80 | 18.0 | 3.55 | 6.52 | 0.52 | 7.05 |
| 10 | 38 | 0.80 | 19.0 | 3.99 | 7.93 | 0.62 | 8.55 |
| 11 | 38 | 0.80 | 20.0 | 4.46 | 9.55 | 0.73 | 10.28 |
| 12 | 38 | 0.80 | 21.0 | 4.97 | 11.41 | 0.86 | 12.27 |
| 13 | 38 | 0.80 | 22.0 | 5.50 | 13.55 | 0.99 | 14.54 |
| 14 | 38 | 0.80 | 23.0 | 6.07 | 15.97 | 1.15 | 17.12 |
| 15 | 38 | 0.80 | 24.0 | 6.68 | 18.72 | 1.31 | 20.03 |
| 16 | 38 | 0.80 | 25.0 | 7.32 | 21.81 | 1.50 | 23.31 |
| 17 | 38 | 0.80 | 26.0 | 8.00 | 25.28 | 1.71 | 26.98 |
| 18 | 38 | 0.80 | 27.0 | 8.71 | 29.15 | 1.93 | 31.08 |
| 19 | 38 | 0.80 | 28.0 | 9.47 | 33.47 | 2.17 | 35.65 |

TABLE    SAND FILTER DESIGNS PROVIDING EQUIVALENT PERFORMANCE AT
         WATER TEMPERATURE 77F. WITH A SAFETY FACTOR OF 2
         CO= 5.0MG/L OF IRON,   C=0.3MG/L OF IRON,   C/CO= 0.06

| ALT. | LENGTH OF RUN T HR | SAND SIZE D MM | SAND DEPTH L IN | FILTR RATE Q GPM/SF | HEAD THRU SAND HT-HO FT | INIT HEAD LOSS HO FT | TOTAL HEAD LOSS HT FT |
|---|---|---|---|---|---|---|---|
| 1 | 46 | 0.80 | 10.0 | 1.02 | 0.96 | 0.08 | 1.05 |
| 2 | 46 | 0.80 | 11.0 | 1.20 | 1.29 | 0.11 | 1.39 |
| 3 | 46 | 0.80 | 12.0 | 1.41 | 1.68 | 0.14 | 1.82 |
| 4 | 46 | 0.80 | 13.0 | 1.63 | 2.17 | 0.17 | 2.35 |
| 5 | 46 | 0.80 | 14.0 | 1.88 | 2.76 | 0.22 | 2.98 |
| 6 | 46 | 0.80 | 15.0 | 2.15 | 3.46 | 0.26 | 3.73 |
| 7 | 46 | 0.80 | 16.0 | 2.44 | 4.30 | 0.32 | 4.62 |
| 8 | 46 | 0.80 | 17.0 | 2.75 | 5.27 | 0.38 | 5.66 |
| 9 | 46 | 0.80 | 18.0 | 3.09 | 6.41 | 0.46 | 6.87 |
| 10 | 46 | 0.80 | 19.0 | 3.45 | 7.73 | 0.54 | 8.27 |
| 11 | 46 | 0.80 | 20.0 | 3.84 | 9.25 | 0.63 | 9.88 |
| 12 | 46 | 0.80 | 21.0 | 4.25 | 10.98 | 0.73 | 11.71 |
| 13 | 46 | 0.80 | 22.0 | 4.69 | 12.95 | 0.85 | 13.79 |
| 14 | 46 | 0.80 | 23.0 | 5.16 | 15.17 | 0.97 | 16.15 |
| 15 | 46 | 0.80 | 24.0 | 5.65 | 17.68 | 1.11 | 18.80 |
| 16 | 46 | 0.80 | 25.0 | 6.17 | 20.50 | 1.27 | 21.76 |
| 17 | 46 | 0.80 | 26.0 | 6.72 | 23.64 | 1.43 | 25.08 |
| 18 | 46 | 0.80 | 27.0 | 7.30 | 27.15 | 1.62 | 28.76 |
| 19 | 46 | 0.80 | 28.0 | 7.91 | 31.03 | 1.82 | 32.85 |
| 20 | 46 | 0.80 | 29.0 | 8.55 | 35.33 | 2.03 | 37.36 |

TABLE    SAND FILTER DESIGNS PROVIDING EQUIVALENT PERFORMANCE AT
WATER TEMPERATURE 77F. WITH A SAFETY FACTOR OF 2
CO= 5.0MG/L OF IRON,  C=0.3MG/L OF IRON,  C/CO= 0.06

| ALT. | LENGTH OF RUN T HR | SAND SIZE D MM | SAND DEPTH L IN | FILTR RATE Q GPM/SF | HEAD THRU SAND HT−HO FT | INIT HEAD LOSS HO FT | TOTAL HEAD LOSS HT FT |
|---|---|---|---|---|---|---|---|
| 1 | 42 | 0.80 | 10.0 | 1.07 | 0.95 | 0.09 | 1.04 |
| 2 | 42 | 0.80 | 11.0 | 1.28 | 1.28 | 0.12 | 1.40 |
| 3 | 42 | 0.80 | 12.0 | 1.50 | 1.70 | 0.15 | 1.84 |
| 4 | 42 | 0.80 | 13.0 | 1.75 | 2.20 | 0.19 | 2.39 |
| 5 | 42 | 0.80 | 14.0 | 2.02 | 2.81 | 0.23 | 3.05 |
| 6 | 42 | 0.80 | 15.0 | 2.32 | 3.55 | 0.29 | 3.84 |
| 7 | 42 | 0.80 | 16.0 | 2.64 | 4.43 | 0.35 | 4.77 |
| 8 | 42 | 0.80 | 17.0 | 2.99 | 5.46 | 0.42 | 5.87 |
| 9 | 42 | 0.80 | 18.0 | 3.37 | 6.66 | 0.50 | 7.16 |
| 10 | 42 | 0.80 | 19.0 | 3.77 | 8.06 | 0.59 | 8.65 |
| 11 | 42 | 0.80 | 20.0 | 4.21 | 9.67 | 0.69 | 10.36 |
| 12 | 42 | 0.80 | 21.0 | 4.67 | 11.52 | 0.80 | 12.32 |
| 13 | 42 | 0.80 | 22.0 | 5.16 | 13.62 | 0.93 | 14.55 |
| 14 | 42 | 0.80 | 23.0 | 5.68 | 16.01 | 1.07 | 17.08 |
| 15 | 42 | 0.80 | 24.0 | 6.24 | 18.70 | 1.23 | 19.93 |
| 16 | 42 | 0.80 | 25.0 | 6.82 | 21.73 | 1.40 | 23.13 |
| 17 | 42 | 0.80 | 26.0 | 7.44 | 25.12 | 1.59 | 26.70 |
| 18 | 42 | 0.80 | 27.0 | 8.10 | 28.89 | 1.79 | 30.69 |
| 19 | 42 | 0.80 | 28.0 | 8.78 | 33.09 | 2.02 | 35.11 |

TABLE    SAND FILTER DESIGNS PROVIDING EQUIVALENT PERFORMANCE AT
WATER TEMPERATURE 77F. WITH A SAFETY FACTOR OF 2
CO= 5.0MG/L OF IRON,  C=0.3MG/L OF IRON,  C/CO= 0.06

| ALT. | LENGTH OF RUN T HR | SAND SIZE D MM | SAND DEPTH L IN | FILTR RATE Q GPM/SF | HEAD THRU SAND HT−HO FT | INIT HEAD LOSS HO FT | TOTAL HEAD LOSS HT FT |
|---|---|---|---|---|---|---|---|
| 1 | 50 | 0.80 | 10.0 | 0.97 | 0.98 | 0.08 | 1.06 |
| 2 | 50 | 0.80 | 11.0 | 1.14 | 1.29 | 0.10 | 1.39 |
| 3 | 50 | 0.80 | 12.0 | 1.33 | 1.68 | 0.13 | 1.81 |
| 4 | 50 | 0.80 | 13.0 | 1.53 | 2.15 | 0.16 | 2.31 |
| 5 | 50 | 0.80 | 14.0 | 1.76 | 2.72 | 0.20 | 2.92 |
| 6 | 50 | 0.80 | 15.0 | 2.00 | 3.39 | 0.25 | 3.64 |
| 7 | 50 | 0.80 | 16.0 | 2.27 | 4.19 | 0.30 | 4.49 |
| 8 | 50 | 0.80 | 17.0 | 2.55 | 5.12 | 0.36 | 5.48 |
| 9 | 50 | 0.80 | 18.0 | 2.86 | 6.20 | 0.42 | 6.62 |
| 10 | 50 | 0.80 | 19.0 | 3.19 | 7.45 | 0.50 | 7.95 |
| 11 | 50 | 0.80 | 20.0 | 3.54 | 8.88 | 0.58 | 9.46 |
| 12 | 50 | 0.80 | 21.0 | 3.91 | 10.51 | 0.67 | 11.19 |
| 13 | 50 | 0.80 | 22.0 | 4.30 | 12.37 | 0.78 | 13.14 |
| 14 | 50 | 0.80 | 23.0 | 4.72 | 14.46 | 0.89 | 15.35 |
| 15 | 50 | 0.80 | 24.0 | 5.17 | 16.81 | 1.02 | 17.83 |
| 16 | 50 | 0.80 | 25.0 | 5.63 | 19.44 | 1.16 | 20.60 |
| 17 | 50 | 0.80 | 26.0 | 6.13 | 22.38 | 1.31 | 23.69 |
| 18 | 50 | 0.80 | 27.0 | 6.65 | 25.65 | 1.47 | 27.12 |
| 19 | 50 | 0.80 | 28.0 | 7.19 | 29.26 | 1.65 | 30.91 |
| 20 | 50 | 0.80 | 29.0 | 7.76 | 33.26 | 1.85 | 35.10 |
| 21 | 50 | 0.80 | 30.0 | 8.36 | 37.66 | 2.06 | 39.71 |

# CHAPTER 7

## MODELING AND SIMULATION OF SLURRY BIOLOGICAL REACTORS

Alonzo Wm. Lawrence

Department of Environmental Engineering
Cornell University
Ithaca, New York

## INTRODUCTION

A unified basis for design and operation of biological waste treatment systems employing slurries or suspensions of microorganisms has been developed from microbial kinetic concepts and continuous culture of microorganisms theory.[1] Biological Solids Retention Time ($\theta_c$), the average time period a unit of biological mass is retained in the system, is employed as the independent parameter in steady-state process design and control. $\theta_c$ is functionally related to microbial net specific growth rate and process performance. In addition, $\theta_c$ is a readily controlled operational parameter. The basic relationships between microbial growth and substrate assimilation employed in developing the steady-state models have also been employed in developing dynamic simulation models of suspended growth biological waste treatment processes.

In this chapter, the biological solids retention time concept is used to describe several aspects of modeling and simulation of slurry biological reactors. Both steady-state and dynamic-loading conditions are considered. Steady-state kinetic models are presented for three process configurations, *i.e.*, completely mixed reactor without solids recycle, completely mixed reactor with solids recycle, and plug flow reactor with solids recycle. Dynamic simulation models are described for the first two process configurations mentioned above. In addition to biological reactor process

221

models, this chapter develops the interrelationships between the biological reactor and the gravity separation process usually employed in process configurations with biological solids recycle. For the steady-state situation, attention is focused on: (1) the relationship between $\theta_c$ and biomass settling characteristics, (2) a design model for secondary clarifiers, and (3) a least-cost design approach for reactor-clarifier systems. For the non-steady-state situation, the recycle process model includes an empirical dynamic model of secondary clarifier operation. The concluding section of the paper surveys the kinetic information available for design and control of various aerobic and anaerobic biological waste treatment processes.

## BASIC KINETIC EQUATIONS

### Microbial Growth and Substrate Utilization

The relationship between biological growth and substrate utilization can be formulated in two basic equations. The first equation describes the relationship between net rate of growth of microorganisms and rate of substrate utilization as

$$dX/dt = Y(dF/dt) - bX \qquad (1)$$

in which $dX/dt$ = net growth rate of microorganisms per unit volume of reactor, mass/volume-time; $Y$ = growth yield coefficient, mass/mass; $dF/dt$ = rate of microbial substrate utilization per unit volume, mass/volume-time; $b$ = microorganism decay coefficient, time$^{-1}$; and $X$ = microbial mass concentration, mass/volume. This equation was developed empirically from waste treatment studies[2,3] and more recently has been shown to apply to pure culture microbial systems as well.[4] Another and perhaps more useful form of Equation 1 is obtained by dividing Equation 1 by $X$, the microbial mass concentration. The resulting equation is

$$\mu = YU - b \qquad (2)$$

in which $\mu = (dX/dt)/X$, the net specific growth rate of microorganisms, time$^{-1}$; and $U = (dF/dt)/X$, specific utilization which is the rate of removal of substrate per unit weight of microorganisms, time$^{-1}$.

The second basic equation relates the rate of substrate utilization both to the concentration of microorganisms in the reactor and to the concentration of substrate surrounding the organisms. The equation, in only a slightly different form from that used by Monod, is

$$\frac{dF}{dt} = \frac{kSX}{K_s + S} \qquad (3)$$

in which k = maximum rate of substrate utilization per unit weight of microorganisms (occurring at high substrate concentration), $time^{-1}$; S = concentration of substrate surrounding the microorganisms, mass/volume; and K = half velocity coefficient, equal to the substrate concentration when U = (dF/dt)/X = (1/2)k, mass/volume. This equation indicates that the functional relationship between substrate utilization rate and substrate concentration is continuous over the total range of substrate concentrations. In the two extreme cases—when S is very high ($S \gg K_s$) and when S is very low ($S \ll K_s$), Equation 3 can be approximated by the following discontinuous functions:

$$dF/dt = kX \qquad S \gg K_s \qquad (4)$$
$$dF/dt = k'XS \qquad S \ll K_s \qquad (5)$$

in which $k' = k/K_s$. Equation 4 is a zero-order reaction with respect to substrate concentration while Equation 5 is first-order. While the merits of the two models have been debated, it should be emphasized that both are empirical and a choice between the two should properly be based more upon convenience and ability to furnish a satisfactory solution than upon any fundamental considerations. Use of a continuous function (Equation 3) is, in many ways, more satisfying and allows comparison of results with studies reported in the microbial literature. For example, by combining Equations 1 and 3, the resulting relationship between growth rate and substrate concentration is identical to the equation proposed by van Uden for pure culture microbial systems.[5] This relationship is

$$\mu = \frac{YkS}{K_s + S} - b \qquad (6)$$

Also, the parameter $K_s$ of the continuous function provides valuable information about the shape and limits of the process efficiency curve.

### Process Design and Control Parameters

There are two operational parameters that have been applied widely in the design and operation of biological treatment systems. Their relationship with each other and also with the basic equations already presented is of interest. The first parameter has variously been called: process loading factor, substrate removal velocity, or food to microorganism ratio. In this chapter, the comparable parameter, termed specific utilization (U), is operationally defined as

$$U \equiv \frac{(\Delta F/\Delta t)_T}{X_T} \qquad (7)$$

in which $(\Delta F/\Delta t)_T$ represents the mass of substrate (waste) utilized over a finite time period (usually a day) by a mass of organisms, $X_T$. The relationship of this parameter to microbial growth rate is shown by Equation 2. The second parameter has been variously termed sludge age, mean cell retention time, or biological solids retention time $(\theta_c)$ and is operationally defined as

$$\theta_c \equiv \frac{X_T}{(\Delta X/\Delta t)_T} \tag{8}$$

in which $X_T$ = total active microbial mass in treatment system, mass; and $(\Delta X/\Delta t)_T$ = total quantity of active microbial mass withdrawn daily, including those solids purposely wasted as well as those lost in the effluent, mass/time. Under steady state conditions, $\theta_c$ is the reciprocal of the microbial net specific growth rate $(\mu)$.

Two particular $\theta_c$ values $(\theta_c^m, \theta_c^d)$ will be defined here and will be related to the basic parameters and to particular treatment systems. $\theta_c^m$ represents the lower value or minimum $\theta_c$ at which complete failure of the biological process will occur. Below $\theta_c^m$, the organisms are removed from the system at a rate greater than their synthesis rate so that eventually no organisms will be left in the system. $\theta_c^d$ represents the $\theta_c$ value to be used for design and must be significantly greater than $\theta_c^m$. The ratio $(\theta_c^d/\theta_c^m)$ gives the safety factor for the system and generally varies from an extreme minimum of 2 to conservation values of 20 or more.

By comparison, U is a measure of the rate of substrate (waste) utilization by a unit mass of organisms, while $\theta_c$ is a measure of the average retention time of organisms in the system. A desired treatment efficiency can be obtained by control of either of these parameters since they are both functionally related to net specific growth rate $(\mu)$. However, in a complex treatment system, $\theta_c$ is the more readily measurable and more easily controlled of the two, and so should generally be the parameter of choice.

**Process Performance**

The efficiency of treatment by a biological process is defined as

$$E \equiv \frac{100(S_0 - S_1)}{S_0} \tag{9}$$

in which E = treatment efficiency, percent; $S_0$ = influent waste concentration, mass/volume; and $S_1$ = effluent waste concentration, mass/volume.

There are two efficiencies of interest, the specific efficiency, $E_s$, and the gross efficiency, $E_g$. The specific efficiency refers to removal of some specific component or group of components in the waste stream. For example, one may speak of the specific removal of phenols, or carbohydrates, of acetic acid, or of "soluble" Biochemical Oxygen Demand (BOD). The gross efficiency refers to the overall removals as measured by some gross parameter. Thus the gross efficiency of BOD removal refers to the removal of all forms of BOD (dissolved plus suspended materials) as determined by BOD measurements on composited influent and effluent samples. Specific removals of soluble components are generally much higher than gross BOD or Chemical Oxygen Demand (COD) removals. Gross effluent BOD, COD, or suspended solids measurements indicate both the remainder of the original waste components plus the synthesized biological solids which happen to escape in the effluent. It is generally known that biological solids synthesized during treatment are often the major contributors to effluent BOD in aerobic processes. The quantity of biological solids in the effluent stream is a function of their settling properties and the design and operation of the final solids separating device. The mathematical models to be developed here apply in a strict sense only to specific removal efficiencies for soluble or readily solubilized components of the waste stream. However, these models have also proved useful when the major interest is in gross removals.

## STEADY-STATE TREATMENT MODELS

Relationships describing steady-state conditions in continuous flow biological waste treatment systems can be developed by performing materials balances on the constituents entering and leaving the system. Relationships presented here will be for three suspended microbial waste treatment processes as illustrated in Figure 7.1: (a) completely mixed reactor without biological solids recycle, (b) completely mixed reactor with biological solids recycle, and (c) plug flow reactor with biological solids recycle. A critical assumption is that microbial growth is limited by the availability of one substance or category of substances. All other growth requirements, *i.e.*, inorganic nutrients and trace organic growth factors, are present in excess amounts. Organic waste constituents are the usual category of growth limiting substances for heterotrophic organisms, while the growth of autotrophic organisms is usually limited by the availability of an inorganic energy source. In infrequent cases, other materials such as nitrogen and phosphorus can be growth-limiting.

Because detailed derivations of these three process models have been presented previously by Lawrence and McCarty,[6] only one derivation is described here for illustrative purposes. The process chosen for discussion is the completely mixed reactor with solids recycle. A tabular summary of the equations for all three process configurations appears at the end of this chapter.

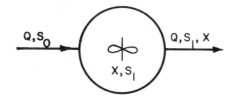

COMPLETELY MIXED - NO SOLIDS RECYCLE

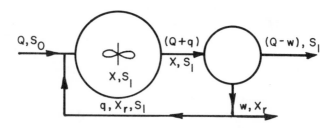

COMPLETELY MIXED - SOLIDS RECYCLE

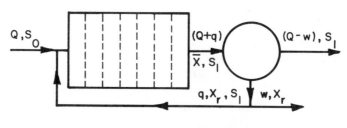

PLUG FLOW - SOLIDS RECYCLE

**Figure 7.1.** Schematic representation of three biological process flow schemes.

### Completely-Mixed Reactor—Biological Solids Recycle

The completely mixed reactor employing post reactor separation of the mixed liquor suspended solids and continuous recycle of a fraction of the separated solids back to the reactor is used with increasing frequency in waste treatment practice. The completely mixed aerobic and anaerobic activated sludge processes belong in this category. The development of the model is based on the following assumptions: (a) liquid waste and recycled solids flow into the reactor at constant rates and are instantaneously and homogenously mixed with the reactor contents, (b) reactor mixed liquor is withdrawn at a rate equal to the inflow rate to maintain the reactor at a constant volume, (c) no active microorganisms are contained in the influent raw waste stream, (d) all waste utilization occurs in the biological reactor, and (e) the total biological mass in the system is equal to the biological mass in the reactor, *i.e.*, the separator volume is small and recycle is continuous.

For this system, $\theta_c$, as defined in Equation 8 is

$$\theta_c = \frac{VX}{wX_r + (Q - w)X_e} \tag{10}$$

in which $Q$ = flow rate of influent waste to the reactor, volume/time; and $V$ = reactor volume; $w$ = flow rate of the fraction of the solids separator underflow which is wasted from the system, volume/time; $X_r$ = biological solids concentration in the underflow from the solids separator, mass/volume; and $X_e$ = biological solids concentration in the clarified overflow from the solids separator, mass/volume.

A materials balance for microbial mass around the entire treatment system gives

(Net rate of change of microbial mass) = (growth rate) - (washout rate)

or

$$\left[\frac{dX}{dt}\right]_N (V) = [Y(dF/dt) - bX]V - (wX_r + QX_e - wX_e) \tag{11}$$

Making a substitution for $(wX_r + QX_e - wX_e)$ from Equation 10 and considering steady state conditions, *i.e.*, $(dX/dt)_N = 0$; the relationship between $\theta_c$ and waste utilization rate will be

$$\frac{1}{\theta_c} = \frac{Y(dF/dt)}{X} - b \tag{12}$$

The relation between $\theta_c$ and effluent waste concentration, found by substituting for $dF/dt$ from Equation 3 is

$$\frac{1}{\theta_c} = \frac{YkS_1}{K_s + S_1} - b \tag{13}$$

Since the right-hand sides of Equations 13 and 6 are equal, it may be formally stated that the steady-state value of $\theta_c$ is equal to the reciprocal of $\mu$, the mean microbial net specific growth rate. The same relationship between $\theta_c$ and $S_1$ is also obtained for the other two process configurations shown in Figure 7.1. Equation 13 can be manipulated to express the effluent waste concentration $(S_1)$ as a function of $\theta_c$ as follows:

$$S_1 = \frac{K_s[1 + b(\theta_c)]}{\theta_c(Yk - b) - 1} \qquad (14)$$

Figure 7.2 illustrates this relationship as well as treatment efficiency for an assumed set of kinetic coefficients and wastewater characteristics. This figure emphasizes that selection and maintenance of an appropriate value of $\theta_c$ determines the effluent quality and hence efficiency of the process. As shown in this figure, process failure (E=0) will occur at some point as the value of $\theta_c$ is progressively reduced. This value of $\theta_c$, defined as $\theta_c^m$, is calculated by setting $S_1 = S_0$ in Equation 13.

$$(\theta_c^m)^{-1} = \frac{YkS_0}{K_s + S_0} - b \qquad (15)$$

In the limiting case when $S_0 \gg K_s$,

$$[\theta_c^m]_{lim} = 1/(Yk - b) \qquad (16)$$

$[\theta_c^m]_{lim}$ is related to the maximum net specific growth rate of microorganisms $(\hat{u})$ and to their minimum generation or doubling time $(t_d)$ as follows:

$$[\theta_c^m]_{lim} = (\hat{u})^{-1} \qquad (17)$$

and

$$t_d = \ln2 [\theta_c^m]_{lim} \qquad (18)$$

These relationships are of value when correlating reported maximum growth values with $\theta_c$.

Generally, the mass of microorganisms removed by controlled wastage $(wX_r)$ is much greater than that lost in the effluent $(Q - w)X_e$. The ability to control organism removal by varying $wX_r$ allows the $\theta_c$ to be maintained independent of hydraulic retention time $(\theta)$. This is the major advantage of a recycle system. Thus, a long $\theta_c$ needed to obtain a low $S_1$ (and hence high treatment efficiency) can be achieved with a short $\theta$ needed for system economy.

However, $\theta_c$ cannot be varied completely independent of $\theta$ because of practical limitations imposed by the settling characteristics of the microbial mass. The relationship between $\theta_c$ and $\theta$, recycle flow rate $(q)$ and clarifier performance as indicated by the underflow solids concentration $(X_r)$ can be evaluated by making a materials balance for microbial mass around the

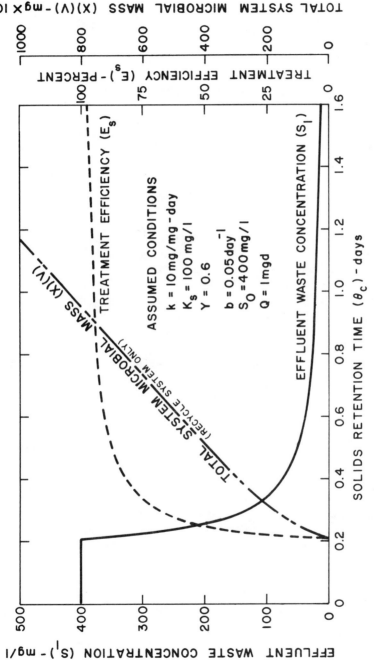

**Figure 7.2.** Steady-state relationships between $\theta_c$, effluent waste concentration, specific treatment efficiency and total system microbial mass for completely mixed process with recycle.

biological reactor alone as follows:

(Net rate of change of microbial mass) = (recycle rate) + (growth rate) - (washout rate)

or

$$\left[\frac{dX}{dt}\right]_n (V) = qX_r + [Y(dF/dt) - bX]V - (q + Q)X \tag{19}$$

in which q = flow rate at which solids separator underflow is recycled to the reactor, volume/time. When steady-state conditions are achieved, $(dX/dt)_n = 0$, the relation of recycle with $\theta_c$, found by combining Equations 12 and 19, is

$$\frac{1}{\theta_c} = \frac{Q}{V}[1 + r - r(X_r/X)] \tag{20}$$

in which r = q/Q, the volumetric recycle ratio. Thus, $\theta_c$ is shown to be functionally related to the volumetric recycle ratio and the ratio $(X_r/X)$. The latter ratio is a function of the performance characteristics of the solid-liquid separator and of the settling characteristics of the biological mass. The relationships between the biological regime and biomass settling characteristics and clarifier design are considered in a separate section.

The total weight of microbial mass in the reactor is of interest, and under steady-state conditions can be determined by noting that $dF/dt = (Q/V)(S_0 - S_1)$ and by substituting this expression for $dF/dt$ in Equation 12

$$(X)(V) = \frac{YQ\theta_c(S_0 - S_1)}{1 + b\theta_c} \tag{21}$$

Once the desired treatment efficiency is stated, a combination of Equations 9, 13 and 21 can be used to calculate the quantity $(X)(V)$, as illustrated in Figure 7.2, for given values of Q, $S_0$ and the coefficients, k, $K_s$, Y, and b. For each value of $\theta_c$, there is a unique value of E, $S_1$, and $(X)(V)$. With these values, Equation 20 can be used to determine the reactor volume and reactor microbial concentration for assumed values of r and $X_r$. This design space is shown in Figure 7.3 for the waste illustrated in Figure 7.2 and for various assumed values of $X_r$ and for values of r in the range of 0 to 2.

The selection of a reactor volume for design will be guided by least cost considerations within certain constraints imposed by considerations related to mixing, pumping, gas (oxygen) transfer, and settling characteristics. Dick and Javaheri have pointed out that an opportunity exists to develop a least capital cost approach to design of a reactor-clarifier system by taking advantage of tradeoffs between reactor volume and clarifier area.[7] This subject is explored later.

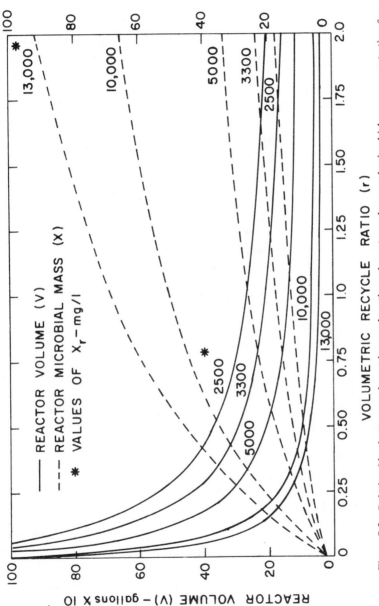

**Figure 7.3.** Relationships between volumetric recycle ratio and reactor volume and microbial mass concentration for various values of $X_I$ and an assumed specific efficiency of 97% in treatment situation given in Figure 7.2.

### Bioprocess Model Summary

Table 7.1 summarizes the equations that constitute the models of the three process configurations shown in Figure 7.1. For a given waste treatment process, these equations facilitate the calculation of the principal process parameters including effluent waste concentration, mixed liquor suspended solids concentration or reactor volume, and excess sludge production.

Appendix A contains a computer program for computing the quantities shown in Table 7.1 for the completely mixed process with recycle. The required data inputs are values of $S_0$, Q and the process coefficients Y, b, k, $K_s$. The program will also generate the values of X and V for various recycle ratios (r) at a specified value of $\theta_c$ and $X_r$. This latter information facilitates construction of design space plots as shown in Figure 7.3. The program is written in PL/1 for use with an IBM 360-65 digital computer. An illustrative print-out is included within the Appendix.

For the system without recycle, the hydraulic retention time is an important parameter since it is equal to the biological solids retention time and hence directly determines process efficiency. However, in the recycle systems, there is considerable opportunity for trade-offs between reactor volume, mixed liquor concentration and clarifier operation. Thus hydraulic retention time $(\theta)$ which equals V/Q is not an independent parameter and hence can assume a spectrum of values without affecting biological process performance. While it appears that undue emphasis has been placed on the role of $\theta$ in process performance, it should be recognized that longer hydraulic retention times do contribute to equalization capacity of the reactor and enhance the stability of the process against hydraulic shock loadings.

The assumptions involved in the derivation of the plug flow process model shown in Table 7.1 are necessarily different from the completely mixed model assumptions and are stated here. The plug flow reactor model assumes that no longitudinal mixing occurs between adjacent elements of fluid. Each element of mixed liquor in such a reactor is analogous to a batch culture of microorganisms which is simply moving along a time axis. Thus, the waste concentration decreases along the length of the tank from the influent to the effluent, while microbial mass concentration increases from its initial value as a result of waste assimilation. However, as operated in practice, the concentration of microorganisms in the reactor effluent is generally not increased significantly over that in the reactor influent, after mixing with the recycle stream (q). Thus, it is valid to make a simplifying assumption that the microbial concentration in the reactor remains constant (at least as long as $\theta_c/\theta > 5$). Also, it should be recognized that the equation in Table 7.1 relating biological solids retention time and influent and effluent quality ($S_0$ and $S_1$) is valid only when the hydraulic recycle ratio (r) is less than one.

Table 7.1. Summary of Steady-State Relationships for Biological Waste Treatment with Suspensions of Microorganisms

| Characteristic | Complete-Mix System | | With Recycle[a] |
| --- | --- | --- | --- |
| | Without Recycle | With Recycle | |
| Specific efficiency | $E_s = \dfrac{100(S_0 - S_1)}{S_0}$ | $E_s = \dfrac{100(S_0 - S_1)}{S_0}$ | $E_s = \dfrac{100(S_0 - S_1)}{S_0}$ |
| Effluent waste concentration | $S_1 = \dfrac{K_s[1 + b(\theta_c)]}{\theta_c(Yk - b) - 1}$ | $S_1 = \dfrac{K_s[1 + b(\theta_c)]}{\theta_c(Yk - b) - 1}$ | b |
| Microorganism concentration in reactor | $X = \dfrac{Y(S_0 - S_1)}{1 + b\theta_c}$ | $X = \dfrac{Y(S_0 - S_1)}{1 + b\theta_c}\left(\dfrac{\theta_c}{\theta}\right)$ | $\bar{X} = \dfrac{Y(S_0 - S_1)}{1 + b\theta_c}\left(\dfrac{\theta_c}{\theta}\right)$ |
| Excess microorganism production rate (mass/time) | $P_X = \dfrac{YQ(S_0 - S_1)}{1 + b\theta_c}$ | $P_X = \dfrac{YQ(S_0 - S_1)}{1 + b\theta_c}$ | $\bar{P} = \dfrac{YQ(S_0 - S_1)}{1 + b\theta_c}$ |
| Hydraulic retention time (V/Q) | $\theta = \theta_c$ | $\theta = \theta_c[1 + r - r(X_r/X)]$ | $\theta = \theta_c[1 + r - r(X_r/\bar{X})]$ |
| Solid retention times general | $(\theta_c)^{-1} = \dfrac{YkS_1}{K_s + S_1} - b$ | $(\theta_c)^{-1} = \dfrac{YkS_1}{K_s + S_1} - b$ | $(\theta_c)^{-1} = \dfrac{Yk(S_0 - S_1)}{K_s\ln(S_0/S_1) + (S_0 - S_1)} - b$ [c] |
| Limiting minimum | $[\theta_c^m]_{lim.} = (Yk - b)^{-1}$ | $[\theta_c^m]_{lim.} = (Yk - b)^{-1}$ | d |

aFor situation in which reactor microbial mass concentration (X) is assumed constant
bNo explicit solution for $\theta_c$
cFor situation in which recycle ratio r < 1.[6]
dNot mathematically defined for this system.

As previously stated the equations shown in Table 7.1 apply directly only to soluble wastes. Thus, the values of volatile solids computed with these equations, *i.e.*, mixed liquor microbial mass concentration (X), clarifier underflow concentration ($X_r$), and waste sludge production rate ($P_x$), represent volatile solids of biological origin only. In municipal waste treatment situations and many industrial waste situations, volatile suspended solids will be introduced into the reactor with the untreated wastewater. If a fraction of these wastewater volatile solids are refractory, they will accumulate to a steady-state level in the treatment system and increase measured volatile solids levels above those calculated by the models presented here. Nonvolatile suspended solids contained in the influent waste will also attain a steady-state level in the treatment system further increasing the total suspended solids in the system as well as the total sludge production rate. These influent suspended solids (nonvolatile and volatile refractory) can be incorporated into treatment process design by including appropriate mass balances for these materials in the treatment process models. In processes employing recycle, the relative proportion of these non-biological suspended solids in the mixed liquor increases with increasing values of $\theta_c$. This is illustrated by the following equation for the steady-state mixed liquor concentration of a biologically refractory suspended material.

$$X^i = X^i_o \, (\theta_c/\theta) \tag{22}$$

where $X^i$ = steady state mixed liquor concentration of a refractory suspended constituent of the influent wastewater, mass/volume and $X^i_o$ = the influent concentration of a refractory suspended constituent of the wastewater, mass/volume.

## Gravity Separation of Mixed Liquor Suspended Solids

In the completely mixed process without solids recycle, neither process efficiency (based on specific removal of soluble waste constituents) nor process stability is influenced by the effectiveness of post-reactor liquid-suspended solids separation. By contrast, the stability and hence both specific and gross efficiencies of solids recycle processes are totally dependent on effective post-reactor separation and recycle of all or a portion of the mixed liquor suspended solids. Dick has appropriately suggested that many reported failures and malfunctions of activated sludge plants are more likely related to ineffective settling of the mixed liquor suspended solids than to biological malfunction.[8] To correct this situation, Dick has suggested that design of secondary clarifiers should receive attention equal to that customarily directed towards the biological regime.

Performance of the secondary clarifier is generally considered to be dependent on the successful accomplishment of two operations, *i.e.*, clarification and solids transmission or thickening. The clarifier surface area required for successful clarification is usually determined by dividing the net forward flow of liquid by the allowable surface overflow rate. This overflow rate is set equal to the zone settling velocity of the reactor mixed liquor. Dick has stated that the second function, *i.e.*, solids transmission will, in many cases, require a greater area than the area for clarification.[8] In such a situation, the clarifier is said to be "flux limited." He has proposed a technique, known as the batch flux method, for determining the design or limiting solids flux ($G_L$). This method is adopted for use here.

The total solids flux in a continuous flow clarifier can be considered equal to the sum of two components, *i.e.*, solids flux due to gravity subsidence and solids flux due to bulk liquid transport via the clarifier underflow. The following equation describes this situation

$$G_i = v_i X_i + u X_i \qquad (23)$$

in which $v_i$ = the subsidence velocity taken to be equal to the zone settling velocity (ZSV) of a suspension having an initial suspended solids concentration equal to $X_i$, length/time; and u = the bulk liquid velocity due to underflow which is equal to the clarifier underflow rate ($q' = q + w$) divided by the clarifier surface area ($A_d$), length/time.

The first step in determining the limiting solids flux ($G_L$) for design is to perform a series of zone settling velocity (ZSV) tests over a range of initial suspended solids concentrations which encompass the anticipated range of clarifier operation in terms of $X_r$ values. It is important to perform these tests on a biological slurry, which has been grown at the same value of biological solids retention time, to be used in the full scale design. This is true because zone settling velocity appears to be a function of $\theta_c$, at least for activated sludge. Figure 7.4 shows a relationship observed by Bisogni and Lawrence for laboratory activated sludge grown on soluble wastes.[9] It is anticipated that a lesser slope in the ZSV versus $\theta_c$ relationship might be observed for a treatment system receiving a waste containing refractory suspended solids. However, while the relationship between ZSV and $\theta_c$ might be less dramatic for some cases than that shown in Figure 7.4, it is important that the settling characteristics of the slurry used in settling tests be the same as the settling characteristics anticipated for the full-scale plant.

Figure 7.5 shows the following quantities plotted against the slurry suspended solids concentration ($X_i$) for a representative activated sludge: (a) zone settling velocity, (b) the two components of the total flux for a given

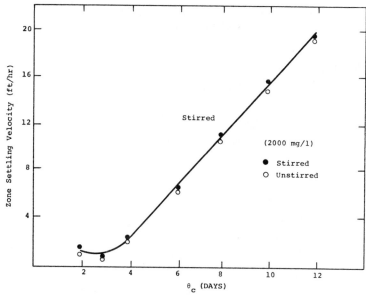

**Figure 7.4.** Zone settling velocity (ZSV) as a function of $\theta_c$ after Bisogni and Lawrence.[9]

value of u, and (c) the total flux for the same value of u. The limiting flux
($G_L$), *i.e.,* that flux requiring the largest area to transmit a given quantity
of solids, is equal to the minimum value of the total flux curve to the right
of the maximum. Further interpretation of the flux curve is beyond the
scope of this discussion. Dick has described a graphical "batch flux"
method for determining the value of $G_L$ that is more useful than the plots
represented by Figure 7.5.[8]  In the graphical batch-flux method, it is not
necessary to draw a new total flux curve for different values of u or $X_r$.

Analytical determinations of the limiting flux ($G_L$) for clarifier design
are possible if the relationship between ZSV and suspended solids concentration ($X_i$) can be expressed mathematically. Vesilind has reported that
plots of the log of ZSV versus initial suspended solids concentration yield
straight lines.[10]  More recently Dick and Young have used log-log plots of
the same variables successfully in evaluation of clarifier performance.[11]
Middleton and Lawrence have found that the semilog relationship suggested
by Vesilind adequately describes the settling characteristics of one and two
stage nitrifying activated sludge pilot systems.[12]  Thus, in this chapter it is
suggested that the ZSV biological slurries can be described by an equation
of the form

$$(ZSV)_i = (a)\exp(-k_e X_i) \qquad (24)$$

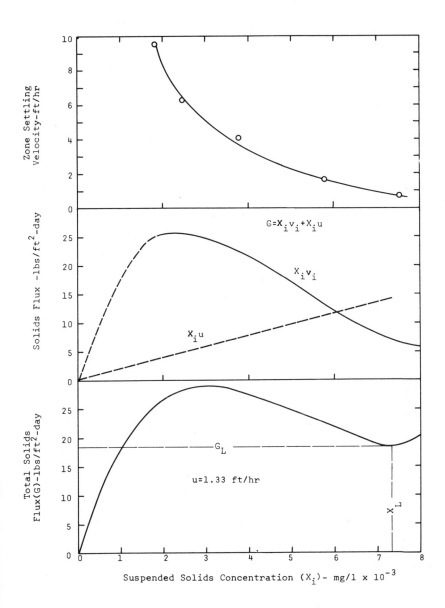

**Figure 7.5.** Zone settling velocity and solids flux versus $X_i$ for representative activated sludge.

in which $(ZSV)_i$ = the zone settling velocity of a biological slurry at initial suspended solids concentration $(X_i)$, length/time; a = slurry specific value of the ordinate intercept (base e); $k_e$ = slope of plot of log (base e) ZSV versus X. Substitution of Equation 24 into Equation 23 and differentiation of the resulting equation with respect to X yields the following expression for the limiting flux $(G_L)$ in terms of the value of $X_i$ associated with the flux limiting condition, *i.e.*, $X_L$

$$G_L = X_L^2 \, ak_e \, \exp(-k_e X_L) \qquad (25)$$

Considerations of hydraulic continuity and suspended solids mass balances around the clarifier lead to the following expression for the flux-limiting suspended solids concentration $(X_L)$ in terms of the desired clarifier underflow solids concentration $(X_r)$:

$$X_L = X_r/2 + 1/2 [(X_r)^2 - 4X_r/k_e]^{\frac{1}{2}} \qquad (26)$$

Thus for a given value of $X_r$, $X_L$ can be computed from Equation 26 and $G_L$ can be computed from Equation 25. The clarifier area required for solids transmission can then be computed from the following equation

$$A_t = L_f/G_L \qquad (27)$$

in which $A_t$ = clarifier area required for transmission of settled solids to the clarifier floor, area; and $L_f$ = the suspended solids load transmitted to the clarifier floor which from mass balance considerations equals $(Q + q)X - (Q - w)X_e$, mass/time.

The design area of the clarifier $(A_d)$ will be the larger of the two areas computed, *i.e.*, the area for clarification $(A_c)$ or the area of solids transmission $(A_t)$.

In summary, use of the solids recycle biological reactor models described in the previous section in conjunction with the above described approach to clarifier design forms the basis for an integrated design of a reactor-clarifier system for steady-state operation. When the value of $\theta_c$ is specified, the performance of the biological component of the system is specified. Attainment of the projected biological performance can be assured through design of an appropriate clarifier by the method described here. The range of clarifier operation possible will be determined by the slurry settling characteristics and desired underflow concentration $X_r$ and/or recycle flow ratio (r). The following section suggests least-cost criteria for rational design selection of a reactor-clarifier combination.

## Toward Least Cost Designs

Design criteria for slurry reactor biological processes or systems will be chosen on the basis of process performance requirements and economic efficiency (least cost). Economic efficiency may also be expressed in terms of maximum volumetric waste loading rates for a biological reactor and, in terms of maximum flux for a gravity clarifier. In waste treatment, requirements for high-removal efficiencies of wastewater constituents will often conflict with the goal of economic efficiency, since effluent quality is inversely related to microbial growth rate and specific utilization. Also, process constraints such as oxygen transfer capabilities, mixing, and biotoxicity must be considered in design. Clearly, the opportunities for trade offs between volumetric waste loading rates, process operating characteristics, and process costs are numerous. Here, the concept of least cost design for a completely mixed process without recycle (anaerobic treatment) and for a completely mixed process with recycle (activated sludge) is considered briefly.

For the completely mixed process without recycle, economic efficiency is most directly enhanced by increasing the waste volumetric loading rate of the reactor, *i.e.,* lbs COD/$10^3$ ft$^3$-day. Increased volumetric loading can be accomplished by increasing the flow rate (Q) of wastewater through the reactor (V) or by maintaining the flow rate (Q) to the process constant and concentrating the waste prior to introduction into the reactor. Since the first method will decrease the hydraulic retention time ($\theta$) and also the biological solids retention time ($\theta_c$), increased volumetric loading is achieved at the expense of effluent quality. O'Rourke has derived the following expression for the value of $\theta_c$ and hence associated with the maximum rate of volumetric waste utilization, $(dF/dt)_m$ for the completely mixed process without recycle in a given waste treatment situation.[13]

$$(\theta_c^*)^{-1} = Yk\ [1 - [K_s/(K_s + S_o)]^{\frac{1}{2}}] - b \tag{28}$$

in which $\theta_c^*$ = value of biological solids retention time at which volumetric waste utilization rate (dF/dt) will be a maximum for a given waste treatment situation. It can be shown that the value of $\theta_c^*$ is quite close to the washout value ($\theta_c^m$) so that it is unlikely that a reactor would be operated at a biological solids retention time equal to $\theta_c^*$. O'Rourke[13] and Lawrence[14] have discussed the benefits associated with prior concentration of wastes, particularly sludges. Since $\theta_c = \theta$ remains constant, it is theoretically possible from a kinetic viewpoint to increase the volumetric loading without limit with no effect on effluent quality. Practically, however, the volumetric loading limit will be determined by an environmental factor such as mixing, gas transfer, or toxic by-products buildup.

Least cost optimization of a biological process with solids recycle is a more complicated exercise because of the introduction of solids separation and recycle as variables. Basic process performance in terms of specific effluent quality $(S_1)$ can be maintained across a wide spectrum of combinations of reactor volume, clarifier area, and recycle policy because the biological solids retention time is no longer equal to the hydraulic retention time. Dick and Javaheri[7] presented a least capital cost approach to design of the activated sludge process—secondary clarifier design as described by Dick.[8] They presented representative results for a given biological regime and one value of clarifier underflow, $X_r$. Cost data were taken from the report by Smith.[15] Using the approach suggested by Dick and Javaheri,[7] Lawrence and Milnes studied the effect of various values of clarifier underflow $(X_r)$ on the least-cost solution for a specified constant value of $\theta_c$.[16] Figure 7.6 shows typical results for one value of $X_r$. Compared to conventional activated sludge design, the results of this study indicated that, based on capital cost of reactor and clarifier only, the least-cost solution favored relatively dilute underflow concentrations; high volumetric recycle ratios; and high mixed-liquor suspended solids concentrations. Jennings and Grady have used a similar approach for evaluating the effect of operational extremes on the design and performance of reactor-clarifier systems.[17] Middleton and Lawrence are presently studying the least-cost design of an expanded process system which includes gravity thickening of waste activated sludge and aerobic digestion, in addition to the activated sludge reactor-clarifier sequence.[18] Operating costs are being included in the present study. Also it will be possible to evaluate the design costs over several values of $\theta_c$ as well as for any given value of $\theta_c$.

Utilizing approaches such as those described here, it is now possible to approach biological process and system design in a more rational manner than was possible heretofore. It is now possible to more closely attain the goal of least-cost design at a specified level of effluent quality.

## DYNAMIC TREATMENT SYSTEM MODELS

While the dominant focus in modeling of biological slurry reactors has been the steady-state condition, considerable justification exists for the development and application of dynamic models in both process design and operation. Most wastewater treatment systems are subjected to time-varying waste loads rather than steady-state waste loads. Thus, it is apparent that only a dynamic model can predict the time varying character of treatment system performance and waste treatment efficiency. Such knowledge is important in operation both to assess the impact of the effluent on the receiving stream and, in the case of water reclamation, to assess the effect

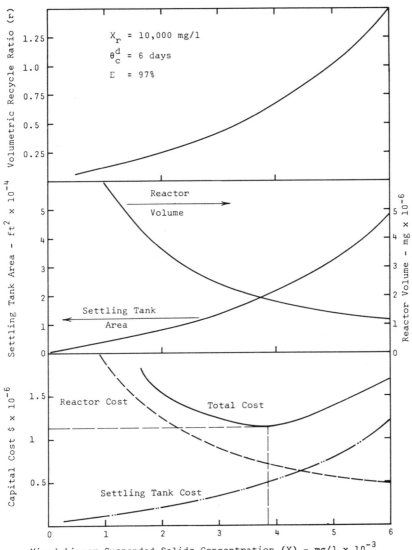

FIGURE 6   LEAST CAPITAL COST RELATIONSHIPS FOR HYPOTHETICAL
ACTIVATED SLUDGE PROCESS DESIGN

**Figure 7.6.** Least capital cost relationships for hypothetical
activated sludge process design.

of variable quality on subsequent advanced treatment steps and reuse. The effectiveness of process control procedures which might be implemented to minimize fluctuations in process performance over time is also dependent on the availability of adequate dynamic models. In design, dynamic models can be used to advantage in studying the response of load intensive elements of the treatment system such as oxygen transfer equipment, pumping and sludge recycle capacity, and clarifier performance. Such studies can be used to establish more rationally based excess capacity or safety factors in design. Thus, while steady-state models provide valuable information with regard to design and performance for time-averaged conditions, and serve as the norm against which process controlled performance is evaluated, dynamic models will also improve our ability to design and properly operate biological slurry reactors in wastewater treatment systems.

A dynamic model of the anaerobic digester has been developed by Andrews.[19,20] The central equation in this model describes microbial growth as a function of substrate concentration. The model includes a term that accounts for possible growth inhibition effects of high concentrations of the substrate. Andrews discussed the applicability of this model for describing transient situations in digester operation, *e.g.*, digester start-up and onset of digester upsets.

The response of the activated sludge process to transient or time varying organic and hydraulic loads has been the subject of numerous computer simulation studies and experimental studies. The majority of these studies have been concerned with the biological response to both continuous time varying organic loads and quantitative organic shock loads. Response of the secondary clarifier either was not considered explicitly or was assumed to be describable by a steady-state model unaffected by transient conditions. Included in this category of effort are the studies of Eckhoff and Jenkins,[21] Ott and Bogan,[22] Burkhead and Wood,[23] and Grady.[24]

Bryant, Wilcox, and Andrews developed a dynamic model of a conventional secondary wastewater treatment plant.[25] Their model included dynamic process models of primary sedimentation, the activated sludge process, and a secondary clarifier. The activated sludge process was simulated by five completely mixed reactors in series. A Monod type saturation function was used to describe microbial waste assimilation and growth. Based on simulation studies and comparisons of simulated performance with data from a full-scale plant, it was concluded that: (a) the secondary clarifier was perhaps the most sensitive unit in the treatment system under dynamic loading conditions, (b) long delays were involved in translation through the process sequence of the effects of loading changes, (c) because of the delays, the conventional wastewater treatment system was more amenable to feed forward control as compared with feed back control, and (d) further model development was needed before the model would function adequately in a feed forward process control capacity.

Milnes and Lawrence described a dynamic model of a nitrifying, completely mixed activated sludge reactor in series with a gravity clarifier-thickener.[26] Figure 7.7 shows a functional block diagram of the system modeled. The biological component of the model was developed from

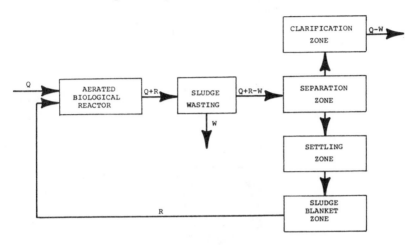

**Figure 7.7.** Functional block diagram: completely mixed activated sludge process after Milnes.[27]

mass balance expressions around the reactor for microbial mass and soluble waste (degradable COD). Monod type microbial growth relationships (Equation 6) were used to describe the growth of the heterotrophic microbial mass, the *Nitrosomonas* type bacterial mass, the *Nitrobacter* type bacterial mass. Microbial oxygen uptake rate equations were also written. An empirical dynamic model was developed to describe the operation of the secondary clarifier. The mathematical form of the clarifier model was based on results of laboratory studies performed on a 150-liter completely mixed reactor coupled to an 8-ft deep, 6-in. inside diameter plexiglass clarifier equipped with sludge scrapers and continuous, metered, pumped, sludge return. The laboratory system was grown on a soluble proteinaceous synthetic waste. Figure 7.8 shows a schematic of the clarifier geometry and the hydraulic retention time assigned to the associated functional blocks. The efficiency of clarification was determined according to an empirical observed relationship between clarifier effluent suspended solids and net forward hydraulic flow rate (Figure 7.9). The sludge thickening zone was modeled on the basis of an observed empirical relationship

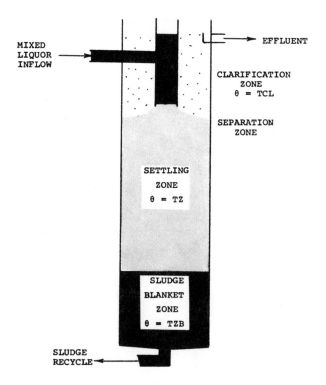

**Figure 7.8.** Physical location of zones in the secondary clarifier, after Milnes.[27]

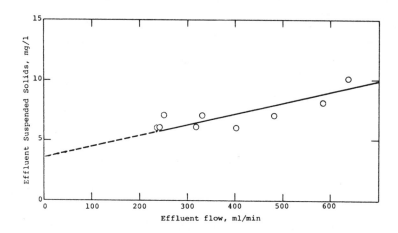

**Figure 7.9.** Regression line fitted to effluent flow and suspended solids data, laboratory activated sludge study after Milnes.[27]

describing clarifier underflow suspended solids concentration as a function of weight of solids in the sludge blanket and the solids retention time in the blanket. The steady-state form of this relationship (Figure 7.10) is expressed by the following equation

$$C_u = K \ (WB) \ (WB/BLKLOD)^m \qquad (29)$$

in which $C_u$ = clarifier underflow suspended solids concentration, mass/volume; K = empirical constant from plot such as shown in Figure 7.10; (WB) = weight of suspended solids in the blanket, mass; (BLKLOD) = mass loading rate to the clarifier sludge blanket, mass/time; m = empirical constant from plot such as shown in Figure 7.10. In the dynamic model version of Equation 29, successively reevaluated time averaged values are used for the quantities (WB) and (BLKLOD).

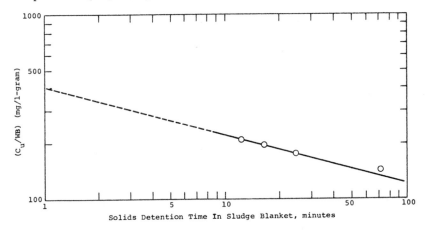

**Figure 7.10.** Empirical basis for sludge blanket model, laboratory activated sludge study, after Milnes.[27]

Table 7.2 presents a summary of the equations which constitute the dynamic model. Symbols used are defined within the table, in notes at the end of the table, and on Figures 7.7 and 7.8. The symbols appearing in Table 7.2 are not redefined in the Nomenclature at the end of this chapter. The model was programmed for solution in the CSMP digital-analog simulation language and run on an IBM Model 360-65 digital computer.[28] The responses of the model to four different time varying loading conditions were studied and compared to observed performance of the laboratory pilot plant operated under the same loading conditions. Based on these results, it was concluded that the dynamic model adequately simulated the observed dynamic performance of the laboratory system. The model

Table 7.2.  Summary of Equations Constituting Dynamic Model[27]

| Functional Block & Dependent Variable | Mass Balance & Transfer Equations | Auxiliary Equations |
|---|---|---|
| **Aerated Biological Reactor (see Figure 7.7)** | | |
| Nondegradable COD, SUDCOD | (V) (dSUDCOD/dt) = (CONST1) (SUBCOD) + (R) (SRUDCD) - (Q+R) (SUDCOD) | |
| Degradable COD, SCOD | (V) (dSCOD/dt) = (SUBCOD) (1 - CONST1) - (Q+R) (SCOD) - (dFC)D/dt) (V) + (R) (SRCOD) | $(dFCOD/dt) = \dfrac{(KCOD)\ (XH)\ (SCOD)}{CODK + SCOD}$ |
| Heterotrophic sludge, XH | (V) (dXH/dt) = (YH) (dFCOD/dt) (V) + (R) (XRH) - (BH) (XH) (V) - (Q+R) (XH) | |
| *Nitrosomonas* sludge, XNS | (V) (dXH/dt) = (YNS) (dFNH3N/dt) (V) + (R) (XRNS) - (BNS) (XH) (V) - (Q+R) (XH) | $(dFNH3N/dt) = \dfrac{(KNH3N)\ (XNS)\ (SNH3N)}{NH3NK + SNH3N}$ |
| Ammonia nitrogen, SNH3N | (V) (dSNH3N/dt) = (CONST2) (dFCOD/dt) (V) + (R) (SRNH3N) - (dFNH3N/dt) (V) - (Q+R) (SNH3N) - 0.109 [(YH) (dFCOD/dt) + (YNS) (dFNH3N/dt) + (YNB) (dFNO2N/dt)] (V) | |
| *Nitrobacter* sludge, XNB | (V) (dXNB/dt) = (YNB) (dFNO2N/dt) (V) + (R) (XRNB) - (BNB) (XNB) (V) - (Q+R) (XNB) | $(dFNO2N/dt) = \dfrac{(KNO2N)\ (XNB)\ (SNO2N)}{NO2NK + SNO2N}$ |
| Nitrite nitrogen, SNO2N | (V) (dSNO2N/dt) = (dFNH3N/dt) (V) + (SRNO2N) (R) - (dFNO2N/dt) (V) - (Q+R) (SNO2N) | |
| Nitrate nitrogen, SNO3N | (V) (dSNO3N/dt) = (dFNO2N/dt) (V) + (SRNO3N) (R) - (Q+R) (SNO3N) | |

Debris sludge, XDB

$$(V) (dXDB/dt) = (G) [(YH) (dFCOD/dt) (V) + (YNS) (dFNH3N/dt) (V) + (YNB) (dFNO2N/dt) (V)] + (XRXDB) (R) - (Q+R) (XDB)$$

$$XB^IO = XH + XNB + XNS$$

Oxygen utilization, O2UPTK

$$O2UPTK = (OENG) (dFCOD/dt) + OEND (XB^IO) + (3.43) (dFNH3N/dt) + (1.14) (dFNO3N/dt)$$

**Sludge Wasting**

$$Waste = (X) (W) (Q+R-W) (X) = (Q+R) (X) - (W) (X)$$

**Separation Zone (see Figure 7.8)**

Effluent solids loss, XCL

$$XCL = CONS1O + MM (Q-W)^a$$

$$OUTX = XCL (t-TCL)$$

Solids to settling zone, ZLOAD

$$ZLOAD = (Q+R-W) (X) - (Q-W) (XCL)$$

**Clarification Zone (see Figure 7.8)**

$$TCL = VCL/(Q-W)$$

Effluent total soluble COD, EFFCDT

$$EFFCDT = EFFCOD + EFFCUD$$

Effluent degradable soluble COD, EFFCOD

$$EFFCOD = SCOD (t-TCL)$$

Effluent nondegradable soluble COD, EFFCUD

$$EFFCUD = SUDCOD (t-TCL)$$

Effluent ammonia nitrogen, EFFNH3

$$EFFNH3 = SNH3N (t-TCL)$$

Effluent nitrite nitrogen, EFFNO2

$$EFFNO2 = SNO2N (t-TCL)$$

Effluent nitrate nitrogen, EFFNO3

$$EFFNO3 = SNO3N (t-TCL)$$

**Settling Zone**

$$BLKLOD = ZLOAD (t-TZ)$$

$$TZ = \frac{MZ}{ZLOAD}$$

**Table 7.2, continued**

| Functional Block & Dependent Variable | Mass Balance & Transfer Equations | Auxiliary Equations |
|---|---|---|
| **Sludge Blanket Zone** | | |
| Sludge blanket weight | $dWB/dt = BLKLOD - (XR)$ (R) | $XR = (K) (WBAVG) \left(\dfrac{WBAVG}{BKLDAG}\right)^m$ [b] |
| Recycle degradable soluble COD, SRCOD | $SRCOD = SCOD$ (t-TZB) | |
| Recycle nondegradable soluble COD, SRUDCD | $SRUDCD = SUDCOD$ (t-TZB) | |
| Recycle ammonia nitrogen, SRNH3N | $SRNH3N = SNH3N$ (t-TZB) | |
| Recycle nitrite nitrogen, SRNO2N | $SRNO2N = SNO2N$ (t-TZB) | |
| Recycle nitrate nitrogen, SRNO3N | $SRNO3N = SNO3N$ (t-TZB) | |

[a]Expression for XCL is derived from regression analysis on data plot such as shown in Figure 7.9.

[b]Expression for XR (dynamic model) = $C_u$ (steady-state condition) is derived from regression analysis on data plot such as shown in Figure 7.10.

Notes:

1. The kinetic coefficients (Y, b, k, $K_s$) used in this model are expressed for heterotropic microbial mass, *Nitrosomonas* microbial mass, and *Nitrobacter* microbial mass as follows: YH, BH, KCOD, CODK, YNS, BNS, KNH3N, NH3KN, and YNB, BNB, KNO2N, NO2NK, respectively.

2. Symbols not identified in the table or Figures 7.7 and 7.8 are defined below:

BKLDAG = time average value (over TB time units) of suspended solids mass loading rate to the sludge blanket (BLKLOD).

CONS1O = ordinate axis intercept of the regression line for effluent suspended solids relationship (Figure 7.9)

CONST1 = fraction of total influent COD, which is nondegradable.

CONST2 = mg of ammonia nitrogen generated per mg of substrate COD assimilated.

G = mg of debris solids (in this model the non-volatile suspended solids) generated per mg of microbial material grown.

MM = slope of regression line for effluent suspended solids relationship (Figure 7.9).

MZ = mass of suspended solids in the settling zone of the secondary clarifier, mass.

OEND = mg of endogenous oxygen required per mg of cell mass per unit of time.

OENG = mg of energy oxygen required in the metabolism of one mg of substrate biodegradable COD.

SUBCOD = mass loading rate of COD substrate to the aeration basin, mass/time.

t = time variable, time.

T = indicates units of time.

TB = nominal detention time of suspended solids in the secondary clarifier sludge blanket, time.

TZ = detention time of suspended solids in the settling zone of the secondary clarifier, time.

WBAVG = time average value (over TB time units) of sludge blanket weight, mass.

appears to have application in simulation studies for design and control. However, the clarifier model *per se* is considered to be system specific and probably would need to be modified in each application to compensate for the prototype clarifier geometry. The detailed derivation of the model, computer program listings, and complete study results have been presented by Milnes.[27] A CSMP program listing of the model for the time varying hydraulic and organic loading condition is available on request from the author of this chapter at Cornell University.

In recent years, considerable effort has been directed towards development of dynamic models of biological slurry processes. Some of these efforts have been described here. It is anticipated that dynamic modeling of biological processes and its application in process control and design will continue to develop rapidly in the coming years. Emerging requirements for reliable production of high quality effluents will provide added incentive for the study of process dynamics.

## APPLICATIONS OF TREATMENT MODELS

Application of steady-state process and systems models of the mathematical form described in this chapter is increasing greatly in process evaluation and design. One indication of this increased acceptance is the inclusion of this type of model in a recently published textbook on wastewater engineering authored by personnel of a major national sanitary engineering consulting firm.[29] In a recent presentation, Stensel and Shell have concluded that a process design approach based on the biological solids retention time concept offers significant advantages over another widely used design concept based on (F/M)[30] While it is anticipated that practical application of dynamic models will increase markedly in the future, usage of dynamic models (1972) is restricted largely to research and development activities. Accordingly, this discussion of applications deals with the steady-state models. This emphasis is consistent with present design practice which usually considers only steady-state performance at projected average, maximum, and minimum organic and hydraulic loading conditions.

The principal information required for using these models are numerical values of the kinetic coefficients ($Y$, $b$, $k$, and $K_s$) which are applicable to the wastewater treatment situation under consideration. Additional information required for certain processes will be gravity settling data and microbial oxygen uptake rates. Values of the coefficients for many processes and types of waste constituents have appeared in the literature in recent years. In previous studies, the author and co-workers have summarized a considerable amount of available kinetic information for activated sludge and anaerobic methanogenic processes.[1,14] These data are not repeated here.

Other sources of data are the report by Eckhoff and Jenkins[21] for activated sludge coefficients and the textbook, *Wastewater Engineering.*[29] McCarty has developed a technique for evaluating the stoichiometry of microbial mediated processes operated at any desired value of $\theta_c$.[31] McCarty has also presented a method for estimating values of Y and K from thermodynamic calculations for a specific microbial mediated reaction.[32] Thus, there is considerable opportunity to find or at least estimate "order of magnitude" values for most biological wastewater treatment processes of interest. It will almost always be necessary before using the kinetic models in final process design to verify or develop values of the coefficients which are system specific for the given wastewater treatment situation. A note of caution is in order regarding dimensional units of reported coefficients. Dimensional units observed in the literature range from electron-equivalent/ electron-equivalent to mg volatile suspended solids/mg COD. The user should verify that he is working with appropriate and dimensionally consistent values of coefficients. In the following section, brief comments are directed to several of the more widely used microbial mediated waste treatment processes.

**Aerobic Processes**

*Activated Sludge*

There are numerous modifications of the activated sludge process ranging from high rate aeration to extended aeration. Characterization of these many process modifications according to representative values of $\theta_c$ contributes to a rational understanding of the principal advantages which each modification offers. The majority of these relative advantages relate to settling characteristics and excess or waste sludge production rates. Both of these parameters have been shown to be functions of $\theta_c$. An excellent description of activated sludge process modifications appears in Chapter 12 of *Wastewater Engineering.*[29]

Since the earlier review of activated sludge applications of kinetic models,[1,14] several published reports have appeared describing uses of models similar to those presented here. Walker used biological solids retention time (defined as in the steady state form of Equation 2) as a control parameter in the operation of 3 mgd activated sludge plant.[33] Sherrard and Schroeder reported that steady-state operation of a laboratory activated sludge system was adequately described by an equation of the form, $(\theta_c)^{-1} = YU - b$ (Equation 2).[34] The values of $\theta_c$ studies ranged from 2 to 18.1 days. Their values of the coefficients Y and b were 0.414 mg V.S.S./mg COD and 0.093 day$^{-1}$ respectively (coefficient of correlation = 0.99). Table 7.3

presents suggested order of magnitude values of the kinetic coefficients for activated sludge systems receiving a municipal type heterogeneous organic waste at approximately 20°C. These values are based on our experience and evaluation of numerous literature citations. Again, it should be emphasized that such generalizations are valid for first approximations only and must be modified for the situation of interest.

Table 7.3. Representative Values of Kinetic Coefficients for Activated Sludge at 20°C

| Coefficient | Unit | Value Range |
|---|---|---|
| Y | mg V.S.S./mg COD | 0.35-0.45 |
| b | $day^{-1}$ | 0.05-0.10 |
| k | mg COD/mg V.S.S.-day | 6-8 |
| $K_s$ | mg/l COD | 25-100 |

A review of reported coefficients for activated sludge reveals that values of Y, b, and k are closely grouped around the representative values shown in Table 7.3. Thus, these three coefficients may be used with a reasonable degree of confidence. While a lesser number of values have been reported for $K_s$, the scatter is quite wide. Meaningful measurement of this coefficient ($K_s$) is very difficult in the mixed organic waste activated sludge process. A nonspecific parameter such as COD is often used as a measure of the "growth limiting" substrate. However, a significant fraction of the effluent soluble organic material is metabolically produced refractory material and should not be included in estimates of the effluent "growth limiting" substrate concentration. Subtraction of a constant value of COD from effluent concentrations before evaluating kinetic coefficients might lessen the system specific nature of reported values of $K_s$. A reasonable estimate for refractory COD in municipal effluents is 30 mg/l. However, the comparable value could be much greater for industrial wastes. Given these difficulties in identifying and measuring the "growth limiting" substrate in mixed organic waste activated sludge systems, it may be preferable to avoid reliance on k and $K_s$ by using the model in the form suggested by Equation 2, *i.e.,* $(\theta_c)^{-1} =$ YU - b. Using this form of the model necessitates system specific pilot plant investigations to determine values of U at different values of $\theta_c$. This is required because U is experimentally measured as waste removal rate per unit weight of volatile suspended solids. Since the active bacteria fraction of volatile suspended solids will vary according to the composition of the wastewater, one would expect to find variation from system to system in the experimentally determined values of U at a constant value of $\theta_c$. This variability of U at the same microbial growth rate $(\theta_c)^{-1}$ is one of the principal disadvantages involved in using $(F/M) \sim U$ in process design and control as compared to the $\theta_c$ based approach. Clarifier design for activated sludge systems should be performed as described earlier. Volumetric microbial oxygen uptake rate for carbonaceous waste removal $(dO_2/dt)_c$ can be estimated from the following equation:

$$\left[\frac{dO_2}{dt}\right]_c = (1 - 1.42 \ Y)(dF/dt) + 1.42 \ (b)(X) \tag{30}$$

For final design, estimates based on the above equation should be verified by experimental measurements.

In summary, high specific-treatment efficiencies are possible with activated sludge processes at $\theta_c$ values of less than one day. However, high gross BOD removal efficiencies at such short $\theta_c$ values are usually not possible as the microorganisms will not flocculate well and the effluent will contain large quantities of oxygen-consuming suspended microbial solids. A $\theta_c$ value of greater than 3 days appears necessary in activated sludge systems to effect the bioflocculation required for a clear effluent and to obtain gross removal efficiencies greater than 85%. The value of $\theta_c$ selected for design should reflect the designer's objective, *i.e.*, short $\theta_c$ for maximum soluble waste removal rates by dispersed growth, medium $\theta_c$ for gravity separation of microbial mass and high gross treatment efficiencies, long $\theta_c$ for minimum net microbial solids production.

*Nitrification*

The studies by Downing, *et al.*[35,36] and Stratton and McCarty[37] have convincingly demonstrated the utility of the kinetic approach for describing the phenomenon of microbial mediated nitrification. Their results (Table 7.4) indicate that the limiting value of biological solids retention time, $[\theta_c^m]_{lim}$ at 20°C is in the range of 1-2 days while at 10°C $[\theta_c^m]_{lim}$ is in the range of 4-6 days. Essentially, complete nitrification (effluent $NH_3$-N $\approx$ 0) should be attained in an activated sludge system whenever the biological solids retention time based on total system volatile suspended solids is maintained at a value greater than $[\theta_c^m]_{lim}$ for the nitrifying bacteria at the appropriate temperature. This statement presumes that the mixed-liquor dissolved-oxygen concentration will be adequate to sustain nitrifying bacteria ($\geqslant$ 2 mg/l D.O.) and that toxic substances will be absent. Oxygen uptake rates can be estimated by multiplying the ammonia removal (nitrate production) rate by the appropriate stoichiometric coefficients.

In recent years, much research and development has been devoted to "two sludge" systems for nitrification, *i.e.*, a carbonaceous reactor-clarifier

Table 7.4. Kinetic Coefficients for Biological Nitrification

| Environment and Reaction | Growth Coefficients | | Waste Removal Coefficients | | Maximum[a] Growth Rate | $[\theta_c^m]$lim. | Temp | Reference |
|---|---|---|---|---|---|---|---|---|
| | Y mg/mg N | b days$^{-1}$ | k mg N/mg-day | $K_s$ mg/l N | $\hat{\mu}$ (day$^{-1}$) | days | °C | |
| Synthetic river water | | | | | | | | |
| NH$_3$-N oxidation | 0.29[b] | 0.05[c] | 2.1 | 3.4 | — | 1.8 | 25 | Stratton and McCarty[37] |
| | 0.29 | 0.05 | 1.8 | 3.6 | — | 2.1 | 20 | |
| | 0.29 | 0.05 | 0.9 | 2.8 | — | 4.8 | 15 | |
| NO$_2$-N oxidation | 0.084[b] | 0.05[c] | 6.8 | 0.3 | — | 1.9 | 25 | |
| | 0.084 | 0.05 | 4.7 | 1.1 | — | 2.9 | 20 | |
| | 0.084 | 0.05 | 3.9 | 0.7 | — | 3.6 | 15 | |
| Activated sludge | | | | | | | | |
| NH$_3$-N oxidation | 0.05 | — | 6.6[d] | 1.0 | 0.33 | 3.0 | 20 | Downing et al.[35] |
| NO$_2$-N oxidation | 0.02 | — | 7.0[d] | 2.1 | 0.14 | 7.2 | ≈20 | |
| Thames estuary water | | | | | | | | |
| NH$_3$-N oxidation | 0.05 | — | 13[d] | 0.6 | 0.65 | 1.55 | 18.8 | Knowles, Downing, Barrett[36] |
| | 0.05 | — | 30[d] | 1.7 | 1.5 | 0.7 | 27 | |
| NO$_2$-N oxidation | 0.02 | — | 42[d] | 1.9 | 0.83 | 1.2 | 18.8 | |
| | 0.02 | — | 100 | 4.7 | 2.0 | 0.5 | ≈29 | |

[a]Calculated on basis of b = o
[b]Calculated from thermodynamic considerations
[c]Determined by "fitting factor" approach
[d]Calculated from relationship $\hat{\mu} = Y k$

system in series with a nitrification reactor-clarifier system. The principal advantage attributed to the two sludge system as compared to a single combined carbonaceous-nitrification reactor-clarifier system (nitrifying activated sludge) is that a greater degree of control over the two microbial processes is possible.[38] If the concept of $\theta_c$ is employed to control both one and two sludge nitrifying systems, it is not readily apparent why a two sludge system should be more controllable. This opinion is further reinforced by a recently completed laboratory-scale pilot study using municipal wastewater in which essentially identical performance was attained with "one sludge" nitrifying activated sludge systems and two sludge separate carbonaceous-nitrification systems operated at 20°C and 8°C.[39] Stankewich described the results of pilot nitrification studies on municipal wastewater using pure oxygen aeration.[40] He reported values of both specific utilization (total volatile suspended solids basis) and $\theta_c$ for one sludge and two sludge nitrifying activated sludges. Over the range of $\theta_c$ values studied (4.8-33 days) essentially complete nitrification was attained in both the one sludge and second stage of the two sludge systems. In many situations, it appears that the choice between one and two sludge nitrifying activated sludge systems can be made on the basis of least cost since comparable performance can be attained with either system.

The advantages associated with a $\theta_c$ based model employing the Monod formulation for substrate assimilation become evident when considering phenomena such as nitrification. It is possible to accurately measure the "growth limiting" substrate, *e.g.*, $NH_3$-N and to determine substrate specific values of U and (dX/dt) from enrichment cultures containing only nitrifying bacteria. Substrate specific kinetic coefficients, Y, b, k, and $K_s$, are determined from this data by appropriate analysis, *e.g.*, Table 7.4. Such coefficients can then be used to calculate limiting values ($\theta_c^m$) for nitrifying systems and to predict performance of nitrification in mixed culture systems such as activated sludge. By contrast, a model based on U or (F/M) is handicapped in mixed culture systems, since it is generally not possible to determine what fraction of the system biomass is metabolizing a given substrate. Moreover, it is not meaningful to calculate specific utilization of ammonia on the basis of total volatile mixed-liquor suspended solids, the major fraction of which are involved in carbonaceous removal and not nitrogen oxidation. For example, it would be possible, at a constant value of $\theta_c$ and constant influent ammonia concentration, for two systems to report widely varying values of U or (F/M) for ammonia oxidation (calculated on the basis of total volatile suspended solids) if the two systems had different influent carbonaceous waste concentrations. Similar difficulties are encountered in attempting to relate nitrification performance to (F/M) for carbonaceous removal. Such problems have led to considerable confusion among practitioners in attempting to develop (F/M) type criteria for design and control of nitrification. Use of $\theta_c$ and the Monod type formulation focus on the central issue, *i.e.*, growth rate of the nitrifying bacteria, and avoid unnecessary confusion. Much the same case can be made for other substrate specific microbial processes; *e.g.*, methanogenesis, sulfate reduction.

## Natural Waters

Pearson, *et al.* have applied a model of the form described here for the completely mixed process without recycle to the continuous culture of algae.[41] Their objective was to evaluate the algal growth potential of wastewater effluents and natural waters under continuous flow conditions. Experiments were conducted to determine values of the kinetic coefficients for nitrogen-limited system and for phosphorus-limited systems. It was concluded that continuous culture techniques and the associated mathematical models are valuable tools in assessing biostimulatory effects of wastewater effluents on natural waters. A number of investigators have used Monod type kinetic models to describe microbial mediated phenomena in natural waters.[42-44]

## Anaerobic Processes

### Methanogenesis

Much of the early work on modeling of biological slurry reactors was done in conjunction with studies of anaerobic digestion. Particular attention has been directed to the methane fermentation step of the process, which is considered to be the process rate-limiting step. The kinetic models presented here have been successfully applied to the anaerobic digestion process and methanogenesis.[1,13,14,19,20,45,46]

### Denitrification

Stensel, Loehr, and Lawrence evaluated steady state kinetics of microbial nitrate respiration (denitrification) using methanol as the sole source of carbon.[47] The biological reactors were operated such that carbonaceous substrate (COD) was the "growth limiting" substrate. Nitrate removal rate was computed from stoichiometric relationships established between carbonaceous removal and nitrate reduction. Table 7.5 shows values of the kinetic coefficients determined in their study. Kinetic coefficients were also evaluated at $10°C$ and $30°C$.

**Table 7.5.** Kinetic Coefficients for Denitrification Using Methanol at $20^{\circ}C$ [47]

| Coefficient | Dimension | Value |
|---|---|---|
| Y | mg S.S./mg COD | 0.18 |
| b | day$^{-1}$ | 0.04 |
| k | mg COD/mg V.S.S.-day | 25, 10[a] |
| $K_s$ | mg/l COD | 73, 9[a] |

[a]Computed using effluent COD adjusted for refractory residual COD of 7 mg/l.

Three system parameters were determined to be necessary for design. These are: (a) quantity of carbon (methanol) that must be added, (b) value of $\theta_c$ for design, and (c) effluent quality in terms of nitrate removal efficiency. The methanol requirement expressed as mg/l COD can be described as

$$M = 3.46 \, (N_0 - N) + 1.5 \, DO \tag{31}$$

in which M = methanol requirements expressed as COD, mass/volume; $N_0$ = influent nitrate nitrogen concentration as N, mass/volume; N = effluent nitrate nitrogen concentration as N, mass/volume; DO = influent dissolved oxygen concentration, mass/volume. The limiting values of biological solids retention time, $[\theta_c^m]_{lim}$ were approximately 0.5 days at $20^{\circ}C$ and $30^{\circ}C$ and 2 days at $10^{\circ}C$. Suggested design values of biological solids retention time $(\theta_c^d)$ are $\geqslant 4$ days at $20^{\circ}C$ and $30^{\circ}C$ and $\geqslant 8$ days at $10^{\circ}C$. The nitrate removal efficiency can be calculated from the following formula

$$N' = \frac{\left[ S_0 - \dfrac{K_s[1 + b(\theta_c)]}{\theta_c(Yk - b) - 1} - 1.5 \, (DO) \right] (10^2)}{3.46 N_0} \tag{32}$$

in which N' = nitrate nitrogen removal efficiency in percent. The relationships developed in this study can be modified for use with carbon sources other than methanol. Design of slurry denitrification systems can proceed on a rational basis using the results described here.

## SUMMARY

An approach to modeling of slurry biological reactors has been described. The models emphasize the utility of considering biological solids retention time or microbial growth rate as the independent parameter for process design and control. The state of knowledge with regard to steady-state modeling of the activated sludge and anaerobic digestion processes is considered to be well-advanced. By coupling the biokinetic process model for recycle systems with the clarifier design model presented, it is possible to explore least-cost design solutions for activated sludge process systems. The application of the models to nitrification and denitrification is less developed to date (1972). However, sufficient fundamental kinetic information is available on these processes to facilitate rapid implementation of kinetic based process design. It is hoped that information on the kinetics of the other microbially mediated processes of interest in water quality control, *e.g.,* sulfur cycle and nitrogen fixation, will be forthcoming.

The state-of-the-art (1972) in dynamic modeling of slurry biological reactors is less well developed than modeling of the steady-state. There are significant problems in process operation, control, and design, which can best be solved when we have dynamic models of sufficient sophistication to reliably predict the time varying nature of biological system performance. The dynamic studies described here are steps in that direction.

In a more general context, it is suggested that the models described here can serve to provide a common format for the conductor of research studies on microbial mediated phenomena in slurry reactors and natural waters and for reporting of study results. Hopefully, process development studies on established processes will also be described in the context of the kinetic models. Use of a common model base and system parameters will contribute immeasurably to increased interchange of information between investigators and practitioners and will serve to move the art of design forward to a universally accepted rational basis.

Rational kinetic models of the type presented here provide the water quality engineer with a powerful and fundamental tool for analysis and design of slurry biological reactors. Increasing acceptance of such models is welcome evidence that water quality control process engineering is coming of age.

## REFERENCES

1. Lawrence, A. W. and P. L. McCarty. "Unified Basis for Biological Treatment Design and Operation," *J. San. Eng. Div., ASCE* **96**(SA3), 757 (1970).
2. Heukelekian, H., H. E. Orford, and R. Manganelli. "Factors Affecting the Quantity of Sludge Production in the Activated Sludge Process," *Sewage Ind. Wastes* **23**, 945 (1951).
3. Weston, R. F. and W. W. Eckenfelder. "Applications of Biological Treatment to Industrial Wastes: I. Kinetics and Equilibria of Oxidative Treatment," *Sewage Ind. Wastes* **27**, 802 (1955).

4. van Uden, N. "Kinetics of Nutrient-Limited Growth," *Ann. Rev. Microbiol.* 23 (Palo Alto, Calif.: Annual Reviews, Inc., 1969), p. 473.
5. van Uden, N. "Transport-Limited Growth in the Chemostat and Its Competitive Inhibition; A Theoretical Treatment," *Arch. Mikrobiol.* 58, 145 (1967).
6. Lawrence, A. W. and P. L. McCarty. "A Kinetic Approach to Biological Wastewater Treatment Design and Operation," *Technical Report No. 23*, Cornell University Water Resources and Marine Sciences Center, Ithaca, New York (December 1969).
7. Dick, R. I. and A. R. Javaheri. Discussion of "Unified Basis for Biological Treatment, Design and Operation" A. W. Lawrence and P. L. McCarty. *J. San. Eng. Div., ASCE* 97, 234 (April 1971).
8. Dick, R. I. "Role of Activated Sludge Final Settling Tanks," *J. San. Eng. Div., ASCE* 96, 423 (April 1970).
9. Bisogni, J. J. and A. W. Lawrence. "Relationships Between Biological Solids Retention Time and Settling Characteristics of Activated Sludge," *Water Res.* 5, 753 (1971).
10. Vesilind, P. A. "Design of Prototype Thickeners from Batch Settling Tests," *Water Sewage Works* 115, 302 (July 1968).
11. Dick, R. I. and K. W. Young. "Analysis of Thickening Performance of Final Settling Tanks," Presented at the 27th Annual Purdue Industrial Waste Conference, Purdue University, May 2-4, 1972.
12. Middleton, A. C. and A. W. Lawrence. Unpublished data, Cornell University, Ithaca, New York (July 1972).
13. O'Rourke, J. T. "Kinetics of Anaerobic Treatment at Reduced Temperatures," Ph.D. Thesis, Stanford University, Stanford California (1968).
14. Lawrence, A. W. "Application of Process Kinetics to Design of Anaerobic Processes," *Anaerobic Biological Treatment Process*, F. G. Pohland, Ed. Adv. Chem. Series No. 105 (Washington, D. C.: American Chemical Society, 1971), p. 163.
15. Smith, R. "Cost of Conventional and Advanced Treatment of Wastewater," *J. Water Poll. Control Fed.* 40, 1546 (1968).
16. Lawrence, A. W. and T. R. Milnes. "Biokinetic Approach to Least Cost Design of Activated Sludge Systems," Presented at the 162nd National Meeting of the American Chemical Society, Washington, D.C. September 12-17, 1971.
17. Jennings, S. L. and C. P. L. Grady. "The Use of Final Clarifier Models in Understanding and Anticipating Performance Under Operational Extremes," Presented at the 27th Purdue Industrial Waste Conference, Purdue University, May 2-4, 1972.
18. Middleton, A. C. and A. W. Lawrence. "Optimal Design of Activated Sludge Systems," In preparation, Cornell University, January-December 1972.
19. Andrews, J. F. "Dynamic Model of the Anaerobic Digestion Process," *J. San. Eng. Div., ASCE* 95, 95 (1969).
20. Andrews, J. F. and S. P. Graef. "Dynamic Modeling and Simulation of the Anaerobic Digestion Process," *Anaerobic Biological Treatment Processes*, F. G. Pohland, Ed. Adv. Chem. Series No. 105 (Washington, D.C.: American Chemical Society, 1971), p. 126.

21. Eckhoff, D. W. and D. Jenkins. "Activated Sludge Systems, Kinetics of the Steady and Transient States," *SERL Report No. 67-12*, Sanitary Engineering Research Laboratory, University of California, Berkeley (December 1967).
22. Ott, C. R. and R. H. Bogan. "Theoretical Analysis of Activated Sludge Dynamics," *J. San. Eng. Div., ASCE* **97**, 1 (February 1971).
23. Burkhead, C. E. and D. J. Wood. "Analog Simulation of Activated Sludge Systems," *J. San. Eng. Div., ASCE* **95**, 593 (1969).
24. Grady, C. P. L. "A Theoretical Study of Activated Sludge Transient Response." Presented at the 26th Purdue Industrial Waste Conference, Purdue University, May 4-6, 1971.
25. Bryant, J. O., L. C. Wilcox, and J. F. Andrews. "Continuous Time Simulation of Wastewater Treatment Plants," Presented at the 69th National Meeting of the American Institute of Chemical Engineers, Cincinnati, Ohio, May 1971.
26. Milnes, T. R. and A. W. Lawrence. "Dynamic Modeling of the Completely Mixed Activated Sludge Process: Laboratory and Mathematical Studies," Presented at the 44th Annual Conf. of the Water Pollution Control Federation, San Francisco, October 5, 1971.
27. Milnes, T. R. "Dynamic Modeling of the Completely Mixed Activated Sludge Process: Laboratory and Mathematical Studies," Ph.D. Thesis Cornell University, Ithaca, New York (1972).
28. IBM. "System/360 Continuous System Modeling Program (360A-CX-16X) User's Manual," *IBM Application Program GH20-0367-3*, IBM, New York (1968).
29. Metcalf and Eddy, Inc. *Wastewater Engineering.* (New York: McGraw-Hill Book Co., 1972).
30. Stensel, H. D. and G. L. Shell. "The Design of Biological Wastewater Treatment Processes—Solids Retention Time or Food: Mass Ratio," Presented at the 45th Annual Conf. of the Water Pollution Control Federation, Atlanta, Georgia, October 11, 1972.
31. McCarty, P. L. "Stoichiometry of Biological Reactions." Presented at the International Conf. "Toward a Unified Concept of Biological Waste Treatment Design," Atlanta, Georgia, October 6, 1972.
32. McCarty, P. L. "Energetics and Bacterial Growth," Presented at the Fifth Rudolph Res. Conf., Rutgers University, New Brunswick, New Jersey, July 2, 1969.
33. Walker, L. F. "Hydraulically Controlling Solids Retention Time in the Activated Sludge Process," *J. Water Poll. Control Fed.* **43**, 30 (1971).
34. Sherrard, J. H. and E. D. Shroeder. "Relationship Between the Observed Yield Coefficient and Mean Cell Residence Time in the Completely Mixed Activated Sludge Process," *Water Res.* **6**, 1039 (1972).
35. Downing, A. L., H. A. Painter, and G. Knowles. "Nitrification in the Activated Sludge Process," *J. Inst. Sewage Purif.*, Part 2, 130 (1964).
36. Knowles, G., A. L. Downing, and M. J. Barrett. "Determination of Kinetic Constants for Nitrifying Bacteria in Mixed Culture, with the Aid of an Electronic Computer," *J. Gen. Microbiol.* **38**, 263 (1965).
37. Stratton, F. E. and P. L. McCarty. "Prediction of Nitrification Effects on the Dissolved Oxygen Balance of Streams," *Environ. Sci. Tech.* **1**, 405 (1967).

38. Mulbarger, M. C. "Nitrification and Denitrification in Activated Sludge Systems," *J. Water Poll. Control Fed.* **43**, 2059 (1971).
39. Lawrence, A. W., C. G. Brown, and J. D. Latona. "Biokinetic Approach to Optimal Design and Control of Nitrifying Activated Sludge Systems," In Preparation, Cornell University, Ithaca, New York (1972).
40. Stankewich, M. J. "Biological Nitrification with the High Purity Oxygenation Process." Presented at the 27th Annual Purdue Industrial Waste Conf., Purdue University, May 2-4, 1972.
41. Pearson, E. A., *et al.* "Kinetic Assessment of Algal Growth," *Proceedings, Eutrophication-Biostimulation Assessment Workshop,* E. J. Middlebrooks, *et al.,* Ed. (Berkeley, Calif.: University of California, Sanitary Engineering Research Laboratory, June 19-21, 1969), p. 56.
42. Jannasch, H. W. "Growth of Marine Bacteria at Limiting Concentrations of Organic Carbon in Seawater," *Limnol. Oceanog.* **12**, 264 (1967).
43. Dugdale, R. C. "Nutrient Limitation in the Sea: Dynamics, Identification, and Significance," *Limnol. Oceanog.* **12**, 685 (1967).
44. DiToro, D. M., D. J. O'Connor, and R. V. Thomann. "A Dynamic Model of the Phytoplankton Population in the Sacramento-San Joaquin Delta," *Nonequilibrium Systems in Natural Water Chemistry,* Adv. Chem. Series No. 106 (Washington, D. C.: American Chemical Society, 1971), p. 131.
45. McCarty, P. L. "Kinetics of Waste Assimilation in Anaerobic Treatment," *Develop. Ind. Microbiol.* **7**, 144, American Institute of Biological Sciences, Washington (1966).
46. Pohland, F. G., Ed. *Anaerobic Biological Treatment Processes,* Adv. Chem. Series No. 105 (Washington, D.C.: American Chemical Society, 1971).
47. Stensel, H. D., R. C. Loehr, and A. W. Lawrence. "Biological Kinetics of the Suspended Growth Denitrification Process,' *J. Water Poll. Control Fed.* **45**, 249 (1973).

## NOMENCLATURE

| Symbol | Description |
|---|---|
| $A_c$ | area of the activated sludge secondary clarifier which satisfies the areal requirement for clarification, area |
| $A_d$ | design area of the activated sludge secondary clarifier taken to be equal to the larger of the quantities, $A_c$ or $A_t$, area |
| $A_t$ | area of the activated sludge secondary clarifier which satisfies the areal requirement for solids transmission (thickening), area |
| a | slurry specific value of the ordinate intercept (base e) in Equation 24, length/time |
| b | microorganism decay coefficient, time$^{-1}$ |
| dF/dt | rate of microbial substrate utilization per unit volume, mass/volume-time |
| E | treatment efficiency, percent |
| $G_i$ | total solids flux associated with a suspended solids concentration $X_i$ in the thickening regime of an activated sludge secondary clarifier under continuous flow steady state conditions, mass/area-time |
| $G_L$ | limiting total solids flux, *i.e.*, the minimum solids flux in the range of $X_i$ values from X to $X_r$ in continuous flow clarifier-thickener, mass/area-time |
| g | subscript which indicates performance evaluation based on a gross parameter, *e.g.*, total BOD or total COD |
| $K_s$ | half velocity coefficient, equal to the substrate concentration when $(dF/dt)/X = (\frac{1}{2})k$, mass/volume |
| k | maximum rate of substrate utilization per unit weight of microorganisms, time$^{-1}$ |
| $k'$ | $k/K_s$ |
| $k_e$ | slope of plot of log (base e) ZSV versus X in Equation 24, volume/mass |
| $L_f$ | suspended solids load transmitted to clarifier floor equal to $[(Q+q)X - (Q-w)X_e]$, mass/time |
| N | subscript which denotes total system, *i.e.*, reactor and solids separator |
| n | subscript which denotes reactor only |
| $P_X$ | excess microorganisms production rate, mass/time |
| Q | "net" flow rate of liquid through the reactor, volume/time |
| q | flow rate at which underflow of solids separator is recycled to the reactor, volume/time |
| $q'$ | q + w, solids separator hydraulic underflow rate, volume/time |
| r | volumetric recycle ratio, *i.e.*, q/Q |
| S | substrate concentration, mass/volume |
| s | subscript which indicates performance evaluation based on removal of a specific waste constituent |
| T | subscript referring to total system |

| | |
|---|---|
| t | time |
| $t_d$ | minimum generation or doubling time of microorganisms, time |
| U | specific utilization equal to $(dF/dt)/X$, $time^{-1}$ |
| u | bulk liquid velocity in a clarifier due to underflow which is equal to $(q'/A_d)$, length/time |
| V | reactor volume, volume |
| v | subsidence velocity of suspension at initial concentration X, taken equal to ZSV for that initial value of X, length/time |
| w | flow rate of that fraction of solids separator underflow which is wasted from the system, volume/time |
| X | microbial mass concentration, mass/volume |
| $X_e$ | microbial mass concentration in the clarified overflow from the solids separator, mass/volume |
| $X_i$ | the ith microbial mass concentration in a settling biomass, mass/volume |
| $X_L$ | the microbial mass concentration associated with the limiting total solids flux $(G_L)$ in the secondary clarifier design, mass/volume |
| $X_r$ | microbial mass concentration in the underflow from the solids separator, mass/volume |
| $X_T$ | total active microbial mass in treatment system, mass |
| $X^i$ | refractory suspended constituent of influent waste origin, mass/volume |
| Y | growth yield coefficient, mass/mass |
| ZSV | zone settling velocity of a suspension determined from batch settling test, length/time |
| o | subscript which denotes reactor effluent stream |
| $\theta$ | reactor mean hydraulic retention time based on raw waste flow, *i.e.*, V/Q, time |
| $\theta'$ | reactor actual mean hydraulic retention time for recycle systems, *i.e.*, V/(Q+q), time |
| $\theta_c$ | biological solids retention time as defined by Equation 8, time |
| $\theta_c^d$ | value of biological solids retention time at which $S_1 = S_0$, time |
| $[\theta_c^m]_{lim}$ | limiting value of $\theta_c^m$ which occurs when $S_0 \gg K_s$, time |
| $\theta_c^*$ | biological solids retention time which achieves the maximum rate of volumetric utilization of waste in a completely mixed system without recycle (Equation 28), time |
| $\mu$ | $(dX/dt)/X$, *i.e.*, net specific growth rate of microorganisms, $time^{-1}$ |
| $\hat{\mu}$ | maximum net specific growth rate of microorganisms, $time^{-1}$ |

The following symbols refer to process model for denitrification only:[47]

| | |
|---|---|
| D.O. | influent dissolved oxygen concentration, mass/volume |
| M | methanol requirements expressed as COD, mass/volume |
| N | effluent nitrate nitrogen concentration as N, mass/volume |
| $N_0$ | influent nitrate nitrogen concentration as N, mass/volume |
| $N'$ | nitrate removal efficiency, percent |

## APPENDIX A

### Listint of PL/1 Computer Program for Completely Mixed Biological Slurry Reactor with Solids Recycle

```
ACM: PROCEDURE OPTIONS (MAIN)  ;
/*   SECTION I :   DECLARATION AND DEFINITION OF VARIABLES            */

/*       A.   BIOKINETIC VARIABLES AND CONSTANTS                      */

         DECLARE(
         K,        /*SUBSTRATE UTILIZATION COEFFICIENT, MG/MG/DA      */
         B,        /* MICROBIAL DECAY COEFFICIENT, 1/DA               */
         KS,       /* HALF VELOCITY COEFFICIENT, MG/L                 */
         Y,        /* YIELD COEFFICIENT, MG/MG                        */
         SO,       /* INFLUENT SUBSTRATE CONCENTRATION, MG/L          */
         Q,        /* VOLUMETRIC FLOW RATE, MGD                       */
         SRT,      /* BIOLOGICAL SOLIDS RETENTION TIME, DA            */
         S1(20),   /* EFFLUENT SUBSTRATE CONCENTRATION, MG/L          */
         E,        /* SUBSTRATE REMOVAL EFFICIENCY, PERCENT           */
         XV(20),   /* TOTAL WEIGHT OF BIOLOGICAL SOLIDS IN THE        */
                   /* REACTOR, 1000 LBS                               */
         PX(20),   /* DAILY RATE OF BIOLOGICAL SOLIDS WASTING,        */
                   /* LBS/DA                                          */
         F_TO_M,   /* FOOD TO MICROORGANISM RATIO,                    */
                   /* LBS BOD 5 / LB BIOLOGICAL SOLID / DA            */
         MU,       /* SPECIFIC GROWTH RATE, 1/DA                      */
         U,        /* SPECIFIC UTILIZATION, 1/DA                      */
         Y_OBS     /* OBSERVED YIELD COEFFICIENT,                     */
                   /* LBS BIOLIGICAL SOLIDS / LB COD REMOVED          */
         )FLOAT DECIMAL (6)  ;


/*       B.  DESIGN VARIABLES                                         */

         DECLARE(
         XR,          /* FINAL CLARIFIER UNDERFLOW                    */
                      /* CONCENTRATION, MG/L                          */
         R,           /* RECYCLE RATIO                                */
         X,           /* MLVSS, MG/L                                  */
         V            /* REACTOR VOLUME, MIL GAL                      */
         )FLOAT DECIMAL (6)  ;


/*    C.  OTHERS                                                      */

         DECLARE (
         WVAR1,
         WVAR2,
         WVAR3,
         WVAR4,
         WVAR5
         )FLOAT DEC(6) INIT(0)  ;

         DECLARE DASHES CHARACTER (90)  ;
/*   SECTION II : INPUT OF CONSTANTS AND PAGE HEADING                 */


/*    A.   INPUT OF CONSTANTS                                         */

         GET LIST (
         K,B,KS,Y,SO,Q,DASHES)  ;
```

```
/*      B.   PAGE HEADING                                                  */

         P = 0  ;
         ON ENDPAGE(SYSPRINT) BEGIN  ;
         P = P + 1  ;
         PUT PAGE  ;
         PUT EDIT('PAGE',P,'***** UNIFIED BASIS FOR BIOLOGICA'||
            'L TREATMENT DESIGN *****')
            (SKIP,COL(52),A,COL(57),F(3),SKIP(2),COL(27),A)  ;
         END  ;


/*   SECTION III :  CALCULATION OF BIOKINETIC VALUES                       */

/*      A.  PAGE AND COLUMN HEADINGS                                       */

         SIGNAL ENDPAGE(SYSPRINT)  ;
         PUT EDIT('BIOKINETIC PARAMETERS AS FUNCTION OF BIOLOGICAL '||
            'SOLIDS RETENTION TIME','INPUT DATA','K  = ',K,' MG/MG/'||
            'DAY','B = ',B,' 1/DAY','KS = ',KS,' MG/L','Y = ',Y,
            ' MG/MG','SO = ',SO,' MG/L','Q = ',Q,' MGD')
            (SKIP(3),COL(21),A,SKIP(3),COL(51),A,SKIP(2),COL(31),A,
            COL(36),F(7,2),COL(43),A,COL(63),A,COL(67),F(6,2),COL(73),
            A,SKIP,COL(31),A,COL(36),F(7,2),COL(43),A,COL(63),A,
            COL(67),F(6,2),COL(73),A,SKIP,COL(31),A,COL(36),F(7,2),
            COL(43),A,COL(63),A,COL(67),F(6,2),COL(73),A)  ;
         PUT EDIT('SRT','S1','E','XV','PX','MU','U','Y','F/M',
            'DAYS','MG/L','PERCENT','1000 LBS','LBS/DAY','1/DAY',
            '1/DAY','OBS','(BOD)')
            (SKIP(3),COL(15),A,COL(24),A,COL(34),A,COL(43),A,COL(54),
            A,COL(64),A,COL(73),A,COL(84),A,COL(93),A,SKIP,COL(15),
            A,COL(23),A,COL(32),A,COL(41),A,COL(53),A,COL(63),A,
            COL(72),A,COL(83),A,COL(92),A)  ;
         PUT EDIT(DASHES,' ')(SKIP,COL(11),A,SKIP,COL(11),A)  ;


/*      B.  CALCULATION OF VALUES                                          */

         WVAR1 = (Y*K)-B ; /*  REPEATEDLY USED CONSTANTS               */
         WVAR2 = Y*Q ;


/*    A DO LOOP IS NOW USED TO CALCULATE VALUES OF BIOKINETIC              */
/*    PARAMETERS FOR  SRT = 1 TO 20 DAYS                                   */

         DO I = 1 TO 20  ;
         WVAR3 = 1 + B*I ; /*  ANOTHER CONSTANT                        */
         S1(I) = KS*WVAR3/(I*WVAR1-1)  ; /*  EFFLUENT SUBSTRATE        */
                                         /*  CONCENTRATION, MG/L       */
         WVAR4 = SO-S1(I)  ;  /*  ANOTHER CONSTANT                     */
         XV(I) = WVAR2*I*WVAR4*8.34E-03/WVAR3  ; /*  TOTAL SOLIDS      */
                                      /*  IN REACTOR,   1000 LBS       */
         E = WVAR4*100/SO ; /*  SUBSTRATE REMOVAL EFFICIENCY, %        */
         PX(I) = XV(I)*1000/I ; /*  DAILY BIOLOGICAL SOLIDS            */
                                /*  WASTAGE RATE, LBS/DAY              */
         MU = 1/I ;  /*  SPECIFIC GROWTH RATE, 1/DAY                   */
         U = (MU+B)/Y ;  /*  SPECIFIC UTILIZATION, 1/DAY               */
         Y_OBS = MU/U ;  /*  OBSERVED YIELD, LBS/LB                    */
         F_TO_M = U*66.67/E ;  /*  ORGANIC LOADING RATE,               */
                            /*  LBS BOD-5/LB BIOLOGICAL SOLIDS         */
                            /*  ASSUMING BOD-5 = 2/3 COD AND           */
                            /*  INPUT  ON A COD BASIS                  */
         PUT SKIP EDIT(I,S1(I),E,XV(I),PX(I),MU,U,Y_OBS,F_TO_M)
            (COL(16),F(2),COL(23),F(5,2),COL(33),F(5,2),  COL(42),
            F(7,2),COL(53),F(5),COL(63),F(5,3),COL(72),F(6,3),
            COL(82),F(6,3),COL(92),F(6,3))  ;
         END  ;


/*   SECTION IV:  CALCULATION OF DESIGN FIELD                             */
```

```
AGAIN:  GET LIST(SRT)   ;
        IF SRT = -1 THEN GOTO FINISH   ;
        SIGNAL ENDPAGE(SYSPRINT)   ;
        PUT EDIT('DESIGN FIELD')(SKIP(3),COL(49),A)   ;
        PUT EDIT('SRT =',SRT,'DA')(SKIP(3),COL(49),A,COL(55),F(2),
            COL(58),A)   ;
        WVAR1,WVAR2 = 0   ;
        WVAR3 = SRT*Q   ;
        WVAR4 = SRT*Q/XV(SRT)/1000*8.34   ;
        DO I = 6 TO 16 BY 2   ;
            WVAR1 = WVAR1 + 1   ;
            XR = 1000*I   ;
            PUT EDIT('XR =',XR,'MG/L','R','X,MG/L','V,MGAL')
                (SKIP(3),COL(47),A,COL(52),F(5),COL(58),A,SKIP(2),COL(40),
                A,COL(52),A,COL(66),A)   ;
            DO R = 0 TO 2 BY 0.25   ;
                V = (WVAR3*(1+R))/(1+WVAR4*(R*XR))   ;
                X = XV(SRT)/V*1000/8.34   ;
                PUT EDIT(R,X,V)(SKIP,COL(39),F(4,2),COL(52),F(6),
                    COL(66),F(7,2))   ;


            END   ;
            IF WVAR1 = 3 & WVAR2 = 0 THEN
            DO   ;
                SIGNAL ENDPAGE(SYSPRINT)   ;
                PUT EDIT('SRT =',SRT,'DA')(SKIP(3),COL(49),A,
                    COL(55),F(2),COL(58),A)   ;
                WVAR1 = 0   ;
                WVAR2 = 1   ;
            END   ;
        END   ;
        GOTO AGAIN   ;
FINISH: PUT PAGE   ;
END ACM   ;
```

##### ***** UNIFIED BASIS FOR BIOLOGICAL TREATMENT DESIGN *****

BIOKINETIC PARAMETERS AS FUNCTION OF BIOLOGICAL SOLIDS RETENTION TIME

INPUT DATA

| | | |
|---|---|---|
| K = 7.00 MG/MG/DAY | | B = 0.05 1/DAY |
| KS = 50.00 MG/L | | Y = 0.40 MG/MG |
| SO = 300.00 MG/L | | Q = 20.00 MGD |

| SRT DAYS | S1, MG/L | E, PERCENT | XV, 1000 LBS | PX, LBS/DAY | MU, 1/DAY | U, 1/DAY | Y, OBS | F/M (BOD) |
|---|---|---|---|---|---|---|---|---|
| 1 | 30.00 | 90.00 | 17.15 | 17156 | 1.000 | 2.625 | 0.380 | 1.944 |
| 2 | 12.22 | 95.92 | 34.91 | 17455 | 0.500 | 1.375 | 0.363 | 0.955 |
| 3 | 7.93 | 97.35 | 50.83 | 16945 | 0.333 | 0.958 | 0.347 | 0.656 |
| 4 | 6.00 | 98.00 | 65.38 | 16346 | 0.250 | 0.750 | 0.333 | 0.510 |
| 5 | 4.90 | 98.36 | 78.75 | 15751 | 0.199 | 0.624 | 0.319 | 0.423 |
| 6 | 4.19 | 98.60 | 91.09 | 15181 | 0.166 | 0.541 | 0.307 | 0.366 |
| 7 | 3.69 | 98.76 | 102.50 | 14643 | 0.142 | 0.482 | 0.296 | 0.325 |
| 8 | 3.33 | 98.83 | 113.10 | 14138 | 0.125 | 0.437 | 0.285 | 0.294 |
| 9 | 3.05 | 98.98 | 122.97 | 13663 | 0.111 | 0.402 | 0.275 | 0.271 |
| 10 | 2.83 | 99.05 | 132.18 | 13218 | 0.099 | 0.374 | 0.266 | 0.252 |
| 11 | 2.64 | 99.11 | 140.79 | 12799 | 0.090 | 0.352 | 0.258 | 0.236 |
| 12 | 2.50 | 99.16 | 148.86 | 12405 | 0.083 | 0.333 | 0.249 | 0.224 |
| 13 | 2.37 | 99.20 | 156.45 | 12034 | 0.076 | 0.317 | 0.242 | 0.213 |
| 14 | 2.26 | 99.24 | 163.59 | 11685 | 0.071 | 0.303 | 0.235 | 0.203 |
| 15 | 2.17 | 99.27 | 170.32 | 11354 | 0.066 | 0.291 | 0.228 | 0.195 |
| 16 | 2.09 | 99.30 | 176.67 | 11042 | 0.062 | 0.281 | 0.222 | 0.188 |
| 17 | 2.02 | 99.32 | 182.69 | 10746 | 0.058 | 0.272 | 0.216 | 0.182 |
| 18 | 1.95 | 99.34 | 188.38 | 10465 | 0.055 | 0.263 | 0.210 | 0.177 |
| 19 | 1.90 | 99.36 | 193.79 | 10199 | 0.052 | 0.256 | 0.205 | 0.172 |
| 20 | 1.85 | 99.38 | 198.92 | 9946 | 0.049 | 0.249 | 0.199 | 0.167 |

PAGE   2

***** UNIFIED BASIS FOR BIOLOGICAL TREATMENT DESIGN *****

DESIGN FIELD

SRT = 10 DA

XR =  6000 MG/L

| R | X,MG/L | V,MGAL |
|------|--------|--------|
| 0.00 | 79 | 200.00 |
| 0.25 | 1263 | 12.54 |
| 0.50 | 2052 | 7.72 |
| 0.75 | 2616 | 6.05 |
| 1.00 | 3039 | 5.21 |
| 1.25 | 3368 | 4.70 |
| 1.50 | 3631 | 4.36 |
| 1.75 | 3846 | 4.11 |
| 2.00 | 4026 | 3.93 |

XR =  8000 MG/L

| R | X,MG/L | V,MGAL |
|------|--------|--------|
| 0.00 | 79 | 200.00 |
| 0.25 | 1663 | 9.52 |
| 0.50 | 2719 | 5.82 |
| 0.75 | 3473 | 4.56 |
| 1.00 | 4039 | 3.92 |
| 1.25 | 4479 | 3.53 |
| 1.50 | 4831 | 3.28 |
| 1.75 | 5119 | 3.09 |
| 2.00 | 5359 | 2.95 |

XR = 10000 MG/L

| R | X,MG/L | V,MGAL |
|------|--------|--------|
| 0.00 | 79 | 200.00 |
| 0.25 | 2063 | 7.68 |
| 0.50 | 3386 | 4.68 |
| 0.75 | 4330 | 3.65 |
| 1.00 | 5039 | 3.14 |
| 1.25 | 5590 | 2.83 |
| 1.50 | 6031 | 2.62 |
| 1.75 | 6392 | 2.47 |
| 2.00 | 6693 | 2.36 |

***** UNIFIED BASIS FOR BIOLOGICAL TREATMENT DESIGN *****

SRT = 10 DA

XR = 12000 MG/L

| R | X,MG/L | V,MGAL |
|------|--------|--------|
| 0.00 | 79 | 200.00 |
| 0.25 | 2463 | 6.43 |
| 0.50 | 4052 | 3.91 |
| 0.75 | 5188 | 3.05 |
| 1.00 | 6039 | 2.62 |
| 1.25 | 6701 | 2.36 |
| 1.50 | 7231 | 2.19 |
| 1.75 | 7665 | 2.06 |
| 2.00 | 8026 | 1.97 |

XR = 14000 MG/L

| R | X,MG/L | V,MGAL |
|------|--------|--------|
| 0.00 | 79 | 200.00 |
| 0.25 | 2863 | 5.53 |
| 0.50 | 4719 | 3.35 |
| 0.75 | 6045 | 2.62 |
| 1.00 | 7039 | 2.25 |
| 1.25 | 7812 | 2.02 |
| 1.50 | 8431 | 1.87 |
| 1.75 | 8937 | 1.77 |
| 2.00 | 9359 | 1.69 |

XR = 16000 MG/L

| R | X,MG/L | V,MGAL |
|------|--------|--------|
| 0.00 | 79 | 200.00 |
| 0.25 | 3263 | 4.85 |
| 0.50 | 5386 | 2.94 |
| 0.75 | 6902 | 2.29 |
| 1.00 | 8039 | 1.97 |
| 1.25 | 8924 | 1.77 |
| 1.50 | 9631 | 1.64 |
| 1.75 | 10210 | 1.55 |
| 2.00 | 10693 | 1.48 |

# CHAPTER 8

# MODELING AND SIMULATION OF FIXED FILM BIOLOGICAL REACTORS FOR CARBONACEOUS WASTE TREATMENT

**Billy H. Kornegay**

Carbon Department
Westvaco
Covington, Virginia

## INTRODUCTION

Fixed-film biological reactors are multi-phase processes in which organic wastes are absorbed and subsequently oxidized by an attached microbial film. Since the ultimate removal is dependent upon microbial growth, fixed-film reactors should be amenable to description by the continuous culture theory. In fact, the validity of this approach has been demonstrated.[1-4] There are, however, several differences that must be considered in the development of models for fixed-film systems. First, the organisms and substrate do not occupy a common volume so any relation between these parameters must be made on a mass basis. Second, the entire mass of attached film is not active in the removal of organics. This latter point has been clearly elucidated by several investigators[1,5-8] and is illustrated in Figure 8.1. It should be noted that the active film thickness, d, constitutes only a portion of the total film thickness, h. The term "active" is used to denote that zone of organisms produced before substrate removal becomes independent of film thickness and is not intended to imply that a distinction can be made between viable and nonviable organisms. Additional restrictions are encountered in the trickling filter model since no simple relation exists between the filter volume, liquid volume and the volume of microbial film.

271

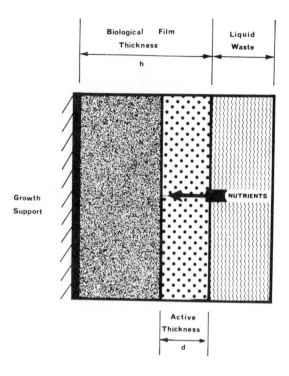

**Figure 8.1.** Biological film section.

Although models have been developed for several fixed-film reactors, only those applicable to trickling filters or rotating biological contactors will be included in this chapter. This will allow more emphasis to be placed on the two most practical and widely used reactor systems. Although some performance trends will be demonstrated with data, no effort will be made to present experimental proof of the models since that may be found elsewhere.[1-4,9,10] This chapter will, instead, emphasize model development, the effect of various design and operating parameters on performance, problem solutions and model limitations and restrictions.

## THEORY

### Trickling Filter Model

In the trickling filter shown in Figure 8.2, a raw waste with a flow rate, $F$, and concentration, $S_o$, is diluted to a concentration, $S_i$, with clarified

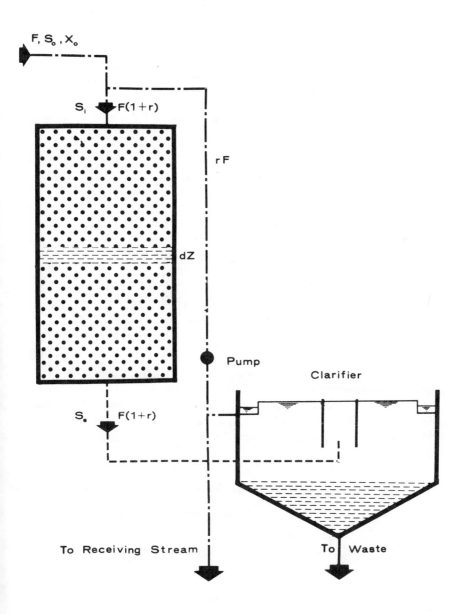

**Figure 8.2.** A trickling filter with recycle.

effluent having a concentration, $S_e$, and flow, $rF$. As this diluted water passes through the filter at a flow rate, $F(1 + r)$, the concentration is reduced from $S_i$ to $S_e$. This substrate removal is produced by the mass of microorganisms which is attached to the surface of the filter media. The removal may be determined from a steady-state substrate balance across the differential element dZ, as

$$\text{Input = Output + Microbial Consumption} \tag{1}$$

The input to the differential element is $F(1 + r)(S + dS)$, the output is $F(1 + r)(S)$, and the substrate removal is equal to the negative value of the microbial growth in the differential element divided by the yield, Y. The substrate balance then becomes

$$F(1 + r)(S + dS) = F(1 + r)(S) - \frac{\mu}{Y}M_o \tag{2}$$

where

$F$ = hydraulic flow rate, $(L^3/T)$
$r$ = hydraulic recycle ratio
$S$ = concentration of the rate controlling nutrient, $(M/L^3)$
$\mu$ = specific growth rate, $(T^{-1})$
$M_o$ = active mass of organisms in the differential element, (M)
$Y$ = yield, (Mass cells produced/Mass of substrate removed)

It has been shown by Kornegay and Andrews[2,3] that the active microbial mass may be expressed as the product of the differential depth, dZ, the cross sectional area, H, the specific surface area of the filter media, a, the unit mass of the biological film, X, and the thickness of the active layer, d. When this relation is substituted into Equation 2 one obtains

$$F(1 + r)(S + dS) = F(1 + r)S - \frac{\mu}{Y}(a)(X)(d)(H)dZ \tag{3}$$

The resulting expression for substrate removal with depth is then

$$-dS/dZ = \frac{\mu(a)(X)(d)(H)}{YF(1 + r)} \tag{4}$$

When the specific growth rate is expressed using the Monod[11] growth relation

$$\mu = \mu_{max} \left( \frac{S}{K_g + S} \right) \tag{5}$$

Equation 4 becomes

$$-dS/dZ = \frac{\mu_{max}(a)(d)(X)(H)}{FY(1 + r)} \left( \frac{S}{K_g + S} \right) \tag{6}$$

where

$\mu_{max}$ = maximum specific growth rate, $(T^{-1})$
$K_g$ = saturation constant, $(M/L^3)$

This expression may be integrated when S goes from $S_i$ to $S_e$ as the depth goes from 0 to Z to produce the expression

$$K_g \ln (S_i/S_e) + (S_i - S_e) = \frac{\mu_{max}}{FY(1 + r)}(a)(X)(d)(H)Z \qquad (7)$$

It may be shown from mass balances on the recycle system that the influent substrate concentration, $S_i$, may be determined by the equation

$$S_i = \frac{S_e r + S_o}{1 + r} \qquad (8)$$

and Equation 7 then becomes

$$K_g \ln \frac{S_o + rS_e}{S_e(1 + r)} + \frac{S_o + rS_e}{1 + r} - S_e = \frac{\mu_{max}}{FY(1 + r)}(a)(X)(d)(H)Z \qquad (9)$$

When there is no recycle, r = 0 and

$$K_g \ln (S_o/S_e) + (S_o - S_e) = \frac{\mu_{max}}{FY}(a)(X)(d)(H)Z \qquad (10)$$

Although Equations 9 and 10 are mathematically correct, they contain a number of elusive kinetic parameters which may lead to either "misuse" or "disuse" by the practicing engineer. However, such parameters as yield, maximum specific growth rate, unit mass of biological film and active film thickness need not be determined on an individual basis for design application. These parameters may be combined into a single term for simplification, but their presence and significance should be thoroughly understood.

Since the product $(1/Y)\mu_{max}(X)(d)$ may be evaluated from pilot studies, it would suggest that a single term, P, be substituted into Equation 9 and 10 such that

$$P = \frac{1}{Y}(\mu_{max})(X)(d) \qquad (11)$$

Equation 9 would then become

$$K_g \ln \frac{S_o + rS_e}{S_e(1 + r)} + \frac{S_o + rS_e}{1 + r} - S_e = \frac{P}{F(1 + r)}(a)(H)Z \qquad (12)$$

and Equation 10 becomes

$$K_g \ln (S_o/S_e) + (S_o - S_e) = \frac{P}{F} (a)(H)Z \qquad (13)$$

The parameters included in the term, P, should be constant for a given set of environmental factors, and may be determined for the region of interest. With the value of P known, the design engineer may then select the flow rate, F, the depth, Z, cross sectional area, H, and specific surface area of the filter media, a, which will most efficiently reduce the influent concentration, $S_o$, to the desired effluent, $S_e$. It must be remembered that the saturation constant, $K_g$, varies with velocity and must be evaluated at each flow rate. In addition, the relation between the flow rate, F, and the saturation constant, $K_g$, must be known to optimize filter design.

The values of the constant, P, may be evaluated in several ways. To obtain a rapid estimate of P, apply the waste to be treated to a pilot filter having a known depth, Z, and cross-sectional area, H. The specific surface area of the wetted filter media must be known and recycle is not practiced. Whenever possible, increase the influent concentration, $S_o$, or decrease the depth, Z, until the ratio $S_o/S_e$ approaches one. As this occurs, the value of $K_g \ln S_o/S_e$ will become insignificant and

$$S_o - S_e \simeq \frac{P}{F} (a)(H)(Z) \qquad (14)$$

From a knowledge of the influent and effluent substrate concentration, flow rate, and filter dimensions, the value of P may be calculated from the approximate expression

$$P \simeq \frac{(S_o - S_e)F}{(a)(H)(Z)} \qquad (15)$$

The process can be more easily understood from an examination of the differential form of the mathematical model. It may be seen that when $S \gg K_g$ the equation

$$-\frac{dS}{dZ} = \frac{\mu_{max}}{FY} (a)(X)(d)(H) \left( \frac{S}{K_g + S} \right) \qquad (16)$$

reduces to

$$-\frac{dS}{dZ} = \frac{\mu_{max}}{FY} (a)(X)(d)(H)(l) \qquad (17)$$

since

$$\left( \frac{S}{K_g + S} \right) = 1 \qquad (18)$$

By proper substitution, Equation 17 becomes

$$-\frac{dS}{dZ} = \frac{P}{F}\,(a)\,(H) \tag{19}$$

Here it is obvious that the rate of substrate removal with depth is constant and P may be estimated from the expression

$$P = \frac{(\Delta S)F}{\Delta Z(a)(H)} \tag{20}$$

With the value of P known, the saturation constant, $K_g$, may be calculated for the flow rate employed.

The above method only provides estimated values of P and $K_g$, but more accurate techniques of evaluation are available. Equation 13 can be written

$$S_O - S_e = \frac{P}{F}\,(a)(H)(Z) - K_g \ln (S_O/S_e) \tag{21}$$

which is similar in form to

$$Y = a + bm \tag{22}$$

Therefore a plot of $(S_O - S_e)$ versus $\ln S_O/S_e$ should produce a line with negative slope, $K_g$, and intercept, $(P/F)(a)(H)(Z)$ at $S_O/S_e = 1$. For ease of presentation of semi-logarithmic paper the value $K_g' \log_{10} S_O/S_e$ may be substituted for $K_g \ln S_O/S_e$, where $K_g' = 2.303\ K_g$. Such a presentation is shown in Figure 8.3.

The product $(P/F)(a)(H)(Z)$ represents the maximum substrate removal that may be expected from a filter having a depth, Z, cross sectional area, H, and filter media with a specific surface area, a. It may, therefore, be considered the maximum filter capacity for the flow rate, F. The term P would then represent the capacity constant for a unit area of filter media and the slope of the line shown in Figure 8.3 represents the negative value of $K_g'$, or the saturation constant for use with logarithms to the base 10.

### Rotating Biological Contactor Model

Depending upon the method of operation the carbonaceous waste removal in rotating biological contactors may be obtained by suspended growth as well as that attached to the discs. Therefore, the formulation of a mathematical model for the system is an extension of both fixed-film and slurry reactor kinetics.

Any substrate removal experienced in the rotating biological contactor shown in Figure 8.4 must be due to the suspended organisms occupying the liquid volume, V, or the microbial film attached to the wetted disc area,

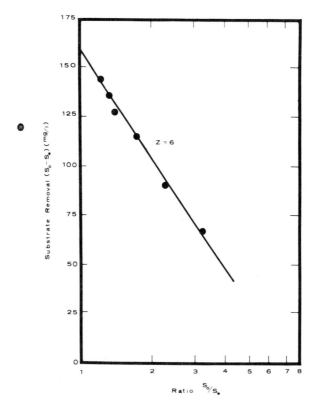

**Figure 8.3.** A graphical determination of the area capacity constant, P, and the saturation constant, $K_g$.

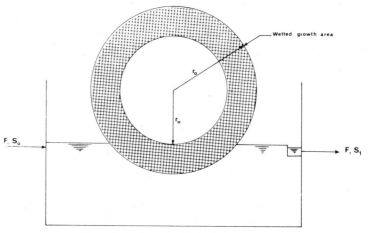

**Figure 8.4.** Rotating biological contactor.

A. The substrate balance across the reactor is then:

Accumulation = Inflow - Outflow - Consumption by Attached Growth -
Consumption by Suspended Growth $\qquad$ (23)

when

1. complete mixing is achieved in the liquid volume,
2. organism decay is neglected; and
3. maintenance energy is not included in explicit terms

Then the substrate balance may be mathematically expressed as

$$(dS/dt) V = FS_0 - FS_1 - \frac{\mu_g}{Y_g} M_0 - \frac{\mu_s}{Y_s} X_s V \qquad (24)$$

where

$V$ = liquid volume ($L^3$)
$\mu_g$ = specific growth rate of the fixed-film organisms ($T^{-1}$)
$Y_g$ = apparent yield of the fixed-film organisms
$\mu_s$ = specific growth rate of the suspended organisms ($T^{-1}$)
$Y_s$ = apparent yield of the suspended organisms
$M_0$ = active mass of fixed-film organisms (M)
$X_s$ = concentration of suspended organisms ($M/L^3$)

The active mass of attached organisms, $M_0$, is equal to the product of the wetted area, A, the active depth, d, and the unit mass of the fixed microbial film, $X_f$. Furthermore, it can be determined from Figure 8.4 that the wetted disc area may be expressed as

$$A = 2N\pi (r_0^2 - r_u^2) \qquad (25)$$

where

$N$ = number of discs
$r_0$ = total disc radius
$r_u$ = unsubmerged disc radius

Equation 24 can, therefore, be written as

$$(dS/dt) V = FS_0 - FS_1 - \frac{\mu_g}{Y_g} 2\pi N (r_0^2 - r_u^2) X_f d - \frac{s}{Y_s} X_s V \qquad (26)$$

When the Monod[11] growth function

$$\mu = \mu_{max} \left( \frac{S}{K + S} \right) \qquad (5)$$

is used to relate the specific growth rate to the substrate concentration, Equation 26 becomes

$$(dS/dt)V = FS_0 - FS_1 - \frac{\mu_{max}}{Y_g} 2N\pi (r_0^2 - r_u^2) X_f d \left(\frac{S}{K_g + S}\right)$$

$$- \frac{\mu'_{max}}{Y_s} X_s V \left(\frac{S}{K_s + S}\right) \tag{27}$$

Equation 27 is a general equation describing the dynamic performance of a completely mixed rotating biological contactor in which the removal by suspended and attached growth is significant. At steady-state conditions, $dS/dt$ is zero and the equation expressing system performance is

$$F(S_0 - S_1) = \frac{\mu_{max}}{Y_g} 2N\pi (r_0^2 - r_u^2) X_f d \left(\frac{S_1}{K_g + S_1}\right) + \frac{\mu'_{max}}{Y_s} X_s V \left(\frac{S_1}{K_s + S_1}\right) \tag{28}$$

In most cases, rotating biological contactors are operated at very short retention times and the suspended organisms are insignificant in comparison to the attached growth. Under these conditions, $X_s$ would be essentially equal to zero and Equation 27 would become

$$(dS/dt)V = FS_0 - FS_1 - \frac{\mu_{max}}{Y_g} 2N\pi (r_0^2 - r_u^2) X_f d \left(\frac{S_1}{K_g + S_1}\right) \tag{29}$$

It should be noted from Equation 29 that the only term containing the volume, V, when suspended growth is negligible, is the "net change" or "accumulation" in the system. Therefore, volume should influence the dynamic performance of the system, but have no influence on the steady-state operation. For example, if the system were converted from a continuous flow to a batch system, F would be zero and

$$dS/dt = \frac{\mu_{max}}{Y_g V} 2N\pi (r_0^2 - r_u^2) X_f d \left(\frac{S_1}{K_g + S}\right) \tag{30}$$

However, at steady-state, continuous flow conditions, $(dS/dt)$ would equal zero and Equation 29 could be expressed as

$$F(S_0 - S_1) = \frac{2\mu_{max}}{Y_g} N\pi X_f d (r_0^2 - r_u^2) \left(\frac{S_1}{K_g + S_1}\right) \tag{31}$$

When reactors are operated in series the total removal, $R_T$, is the sum of the removal in each reactor or

$$R_T = R_1 + R_2 + R_3 + \dots R_{n-1} + R_n \tag{32}$$

If the reactors are identical, the substrate concentration, S, is the only variable from reactor to reactor and the total removal in a series operation is then

$$F(S_O - S_n) = \frac{2\mu_{max}}{Y_g} N\pi X_f d \; (r_O^2 - r_u^2) \; \sum_1^n \left( \frac{S}{K_g + S} \right) \tag{33}$$

It is apparent that Equations 31 and 33 contain the same kinetic parameters that were consolidated into the area capacity constant, P, for trickling filters. That is,

$$P = \frac{1}{Y_g} \; \mu_{max} X_f d \tag{11}$$

When this expression is substituted into Equation 31, the removal equation for a single stage, biological contactor in which suspended growth is negligible becomes

$$F(S_O - S_1) = 2PN\pi \; (r_O^2 - r_u^2) \; \left( \frac{S_1}{K_g + S_1} \right) \tag{34}$$

The equation for a multi-stage system is

$$F(S_O - S_n) = 2PN\pi \; (r_O^2 - r_u^2) \; \sum_1^n \left( \frac{S}{K_g + S} \right) \tag{35}$$

Under extreme loading conditions Equations 34 and 35 may be simplified even further. At very high loading conditions

$$\left( \frac{S}{K_g + S} \right) \approx 1 \tag{36}$$

and Equation 34 becomes

$$F(S_O - S_1) = 2PN\pi \; (r_O^2 - r_u^2) \tag{37}$$

while the equation for n series reactors becomes

$$F(S_O - S_n) = n2PN\pi \; (r_O^2 - r_u^2) \tag{38}$$

If the loading is very light, then it would be anticipated that the substrate concentration would be very low and

$$\left( \frac{S}{K_g + S} \right) \approx \frac{S}{K_g} \tag{39}$$

At these low substrates concentrations, Equation 34 would be

$$F(S_O - S_1) = 2PN\pi \; (r_O^2 - r_u^2) \; \frac{S_1}{K_g} \tag{40}$$

and the fraction of soluble organics remaining in the first reactor would be

$$\frac{S_1}{S_o} = \frac{1}{\left[1 + \frac{2PN\pi}{FK_g} (r_o^2 - r_u^2)\right]} \tag{41}$$

The soluble organic fraction remaining in the $n^{th}$ reactor operating at very low substrate concentrations would be

$$\frac{S_n}{S_o} = \frac{S_1}{S_o} \times \frac{S_2}{S_1} \times \frac{S_3}{S_2} \cdots \frac{S_{n-1}}{S_{n-2}} \times \frac{S_n}{S_{n-1}} \tag{42}$$

or

$$\frac{S_n}{S_o} = \frac{1}{\left[1 + \frac{2PN\pi}{FK_g} (r_o^2 - r_u^2)\right]_1} \times \frac{1}{\left[1 + \frac{2PN\pi}{FK_g} (r_o^2 - r_u^2)\right]_2} \cdots \text{etc.} \tag{43}$$

For identical reactors and operating conditions, this would be

$$\frac{S_n}{S_o} = \frac{1}{\left[1 + \frac{2PN\pi}{FK_g} (r_o^2 - r_u^2)\right]^n} \tag{44}$$

It should be cautioned once again that Equations 37 and 38 are for very high substrate concentrations while Equations 40, 41, 43, and 44 are restricted to substrate concentrations that are much lower than the saturation constant, $K_g$.

The kinetic parameters for the suspended and attached microorganisms must be determined in separate tests for rotating biological contactors, in which both contribute to substrate removal. In the first series of tests, the retention time should be decreased until "washout" occurs as described by Herbert.[12] Removal by suspended growth is then negligible compared to the removal by fixed-film organisms and the steady-state performance may be expressed by the equation

$$F(S_o - S_1) = 2PN\pi (r_o^2 - r_u^2) \left(\frac{S_1}{K_g + S}\right) \tag{34}$$

Equation 34 may be rearranged to give

$$\frac{K_g}{P} \frac{1}{S_1} + \frac{1}{P} = \frac{2N\pi (r_o^2 - r_u^2)}{F(S_o - S_1)} \tag{45}$$

which plots as a straight line with a slope of $K_q/P$ and intercept $1/P$ when $1/S_1$ is plotted against the term on the right side of Equation 45. This provides a method of evaluating $K_g$ and $P$ for the fixed-film organisms as shown in Figure 8.5. If desired, the active depth, d, unit weight of biological film, $X_f$, and yield, Y, may be measured and $\mu_{max}$ calculated from the area capacity constant, P. This will provide the kinetic parameters necessary to describe the substrate removal by the attached films in rotating biological contactors (RBC). If the reactor is to be operated at low retention times, then removal by suspended organisms should be insignificant and this would be the only data required for design. However, if the reactor is to operate at retention times greater than the reciprocal of the specific growth rate, the kinetic constants for suspended organisms must also be obtained. A

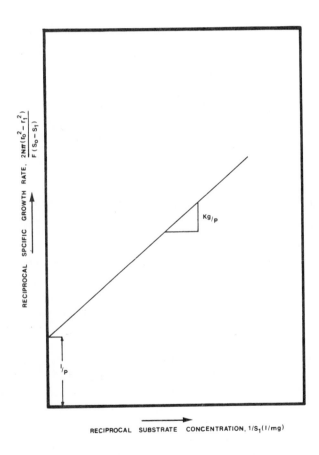

**Figure 8.5.** The evaluation of kinetic constants $K_g$ and P for fixed-films.

rotating biological contactor containing both suspended and attached organisms may be described by the equation

$$F(S_0 - S_1) = 2PN\pi\ (r_0^2 - r_u^2)\ \left(\frac{S_1}{K_g + S}\right) + \frac{\mu'_{max}}{Y_s}\ X_sV\left(\frac{S_1}{K_s + S}\right) \qquad (28)$$

The substrate removal by the suspended growth is the difference between the total removal and that due to the fixed-film organisms or

$$F(S_0 - S_1) - 2PN\pi\ (r_0^2 - r_u^2)\ \left(\frac{S_1}{K_g + S_1}\right) = \frac{\mu'_{max}}{Y_s}\ X_sV\left(\frac{S}{K_s + S}\right) \qquad (46)$$

Equation 46 may be arranged to provide the linear form

$$\frac{X_sV}{Y_s\left[F(S_0 - S_1) - 2PN\pi\ (r_0^2 - r_u^2)\ \dfrac{S_1}{K_g + S_1}\right]} = \frac{K_s}{\mu'_{max}}\frac{1}{S_1} + \frac{1}{\mu'_{max}} \qquad (47)$$

Therefore, a plot of $1/S_1$ versus the term on the right side of Equation 47 should produce a line with a slope of $K_s/\mu'_{max}$ and intercept of $1/\mu'_{max}$ as shown in Figure 8.6. This will provide the data required to describe the removal by the suspended growth.

## THE EFFECTS OF DESIGN AND OPERATING VARIABLES ON PERFORMANCE

### General

Before discussing the effect of each variable in detail, it would be constructive to make several general observations concerning the performance of fixed-film systems. It may be ascertained from the equations previously developed that the effluent concentration from reactors containing only fixed-film organisms will increase, and the efficiency decrease, as the influent concentration or flow rate is increased. However, the rate of organic removal will actually increase with an increased flow rate or concentration. Since the rate of removal, $F(S_0 - S_e)$, and the fraction of organics remaining, $(S_0 - S_e)/S_0$, are both used to express reactor performance, it is imperative to realize that these two performance parameters normally respond in opposite directions to changes in flow rate of influent concentration. However, when other design or operating variables are changed, the efficiency and rate of removal respond in the same direction.

It should also be apparent that the rate of removal, $F(S_0 - S_e)$, will approach a limiting value while the efficiency, $(S_0 - S_e)/S_0$, will continue

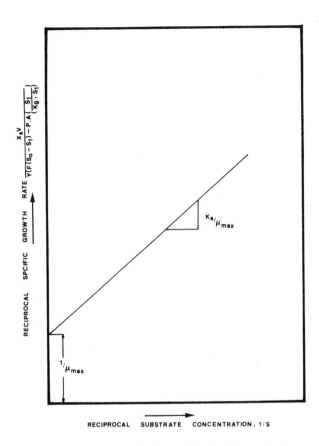

**Figure 8.6.** The evaluation of the kinetic constants,
$K_s$ and $\mu_{max}$ for suspended organisms.

to decrease as $S_o$ is increased and $[S/(Kg+S)]$ approaches unity. Since the area available for biological growth determines the maximum microbial mass in fixed-film systems, a maximum rate of removal must be reached when this microbial mass is responding at the maximum specific growth rate, $\mu_{max}$.

It is also obvious that the performance of systems containing only fixed microbial films is highly sensitive to the area available for biological growth while the liquid volume has no direct effect on the steady-state performance.

One factor that is not readily apparent from the equations presented is the effect of the liquid residence time distribution on reactor performance. However, it can be determined from both theoretical calculations and experimental data that both the efficiency and rate of organic removal increase with a departure from complete mixing toward plug-flow conditions. The

theoretical comparison in Figure 8.7 shows the ratio of the area required
in a completely mixed reactor to that required in a plug-flow reactor for
various influent concentrations and levels of removal. It should be noted
that as the final effluent is reduced to lower concentrations, the ratio
$A_{CM}/A_{PF}$ increases. Figure 8.8 compares the rate of removal per unit
area for completely mixed and plug-flow annular reactors and is based on
the data of Kornegay and Andrews.[3] The advantages of plug-flow reactors
are greater than indicated in Figure 8.8 since the two reactors are not pro-
ducing an equal effluent. For example, when both reactors produce a
31 mg/l effluent from a 100 mg/l influent, the rate of removal per unit area
in the plug-flow reactor is 1.8 times that in the completely mixed reactor.
This is slightly higher than the predicted area ratio of 1.6 from Figure 8.7
since the value of $K_g$ in the work of Kornegay and Andrews[3] was 121 mg/l
compared to the assumed value of 100 mg/l used in Figure 8.7.

**Trickling Filter Variables**

The steady-state performance of trickling filters may be described by the
equation

$$K_g \ln \frac{S_o + rS_e}{S_e(1 + r)} + \left[ \left( \frac{S_o + rS_e}{1 + r} \right) - S_e \right] = \frac{P}{F(1 + r)} (a)(H)Z \qquad (9)$$

or when recycle is not practiced,

$$K_g \ln S_o/S_e + (S_o - S_e) = \frac{P}{F} (a)(H)Z \qquad (13)$$

An examination of Equations 9 and 13 would suggest that the area capacity
constant, P, specific surface area, a, cross sectional area, H, and depth Z,
would produce a similar effect on filter performance if they were independently varied. Figures 8.9, 8.10, 8.11, and 8.12 demonstrate the effect of
P, a, H and Z on filter performance respectively and confirm the fact that
these variables do in fact have a similar effect on performance. In each
case, filter efficiency increases rapidly as these variables are increased, but
the value of each variable then approaches infinity as the efficiency approaches 100%. This can be demonstrated from an examination of the
equation

$$\frac{S_o - S_e}{S_o} = \frac{P}{S_o F} (a)(H)Z - K_g/S_o \ln S_o/S_e \qquad (48)$$

When the ratio, $S_o/S_e$, is approximately one, the term $K_g/S_o \ln S_o/S_e$ is
insignificant and fractional efficiency $(S_o - S_e)/S_o$ is approximately a

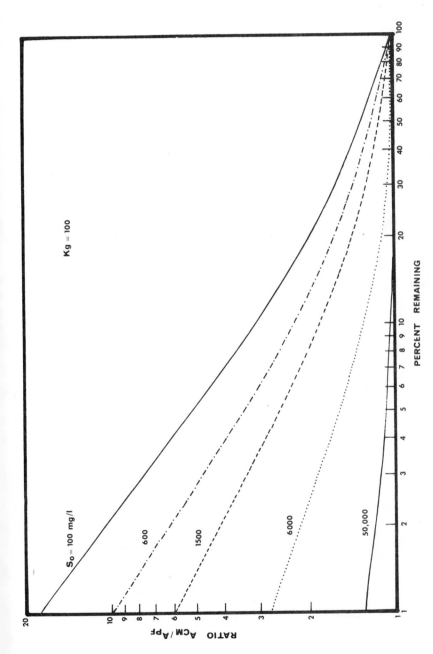

**Figure 8.7.** Comparative area requirements for completely mixed and plug flow fixed-film reactors.

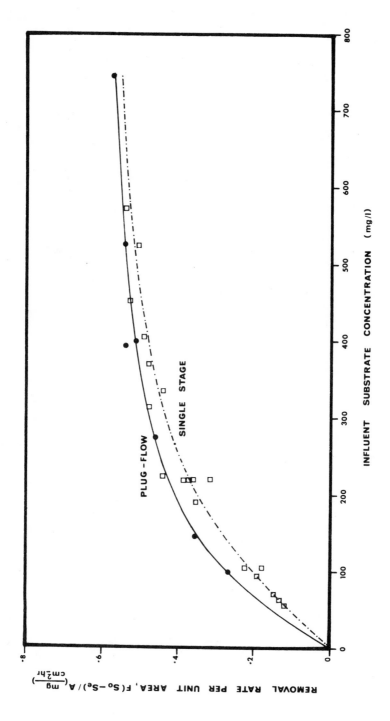

Figure 8.8.  A comparison of single-stage and plug-flow fixed-film reactors.[3]

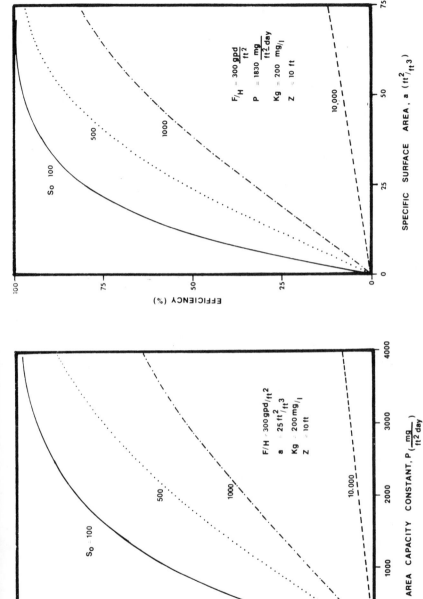

**Figure 8.10.** The effect of specific surface area on efficiency.

**Figure 8.9.** The effect of the area capacity constant on efficiency.

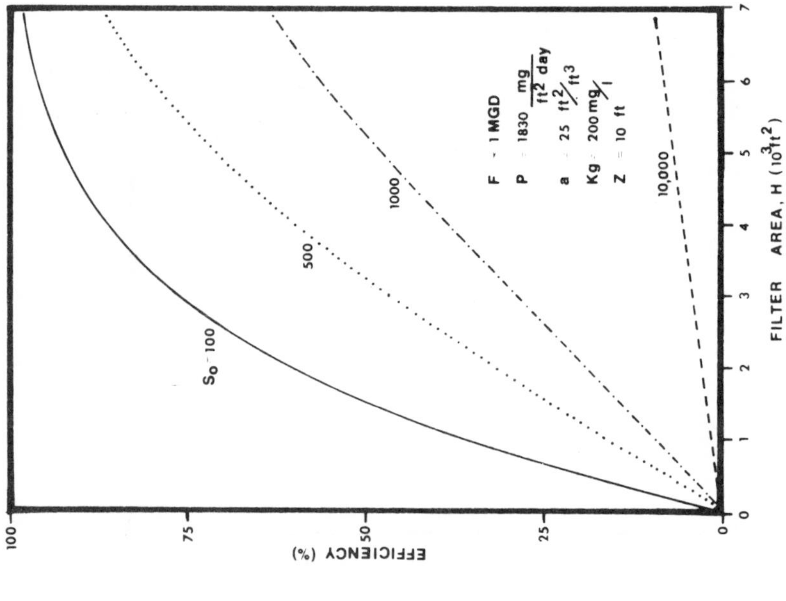

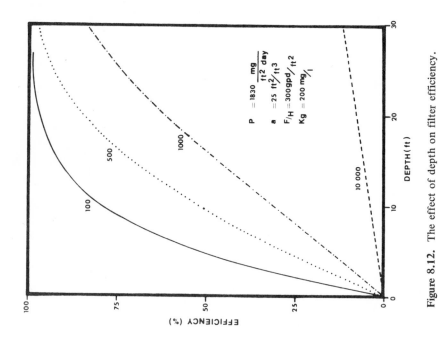

**Figure 8.12.** The effect of depth on filter efficiency.

**Figure 8.11.** The effect of filter area on efficiency.

linear function of P, a, H and Z since

$$\frac{(S_o' - S_e)}{S_o} \approx \frac{P}{S_o F} (a)(H)Z \qquad (49)$$

However, when the influent concentration, $S_o$, is low and high removal efficiencies are obtained, the ratio $S_o/S_e$ increases rapidly and $K_g/S_o$ ln $S_o/S_e$ approaches the value of the term $[(P/FS_o)(a)(H)Z - 1]$.

Based on the previous discussion and Figures 8.9, 8.10, 8.11, and 8.12, it is apparent that filter efficiencies are directly proportional to the area capacity constant, P, specific surface area, a, cross sectional area, H, and filter depth, Z, only when the influent is high and the efficiency is low. This is demonstrated by the 10,000 mg/l influent curve in Figures 8.9, 8.10, 8.11 and 8.12, where the efficiency continues to be a linear function of these four variables throughout the range investigated. At lower influent concentrations, the ratio $S_o/S_e$ increases rapidly with changes in P, a, H or Z and removal ceases to be a linear function of these variables. Fleming and Cook presented experimental evidence similar to the theoretical plot in Figure 8.10 and concluded that a point is reached at which additional increases in specific surface area cannot be justified.[13] For the range of values investigated during that study, 27 $ft^2/ft^3$ provided the optimum area. However, it should be noted that the point of departure between efficiency and the magnitude of these variables depends upon the influent concentration, flow rate and value of the saturation constant, $K_g$. In addition, the important parameter is not the individual value of P, a, H, or Z, but the value of their product. Therefore, a decrease in the cross sectional area, H, or depth, Z, could be achieved by an increase in the specific surface area, a, and maintain the same efficiency.

As previously stated, an increase in the influent substrate concentration increases the rate of removal, $F(S_o - S_e)$, but reduces the fractional efficiency, $(S_o - S_e)/S_o$. This is demonstrated in Figure 8.13 where the filter effluent, $S_e$, rate of removal per unit of surface area, $F(S_o - S_e)/A$, and efficiency are shown as a function of the initial waste concentration, $S_o$. When $S_o$ is low, the effluent and rate of removal are low, but the efficiency is high. As the influent substrate concentration increases, the effluent and rate of removal increase while the efficiency decreases. It should be noted that the rate of removal per unit area approaches the area capacity constant, P, asymptotically as $[S/(K_g + S)]$ approaches unity. As the influent increases, $K_g/S_o$ ln $(S_o/S_e)$ decreases and approaches zero as $S_o$ approaches infinity. At this point, the rate of removal reaches a limiting value as given by the equation

$$F(S_o - S_e) \approx P(a)(H)Z \qquad (50)$$

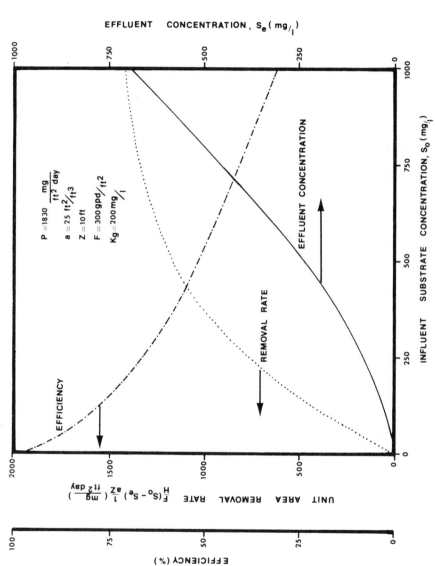

**Figure 8.13.** The effect of influent substrate concentration or efficiency, rate of removal and effluent concentration.

Since the product, P(a)(H)Z, has a fixed value, the efficiency must approach zero as the influent approaches infinity as given by the equation

$$\frac{S_o - S_e}{S_o} = \frac{P}{FS_o} (a)(H)Z \qquad (49)$$

Equation 49 also demonstrates that the performance of low efficiency systems varies inversely as the organic loading rate, $FS_o$, regardless of whether the change is in the flow rate, F, or influent concentration, $S_o$. This indicates that the efficiency of heavily loaded filters varies as $1/F^n$ where $n = 1$. This is in close agreement with data which showed that n had values from 0.91 to 0.92 for heavily loaded systems.[7] However, as the filter loadings are decreased, the term, $K_g \ln S_o/S_e$, becomes significant and filter efficiency is no longer inversely proportional to flow, and the efficiency for a given loading, $FS_o$, depends upon whether $S_o$ or F is varied. Since the magnitude of $K_g \ln S_o/S_e$ is a direct function of $K_g$, it would be expected that the deviation in efficiency for a given organic loading obtained at various flow rates would depend upon $K_g$. This is demonstrated in Figures 8.14 and 8.15 for $K_g$ values of 20 and 200 mg/l respectively. However, it should be

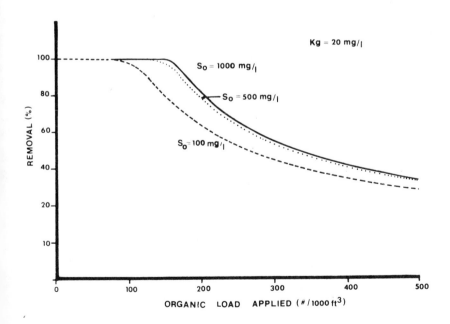

**Figure 8.14.** Efficiency vs. organic load for $K_g = 20$.

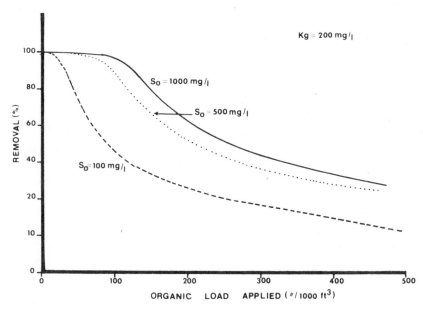

**Figure 8.15.** Efficiency vs. organic load for $K_g = 200$.

cautioned that the difference in efficiency obtained by varying $S_o$ and F to obtain a given organic loading will not be as great as indicated in Figures 8.14 and 8.15. In the equation

$$K_g \ln S_o/S_e + (S_o - S_e) = \frac{P}{F} (a)(H)Z \qquad (13)$$

the effect of flow rate is manifest in two ways. First, the flow rate is expressed in the term, F. Second, the saturation constant, $K_g$, is a function of velocity, and therefore of flow rate, as demonstrated in Figure 8.16. Therefore, an increase in F will cause a reduction in $K_g$ and the curves in Figures 8.14 and 8.15 would be closer than indicated.

In view of this variance of $K_g$ with flow rate, deep filters operated at a higher hydraulic loading rate should be more efficient than an equal volume filter at a low flow rate. This is similar to the conclusions of Sorrels and Zeller[14] and the NRC study,[15] which showed that filters in series are more efficient than filters in parallel. A comparison of the volume and depth of filter required for an assumed condition is shown in Figure 8.17 when $K_g$ varies with hydraulic loading as shown in Figure 8.16. It should be noted that the filter volume required to reduce an influent concentration of 300 mg/l to 30 mg/l decreases from 79,000 ft³ tp 32,000 ft³ as the hydraulic loading rate, F/H, is increased from 100 to 600 gpd/ft². The filter volume requirements approach a value of approximately 32,000 ft³ as the saturation

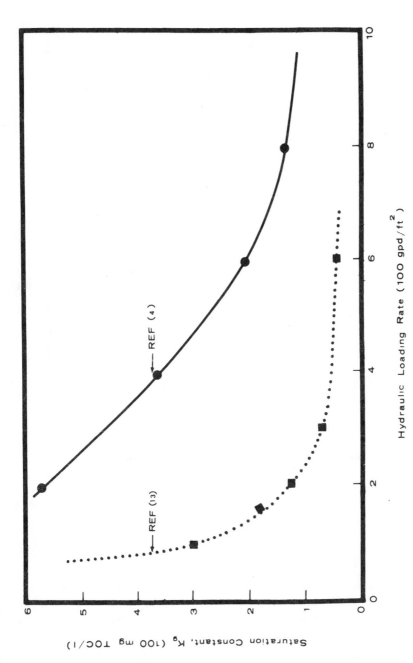

**Figure 8.16.** The effect of hydraulic loading on the saturation constant, $K_g$.

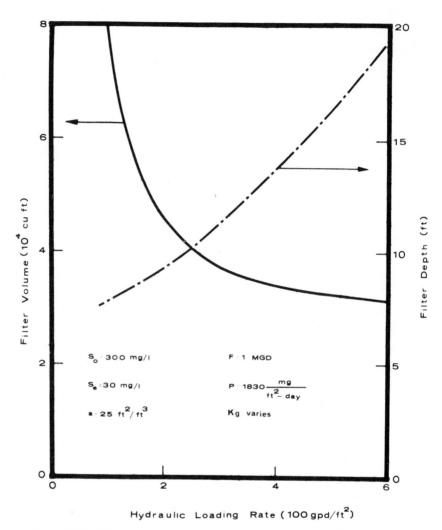

**Figure 8.17.** Filter volume and depth required at various hydraulic loading rates.

constant, $K_g$, approaches a limiting value of 50 mg/l. The question then arises as to the value of recycle. Would it be just as beneficial to provide the increased flow by recycle? The answer is no. Recycle accomplishes two things—one is favorable, the other is not. Recycle increases the hydraulic flow rate and, thereby, reduces the resistance to diffusion as reflected in a reduction in $K_g$. This, of course, is favorable. However, recycle also dilutes the incoming waste, or tends to cause the process to deviate from plug-flow to completely mixed conditions as recycle is increased. This, of course, is detrimental except for the treatment of inhibitory wastes. The data presented in Figure 8.18 indicates that the lowest filter volume for this waste would be obtained at a hydraulic loading rate of 600 gpd/ft$^2$ with no recycle. In this case, recycle will improve the performance of filters designed to operate below this point until the sum of the inflow and recycle approaches 600 gpd/ft$^2$ at which point it will become detrimental.

The most beneficial method of filter design and operation will vary with waste, environmental conditions, and type of media. However, the necessary

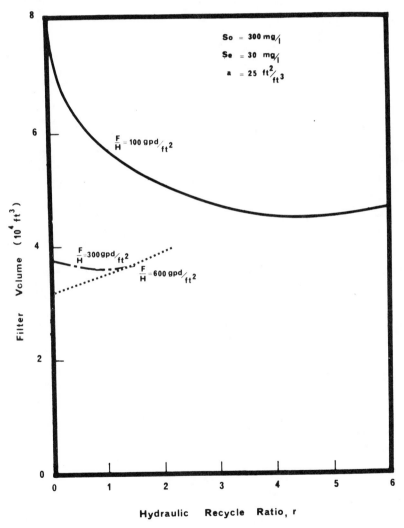

**Figure 8.18.** The effect of hydraulic recycle on filter volume requirements.

data can be provided from pilot data to optimize the design and operation of trickling filters for the treatment of carbonaceous wastes based on the mathematical models presented.

### Rotating Biological Contactor Variables

The performance of rotating biological contactors operated at steady-state conditions can be described by the equation

$$F(S_O - S_1) = 2PN\pi (r_O^2 - r_u^2) \left( \frac{S}{K_g + S} \right) + \frac{\mu'_{max}}{Y_s} X_s V \left( \frac{S}{K_g + S} \right) \qquad (28)$$

Although rotating biological contactors have been investigated at retention times as high as 16 hr,[16] most systems operate at retention times less than one hour.[9] Under these conditions, suspended organisms are "washed out" and the liquid volume has a negligible effect in the normal operating range as demonstrated in Figure 8.19. Since removal by suspended growth is normally negligible and involves factors previously described for slurry reactor systems, the parameters governing removal by suspended organisms will not be discussed. However, it should be pointed out that several investigators have presented equations which contain the retention time, V/F,

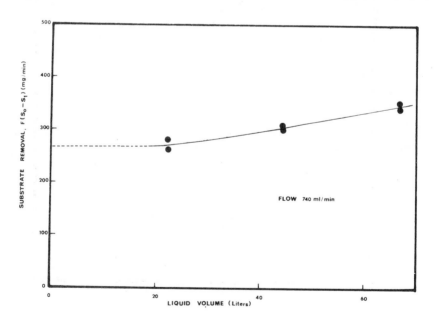

**Figure 8.19.** The effect of reactor volume on substrate removal in a RBC unit.[9]

as a parameter.[17-19] This would erroneously suggest that a change in V would produce the same effects as a change in F for a given retention time. This, of course, is not so. Investigations by Martin[9] and Weng and Molof[20] demonstrated that volume has a negligible effect on reactor performance at retention times low enough to produce "washout" of the suspended cells. In previous studies, the flow was varied and retention times calculated,[17-19] whereas Martin[9] and Weng and Molof[20] actually varied the volume while holding other operating conditions constant.

When the retention time is short enough to produce "washout" of the suspended organisms, rotating biological contactor performance may be expressed as

$$F(S_0 - S_1) = 2PN\pi (r_0^2 - r_u^2) \left( \frac{S_1}{K_g + S_1} \right) \tag{31}$$

and the efficiency of the system is then

$$\frac{S_0 - S_1}{S_0} = \frac{2PN\pi}{FS_0} (r_0^2 - r_u^2) \left( \frac{S_1}{K_g + S} \right) \tag{51}$$

Equation 51 indicates that the efficiency of an RBC unit is directly proportional to the area capacity constant, P, and number of discs, N, only when $S_0$ is high enough to insure that

$$\left( \frac{S}{K_g + S} \right) \approx 1 \tag{18}$$

At this point the maximum rate of removal is reached and may be calculated by the equation

$$F(S_0 - S_1) = 2PN\pi (r_0^2 - r_u^2) \tag{52}$$

At lower influent concentrations, an increase in the number of discs would result in a reduction of $S_1$. However, this reduction in $S_1$ would produce a lower specific growth rate and the removal per unit area would then be reduced and result in the relation shown in Figure 8.20. A similar relation is achieved for changes in P as shown in Figure 8.21.

Equation 51 also demonstrates that rotating biological contactor efficiency is an inverse function of the flow rate and influent concentration only at high substrate concentrations. At lower substrate concentrations, the effect of the influent concentration on efficiency is similar to that previously described for trickling filters. An important similarity of operation is that a change in flow rate will not produce the same efficiency as a change in influent concentration. This is demonstrated in Figure 8.22 where the

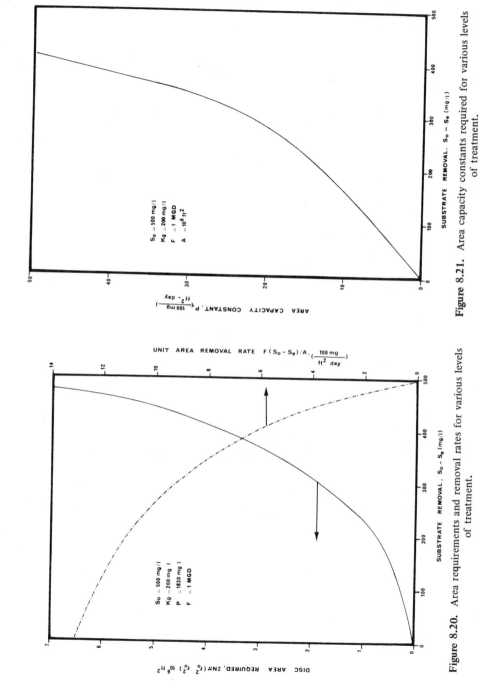

**Figure 8.21.**  Area capacity constants required for various levels of treatment.

**Figure 8.20.**  Area requirements and removal rates for various levels of treatment.

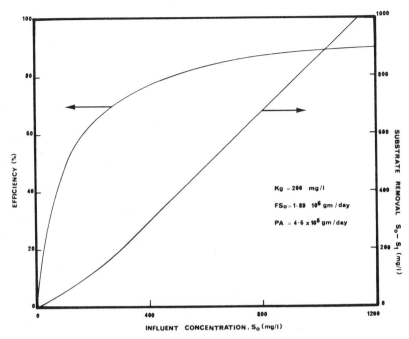

**Figure 8.22.** The effect of the influent substrate concentration on RBC performance for a constant organic loading.

efficiency for an influent concentration of 100 mg/l and flow of 5 mgd is 48% while that for an influent of 500 mg/l and 1 mgd flow is 80%.

The total and unsubmerged disc radii are obviously significant variables in the performance of RBC treatment units. When $r_o = r_u$, there is no wetted area and no removal is achieved. As $r_u$ approaches zero, the removal increases until 50% submergence is obtained and then remains constant as shown in Figure 8.23. The upper curve in Figure 8.23 represents the theoretical removal for an influent substrate concentration of 400 mg/l and a disc area of 1830 cm$^2$. Sloughing occurred just before sampling and required that the test area be reduced to 900 cm$^2$. The lower curve shows the observed results and compares reasonably well with the theoretical values. The efficiency increased with submergence depth to a d/D ratio of 0.5 and then approached a constant efficiency of 22% compared to the theoretical value of 24%. Systems are normally operated at 40-60% submergence and submergence ratios greater than 0.5 appear to be effective only when long retention times are used and removal by the suspended growth becomes significant.

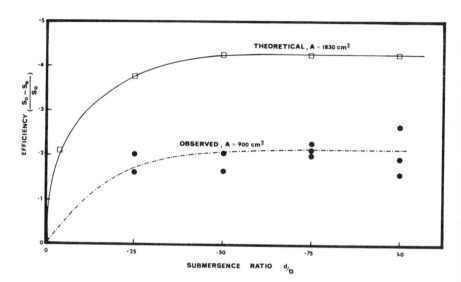

**Figure 8.23.** The effect of the submergence ratio on efficiency.[10]

The mixing speed is another important parameter in RBC operations as shown in Figure 8.24. The effect of mixing is to increase the turbulence and therefore reduce $K_g$. In this respect it serves the same function as flow rate and recycle in trickling filters. However, flow rate and recycle have a minimal effect on turbulence in RBC units compared to that produced by the mechanical rotation. Therefore, flow rate and recycle have no significant effect on $K_g$ in rotating biological contactors. In fact, recycle has no effect on single stage RBC systems when $X_s$ is negligible and simply provides an additional mode of mixing. In series RBC systems, recycle may be detrimental since it produces a deviation from a plug-flow toward a completely mixed mode. However, when the retention times are longer and the suspended organism concentration is significant, recycle can provide a beneficial effect by increasing $X_s$ and improving removal by suspended growth.

The equations for series RBC units have not been specifically discussed, but it should be mentioned that series operations provides a definite advantage. This should be apparent from the general discussion, and will be demonstrated in the solution of subsequent problems. The area requirements for series fixed-film systems lies between that required for completely mixed and plug-flow treatment systems.

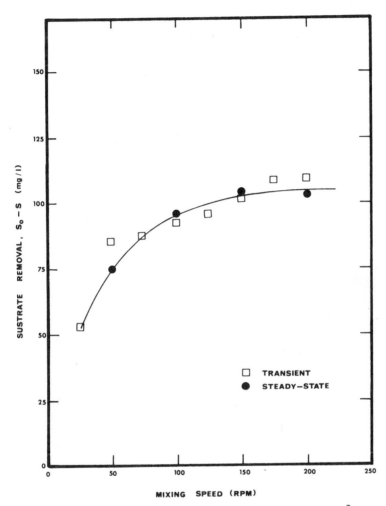

**Figure 8.24.** The effect of mixing speed on substrate removal.[3]

## SUMMARY AND CONCLUSIONS

Equations were developed for trickling filters and rotating biological contactors and the method of evaluating the biological constants described. Organism decay, maintenance energy and feedback may also be included in these models but would require a computer solution for some equations and may obscure the effect of major design or operating variables in others. It was, therefore, decided that the basic models should be presented and their use and significance discussed. Although time would not permit an extensive review of published data, the equations presented provide considerable insight into the operation of both trickling filters and rotating biological contactors and serve to clarify much of the controversy in the published literature.

## REFERENCES

1. Kornegay, B. H. and J. F. Andrews. "Kinetics of Fixed-Film Biological Reactors," *J. Water Poll. Control Fed.* **4**, R460 (1968).
2. Kornegay, B. H. and J. F. Andrews. "Application of the Continuous Culture Theory to the Trickling Filter Process," *Proceedings, 24th Industrial Waste Conference* (Lafayette, Ind.: Purdue University, 1969), p. 1398.
3. Kornegay, B. H. and J. F. Andrews. "Characteristics and Kinetics of Fixed-Film Biological Reactors," Final Report, Federal Water Pollution Control Administration Research Grant No. WP 01181, Clemson University, Clemson, South Carolina (August 1969).
4. Kornegay, B. H. and M. L. Massey. "A Rational Approach to Trickling Filter Design," Presented at the 162nd Nat. Amer. Chem. Soc. Meeting, Washington, D.C. (September 1971).
5. Sanders, W. M., III. "The Relationship Between the Oxygen Utilization of Heterotrophic Slime Organisms and the Wetted Perimeter," Ph.D. Thesis, The Johns Hopkins University, Baltimore, Maryland (January 1964).
6. "Relation Between Weight of Film, Retention Characteristics and Performance of a Percolating Filter," In *Water Pollution Research for 1963.* (London, England: Her Majesty's Stationery Office, 1964).
7. Hoehn, R. C. "The Effects of Thickness on the Structure and Metabolism of Bacterial Films," Ph.D. Thesis, University of Missouri, Columbia (1970).
8. Grieves, C. G. "Dynamic and Steady State Models for the Rotating Biological Disc Reactor," Ph.D. Thesis, Clemson University, Clemson, South Carolina (August 1972).
9. Martin, D. T. "Determination of the Effect of Volume and Biomass Area Variation on the Performance of a Fixed-Film Rotating Disc Biological Treatment Process." Special Research Problem, School of Civil Engineering, Georgia Institute of Technology, Atlanta (May 1972).
10. Yunus, S. M. "Determination of the Effect of Flow Rate and Substrate Concentration on Substrate Removal as Applied to the Operation of Bio-Discs in Biological Treatment Processes," Special Research Problem, Georgia Institute of Technology, Atlanta (August 1972).
11. Monod, J. "La Techniques de Culture Continue, Theorie et Applications," *Annal. Inst. Pasteur* **79**, 390 (1950).
12. Herbert, D. "Multistage Continuous Culture," *Continuous Cultivation of Microorganisms,* I. Malek, K. Beran and J. Hospodka, Eds. (New York: Academic Press, 1964), p. 23
13. Fleming, M. and E. C. Cook. "The Effect of the Specific Area Provided by a Synthetic Medium on the Performance of a Trickling

Filter." Presented at the 27th Annual Purdue Ind. Waste Conf., Purdue University, Lafayette, Ind. (May 1972).

14. Sorrels, J. H. and P. S. A. Zeller. "Two-Stage Trickling Filter Performance," *Sewage Ind. Waste* **28**, 943 (1956).

15. NRC SubCommittee Report. "Sewage Treatment at Military Installations," *Sewage Works J.* **18**, 897 (1946).

16. Pretorius, W. A. "Some Operational Characteristics of a Bacterial Disc Unit," In *Water Research* (New York: Pergammon Press, 1971), p. 1141.

17. Antonie, R. L. and F. M. Welch. "Preliminary Results of a Novel Biological Process for Treating Dairy Waste," *Proceedings 24th Annual Purdue Industrial Waste Conference* (Lafayette, Ind.: Purdue University, 1969), p. 115.

18. Joost, R. H. "Systemation in Using the Rotating Biological Surface (RBS) Waste Treatment Process," *Proceedings 24th Annual Purdue Industrial Waste Conference* (Lafayette, Ind.: Purdue University, 1969), p. 365.

19. Hartman, H. *Investigation of the Biological Clarification of Wastewater Using Immersion Drip Filters*, Stuttgart, Report of City Water Economy (Munich, West Germany: Oldenbourg, 1960).

20. Weng, Cheng-Nan and A. H. Molof. "Nitrification in the Biological Fixed-Film Rotating Disk System," Presented at 45th Annual Water Poll. Control Federation Conference, Atlanta, Georgia (October 1972).

## NOMENCLATURE

| Symbol | Description |
|---|---|
| A | Surface area of biological film $(L^2)$ |
| a | Specific surface area of filter media $(L^2/L^3)$ |
| d | Active thickness of biological film (L) |
| E | Removal efficiency (%) |
| F | Hydraulic flow rate, $(L^3/T)$ |
| H | Cross sectional area of a trickling filter $(L^2)$ |
| h | Total thickness of biological film (L) |
| K | A constant $(M/L^3)$ |
| $K_g$ | Essential nutrient or substrate concentration where $\mu = \frac{1}{2}\mu_{max}$ in fixed-film systems $(M/L^3)$ |
| $K_s$ | Essential nutrient or substrate concentration where $\mu = \frac{1}{2}\mu'_{max}$ in slurry systems $(M/L^3)$ |
| $M_o$ | Mass of organisms in the active layer (M) |
| $M_s$ | Mass of substrate or essential nutrient (M) |
| N | Number of rotating discs |
| n | Number of reactors in series |
| P | Area capacity constant and equal to $\frac{1}{Y}(\mu_{max})$ (d)(X) $(M/L^2T)$ |
| $r_o$ | Rotating disc radius |
| $r_u$ | Unsubmerged disc radius |

| | |
|---|---|
| $S_o$ | Influent substrate concentration ($M/L^3$) |
| $S_e$ | Effluent substrate concentration from a plug-flow reactor ($M/L^3$) |
| $t$ | Time (T) |
| $\mu$ | Specific growth rate ($T^{-1}$) |
| $\mu_{max}$ | Maximum specific growth rate of fixed films ($T^{-1}$) |
| $\mu'_{max}$ | Maximum specific growth rate of suspended organisms ($T^{-1}$) |
| $V$ | Reactor volume ($L^3$) |
| $X_f$ | Unit mass of biological film on a dry weight basis ($M/L^3$) |
| $X_s$ | Concentration of suspended organisms ($M/L^3$) |
| $Y_g$ | Apparent growth yield of fixed-film systems |
| $Y_s$ | Apparent growth yield of suspended organisms |
| $Z$ | Depth of trickling filter (L) |

## PROBLEMS

1. Calculate the disc area required in a single stage, biological contactor to reduce an influent concentration of 200 mg/l to 20 mg/l. The flow is 10,000 gpd, $K_g$ is 100 mg/l and P is 2000 mg/ft$^2$-day. How many 10-ft diameter discs would be required if $r_u \approx 0$?

*Solution:*

$$F(S_0 - S_1) = P.A. \left( \frac{S_1}{K_g + S_1} \right)$$

where

$$A = 2N\pi (r_0^2 - r_u^2)$$

$$A = \frac{F(S_0 - S_1)}{P} \left( \frac{K_g + S}{S} \right)$$

$$A = \frac{(10,000 \text{ gal/day} \times 3.785 \text{ L/gal})(200 \text{ mg/l} - 20 \text{ mg/l})}{2000 \text{ mg/ft}^2\text{-day}} \quad \frac{100 + 20}{20}$$

$$A = 20,440$$

$$A = 2N\pi (10 \text{ ft})^2 - 0^2 = 20,440 \text{ ft}^2$$

$$N = 20,440 \text{ ft}^2/(2)(3.14)(100 \text{ ft}^2)$$

$$N = 20,440/628 = 32.6$$

Use 33

2. Calculate the total area and number of discs required in Problem 1 for two reactors in series. Solve by trial and error. Assume $A_T = 12,000$ $ft^2$ and $A_1 = A_2 = 6,000$ $ft^2$. The influent to the second reactor is then:

$$S_1 = \frac{(6000 \text{ ft}^2)(2000 \text{ mg/ft}^2\text{-day})(20/120)}{10,000 \text{ gal/da6 X } 3.785 \text{ L/gal}} + S_2$$

$S_1 = 52.8 + 20 = 72.8$ mg/l

$$S_o = \frac{(6000 \text{ x } 2000)}{10,000 \text{ x } 3.785} \left(\frac{72.8}{172.8}\right) + S_1$$

$\quad = 134 + 72.8 = 206.8$

$\quad 206.8 > 200$ try a total area of 11,750 $ft^2$

Then:

$S_1 = 51.4 + 20 = 71.4$ mg/l

$S_o = 129.2 + 71.4 = 200.6$ mg/l

This value is close enough to 200 mg/l so $A_T = 11.750$ $ft^2$

$$N = \frac{11,750 \text{ ft}^2}{628 \text{ ft}^2/\text{disc}} = 18.7 \quad \text{Use 19 discs}$$

3. Calculate the total area required in Problem 1 for a plug flow reactor.

For plug flow reactors:

$$K_g \ln S_o/S_e + (S_o - S_e) - \frac{P}{F} (A)$$

$$A = \frac{F}{P} [K_g \ln S_o/S_e + (S_o - S_e)]$$

$$A = \frac{(10,000 \text{ gal/day})(3.785 \text{ L/gal})}{2000 \text{ mg/ft}^2\text{-day}} \left(100 \ln \frac{200}{20} + (200 - 20)\right)$$

$A = 7,764$ $ft^2$

4. Make a plot of area requirements versus number of reactors in series from Problems 1, 2 and 3. Remember that plug-flow can be approximated by an infinite number of stages in series.

*Solution:*

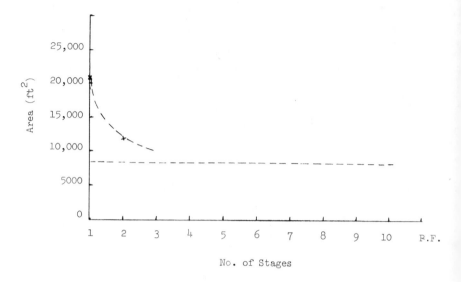

No. of Stages

5. Based on the plot in Problem 4 try to estimate the number of discs (628 $ft^2$/disc) required for a three-stage unit.

*Solution:*

The required area appears to lie between 8000 $ft^2$ (12.8 discs) and 10,000 $ft^2$ (15.8 discs). Try 15 total with 5 in each reactor.

$$S_O = 200 \text{ mg/l}$$

$$S_3 = 20 \text{ mg/l}$$

$$S_2 - S_3 = \frac{P}{F} A_3 \left( \frac{S_3}{K_3 + S_3} \right)$$

$$S_2 = \frac{P}{F} A_3 \left( \frac{S_3}{K_g + S_3} \right) + S_3$$

$$S_2 = \frac{(2000)}{3.78 \times 10,000} (5 \times 628) \left( \frac{20}{120} \right) + 20$$

$$S_2 = 27.7 + 20 = 47.7 \text{ mg/l}$$

$$S_1 = \frac{P}{F} A_2 \left( \frac{S_2}{K_g + S_2} \right) + S_2$$

$$= \frac{2000}{(10{,}000 \times 3.785)} (5 \times 628) \left( \frac{47.7}{147.7} \right) + 47.7$$

$$= 53.6 + 47.7 = 101.3$$

$$S_0 = \frac{P}{F} A_1 \left( \frac{S_1}{K_g + S_1} \right) + S_1$$

$$\frac{S}{S_0} = \frac{2000}{(10{,}000 \times 3.785)} (5 \times 628) \left( \frac{101.3}{201.3} \right) + 101.3$$

$$= 84.2 + 101.3 = 185.5$$

Use 6 in reactor 1

then

$$S_0 = 101 + 101.3 = 202.3$$

## APPENDIX A

### Trickling Filter Model

```
BURROUGHS B-5500 ALGOL COMPILER LEVEL   4

$COMPILE BTN0216/CE1      ALGOL                                      0000
$PROCESS.=  00000015;IN=   00n00003.         .13S800007             0000
$DATA.                                                              0000
$DATA MASIN.                                                        0000
$                                                                   0000
BEGIN                                                               0000
                             ******** START OF SEGMENT ********     0000
  FILE IN MASIN (2,10);                                            0000
  FILE OUT MASOUT 6(2,15);                                         0003
  INTEGER I,N;                                                      0007
  REAL F,X,D,A,Y,MUBAR,KS,SO,SE,G;                                  0007
  FORMAT IN1 (5F15.5), OUT1 (X8,"F",X13,"X",X13,"D",X13,"A",X13,"Y  0007
  X13,"MURAR",X9,"KS"), OUT2 (7F15.5), OUT3 (X5,"SO",X16,"SE",X16,
                             ******** START OF SEGMENT ********
  "SO-SE"), OUT4 (3(F15.5,X3)), OUT5 (///), IN2 (2F15.5);           0007
                                                                    0007
                                    3 IS   51 LONG, NEXT
  WRITE (MASOUT [NO]);                                              0007
  READ (MASIN, IN1, F,X,D,A,Y);                                     0011
  READ (MASIN, IN2,MUBAR,KS);                                       0025
  WRITE (MASOUT, OUT1);                                             0036
  WRITE (MASOUT,OUT2,F,X,D,A,Y,MUBAR,KS);                           0039
  WRITE (MASOUT, OUT5);                                             0057
  WRITE (MASOUT, OUT3);                                             0060
  FOR SE + 25 STEP 25 UNTIL 200 DO                                  0063
  BEGIN                                                             0064
  SO + SE + MUBAR x A x D x x x SF / ((Y x F) x (KS + SE)));        0064

  G + (SO-SE);                                                      0069
  WRITE (MASOUT, OUT4,SO,SE,G);                                     0070
  END;                                                              0083
  FOR SE + 400 STEP 400 UNTIL 4000 DO                               0085
  BEGIN                                                             0087
  SO + SE + MUBAR x A x D x x x SF / ((Y x F) x (KS + SE)));        0087
  G + (SO-SE);                                                      0092
  WRITE (MASOUT, OUT4,SO,SE,G);                                     0093
  END;                                                              0106
                                                                    0108
                                   2 IS  113 LONG, NEXT SEG
```

```
OUTPUT(W) IS SEGMENT NUMBER 0004,PRT ADDRESS IS 0054
BLOCK CONTROL IS SEGMENT NUMBER 0005,PRT ADDRESS IS 0005
INPUT(W) IS SEGMENT NUMBER 0006,PRT ADDRESS IS 0050
ALGOL WRITE   IS SEGMENT NUMBER 0007,PRT ADDRESS IS 0014
ALGOL READ    IS SEGMENT NUMBER 0008,PRT ADDRESS IS 0015        1 IS    2 LONG,
ALGOL SELECT IS SEGMENT NUMBER 0009,PRT ADDRESS IS 0016        10 IS   69 LONG,

NUMBER OF SYNTAX ERRORS DETECTED = 0. NUMBER OF SEQUENCE ERRORS DETECTED =    0
COMPILER TIMES: PROCESSOR = 6 SECONDS; IO = 17 SECONDS; ELAPSED = 32 SECONDS.
PRT SIZE = 54; TOTAL SEGMENT SIZE = 235 WORDS; DISK SIZE = 15 SEGS; NO. PGM. SEGS = 10
ESTIMATED CORE STORAGE REQUIREMENT = 931 WORDS.
```

| F | X | D | A | Y | MUBAR | KS |
|---|---|---|---|---|---|---|
| 8040.00000 | 95000.00000 | 0.00700 | 2000.00000 | 0.26000 | 0.28000 | 121.00000 |

| SO | SE | SO-SE |
|---|---|---|
| 55.50475 | 25.00000 | 30.50475 |
| 102.08998 | 50.00000 | 52.08998 |
| 143.16877 | 75.00000 | 68.16877 |
| 180.60983 | 100.00000 | 80.60983 |
| 215.52222 | 125.00000 | 90.52222 |
| 248.60575 | 150.00000 | 98.60575 |
| 280.32382 | 175.00000 | 105.32382 |
| 310.99547 | 200.00000 | 110.99547 |
| 536.77368 | 400.00000 | 136.77368 |
| 954.74286 | 800.00000 | 154.74286 |
| 1361.82988 | 1200.00000 | 161.82988 |
| 1765.62252 | 1400.00000 | 165.62252 |
| 2167.98465 | 2000.00000 | 167.98465 |
| 2569.59720 | 2400.00000 | 169.59720 |
| 2970.76810 | 2800.00000 | 170.76810 |
| 3371.65694 | 3 00.00000 | 171.65694 |
| 3772.35469 | 3600.00000 | 172.35469 |
| 4172.91698 | 4000.00000 | 172.91698 |

```
/*FIXED FILM REACTORS*/


/*FIXED FILM REACTORS*/
/*FFR2 PROGRAM FOR REACTORS IN SERIES W/O RECYCLE & UNIFORM MAS
1      FFR1: PROCEDURE OPTIONS(MAIN);
2          ON ERROR GO TO DONE;
4              DCL KS FLOAT; DCL MOT FLOAT; DCL MUH FLOAT; DCL MO FLOA
8              ST1: GET DATA(N,MUH,Y,A,X,D,F,KS,SO);
9                   PUT SKIP(3) DATA(N,MUH,Y,A,X,D,F,KS,SO);
10                      SOO=SO;
11                      MOT=A*X*D;
12                      E=MUH/(F*Y);
13                          DO I=1 TO N;
14                              MO=MOT/N;
15                              C=E*MO;
16                              RAD=(SO-C-KS)**2+(4*SO*KS);
17                              SI=((SO-C-KS)+SQRT(RAD))/2;
18                              RAI=F*(SO-SI)/(A/N);
19                              RAT=F*(SOO-SI)/(A*I/N);
20                              PUT SKIP DATA(I,MO,SI,RAI,RAT);
21                              SO=SI;
22                          END;
23                      GO TO ST1;
24                  DONE: END;
```

NO ERROR DETECTED, ANY WARNINGS ARE NOT PRINTED.

COMPILE TIME        .49 MINS

```
IEF285I    SYSOUT                                               SYSOUT
IEF285I    VOL SER NOS=            .
IEF285I    SYS68148.TOR1454.RPOO8.CU.R0000141                   PASSED
IEF285I    VOL SER NOS= 222222.
IEF285I    SYS68148.TOR1454.RPOO8.CU.R0000142                   DELETED
IEF285I    VOL SER NOS= 333333.
IEF236I    ALLOC. FOR CU        L         STEP1
IEF237I    SYSLIB    ON 190
IEF237I    SYSLIN    ON 191
IEF237I    SYSLMOD   ON 191
IEF237I    SYSUT1    ON 192
IEF285I    SYS1.PL1LIB                                          KEPT
IEF285I    VOL SER NOS= 111111.
IEF285I    SYS68148.TOR1454.RPOO8.CU.R0000141                   PASSED
IEF285I    VOL SER NOS= 222222.
IEF285I    SYS68148.TOR1454.RPOO8.CU.PDS                        PASSED
IEF285I    VOL SER NOS= 222222.
IEF285I    SYS68148.TOR1454.RPOO8.CU.R0000144                   DELETED
IEF285I    VOL SER NOS= 333333.
IEF236I    ALLOC. FOR CU        G         STEP1
IEF237I    PGM=*.DD ON 191
IEF237I    SYSIN     ON 00A
```

```
N=   6                    MIH= 2.79999E-01          Y= 2.59999E-01            A=  2.00000E+03
D= 1.99999E-03            F=  8.00000E+00           KS= 1.21000E+02           SO= 5.00000E+01;
I=   1                    MD= 6.33332E+00           SI= 4.75932E+01           RAI= 5.77620E-02
I=   2                    MD= 6.33332E+01           SI= 4.57719E+01           RAI= 5.57120E-02
I=   3                    MD= 6.33332E+01           SI= 4.30351E+01           RAI= 5.36817E-02
I=   4                    MD= 6.33332E+01           SI= 4.08820E+01           RAI= 5.16742E-02
I=   5                    MD= 6.33332E+01           SI= 3.88115E+01           RAI= 4.96926E-02
I=   6                    MD= 6.33332E+01           SI= 3.68223E+01           RAI= 4.77400E-02

N=   6                    MIH= 2.79999E-01          Y= 2.59999E-01            A=  2.00000E+03
D= 1.99999E-03            F=  8.00000E+00           KS= 1.21000E+02           SO= 1.00000E+02;
I=   1                    MD= 6.33332E+01           SI= 9.62234E+01           RAI= 9.06382E-02
I=   2                    MD= 6.33332E+01           SI= 9.25289E+01           RAI= 8.86666E-02
I=   3                    MD= 6.33332E+01           SI= 8.89176E+01           RAI= 8.66722E-02
I=   4                    MD= 6.33332E+01           SI= 8.53902E+01           RAI= 8.46562E-02
I=   5                    MD= 6.33332E+01           SI= 8.19477E+01           RAI= 8.26212E-02
I=   6                    MD= 6.33332E+01           SI= 7.85906E+01           RAI= 8.05696E-02

N=   6                    MIH= 2.79999E-01          Y= 2.59999E-01            A=  2.00000E+03
D= 1.99999E-03            F=  8.00000E+00           KS= 1.21000E+02           SO= 2.00000E+02;
I=   1                    MD= 6.33332E+01           SI= 1.94741E+02           RAI= 1.26202E-01
I=   2                    MD= 6.33332E+01           SI= 1.89537E+02           RAI= 1.24892E-01
I=   3                    MD= 6.33332E+01           SI= 1.84389E+02           RAI= 1.23547E-01
I=   4                    MD= 6.33332E+01           SI= 1.79299E+02           RAI= 1.22170E-01
I=   5                    MD= 6.33332E+01           SI= 1.74267E+02           RAI= 1.20770E-01
I=   6                    MD= 6.33332E+01           SI= 1.69295E+02           RAI= 1.19329E-01

N=   6                    MIH= 2.79999E-01          Y= 2.59999E-01            A=  2.00000E+03
D= 1.99999E-03            F=  8.00000E+00           KS= 1.21000E+02           SO= 3.00000E+02;
I=   1                    MD= 6.33332E+01           SI= 2.93960E+02           RAI= 1.44955E-01
I=   2                    MD= 6.33332E+01           SI= 2.87957E+02           RAI= 1.44076E-01
I=   3                    MD= 6.33332E+01           SI= 2.81991E+02           RAI= 1.43179E-01
I=   4                    MD= 6.33332E+01           SI= 2.76063E+02           RAI= 1.42265E-01
I=   5                    MD= 6.33332E+01           SI= 2.70174E+02           RAI= 1.41328E-01
I=   6                    MD= 6.33332E+01           SI= 2.64326E+02           RAI= 1.40367E-01
```

```
                    ****CONTINUOUS SYSTEM MODELING PROGRAM****

              ***PROBLEM INPUT STATEMENTS***

        TITLE       FIXED FILM REACTORS
        *
        INIT
                    V=1./VP
        PARAM       K= 1432., F= 48., KS= 121., VP=1.
        PARAM       SO= 400.
        *
        DYNAM
                    M1DOT= F*(SO-S1)-K*(S1/(KS+S1))
                    S1= V*INTGRL(500.97,M1DOT)
                    M2DOT= F*(S1-S2)-K*(S2/(KS+S2))
                    S2= V*INTGRL(477.17,M2DOT)
                    M3DOT= F*(S2-S3)-K*(S3/(KS+S3))
                    S3= V*INTGRL(453.61,M3DOT)
                    M4DOT= F*(S3-S4)-K*(S4/(KS+S4))
                    S4= V*INTGRL(430.33,M4DOT)
                    M5DOT= F*(S4-S5)-K*(S5/(KS+S5))
                    S5= V*INTGRL(407.33,M5DOT)
                    M6DOT= F*(S5-S6)-K*(S6/(KS+S6))
                    S6= V*INTGRL(384.63,M6DOT)
        *
        *
        METHOD      MILNE
        TIMER       DELT=0.10, FINTIM=1.5
        PRINT       S1,S2,S3,S4,S5,S6
        PREPAR      S1,S2,S3,S4,S5,S6
        END
        STOP

OUTPUT VARIABLE SEQUENCE
V        S1       M1DOT   ZZ0001 S2        M2DOT   ZZ0003 S3        M3DOT   ZZ0005
S4       M4DOT    ZZ0007 S5        M5DOT   ZZ0009 S6        M6DOT   ZZ0011

OUTPUTS      INPUTS      PARAMS     INTEGS + MEM BLKS    FORTRAN   DATA CDS
23(500)      37(1400)    7(400)     6+  0=  6(300)       20(600)      8
```

```
FORTRAN IV G LEVEL 1, MOD 1                    DATA

0001          SUBROUTINE DATA
0002          DIMENSION LC(1)
0003          COMMON DDUM1(64),C(8000),NALARM,KPOINT,DDUM2(1732)
0004          COMMON NOINTG,NOSYMB,SYMB(56)
0005          EQUIVALENCE (C(1),LC(1))
0006          KPOINT =  38
0007          NOINTG =   6
0008          NOSYMB =  37
0009          C(19) = 500.97
0010          C(20) = 477.17
0011          C(21) = 453.61
0012          C(22) = 430.33
0013          C(23) = 407.33
0014          C(24) = 384.63
0015          READ(5,1)(SYMB(I),I=1,56)
0016        1 FORMAT(18A4)
0017          RETURN
0018          END

FORTRAN IV G LEVEL 1, MOD 1                    DATA

                         COMMON BLOCK  /            /  MAP SIZE    9400
SYMBOL   LOCATION   SYMBOL   LOCATION   SYMBOL   LOCATION   SYMBOL   LOCATION
DDUM1    0          C        100        LC       100        NALARM   7E00       KPOINT   7E00
DDUM2    7E08       NOINTG   9918       NOSYMB   991C       SYMB     9920

                         SCALAR MAP
SYMBOL   LOCATION   SYMBOL   LOCATION   SYMBOL   LOCATION   SYMBOL   LOCATION
I        A0

                         SUBPROGRAMS CALLED
SYMBOL   LOCATION   SYMBOL   LOCATION   SYMBOL   LOCATION   SYMBOL   LOCATION
IBCOM#   A4

                         FORMAT STATEMENT MAP
SYMBOL   LOCATION   SYMBOL   LOCATION   SYMBOL   LOCATION   SYMBOL   LOCATION
1        A8
```

```
FORTRAN IV G LEVEL 1, END 1                                    UPDATE

0001          SUBROUTINE UPDATE
0002          COMMON DDUM1(64)
0003          COMMON TIME
             1,DELT  ,DELMIN,FINTIM,PRDEL ,OUTDEL,ZZ0001,ZZ0003,ZZ0005,ZZ0007
             1,ZZ0009,ZZ0011,M1DT ,M2DT ,M3DT ,M4DT ,M5DT ,M6DT ,ZZ0002
             1,ZZ0004,ZZ0006,ZZ0008,ZZ0010,ZZ0012,ZZ0000,K   ,KS
             1,VP  ,S0 ,V ,S1 ,S2 ,S3 ,S4 ,S5 ,S6
0004          COMMON DDUM2(7963),NALARM,DDUM3(418),KEEP
0005          REAL
             1,M1DT ,M2DT ,M3DT ,M4DT ,M5DT ,M6DT ,K   ,KS
                A1DT
0006          IF(ZZ0000) 39996,39997,39998
0007    39996 ZZ0000 = 0.0
0008          V=1./VP
0009    39997 CONTINUE
0010          S1=V*ZZ0001
0011          770001
0012          M1DT=F*(S0-S1)-K*(S1/(KS+S1))   =INTGRL  (ZZ0002  ,M1DT  )
0013          770002
              S2=V*ZZ0003
              770003
0014          M2DT=F*(S1-S2)-K*(S2/(KS+S2))   =INTGRL  (ZZ0004  ,M2DT  )
0015          770004
              S3=V*ZZ0005
              770005
0016          M3DT=F*(S2-S3)-K*(S3/(KS+S3))   =INTGRL  (ZZ0006  ,M3DT  )
0017          770006
              S4=V*ZZ0007
              770007
0018          M4DT=F*(S3-S4)-K*(S4/(KS+S4))   =INTGRL  (ZZ0008  ,M4DT  )
0019          770008
              S5=V*ZZ0009
              770009
0020          M5DT=F*(S4-S5)-K*(S5/(KS+S5))   =INTGRL  (ZZ0010  ,M5DT  )
0021          770010
              S6=V*ZZ0011
              770011
0022          M6DT=F*(S5-S6)-K*(S6/(KS+S6))   =INTGRL  (ZZ0012  ,M6DT  )
0023    39998 CONTINUE
0024          RETURN
              END
```

```
FORTRAN IV G LEVEL 1, MOD 1                    UPDATE

                COMMON BLOCK /          / MAP SIZE  8490
```

| SYMBOL | LOCATION | SYMBOL | LOCATION | SYMBOL | LOCATION | SYMBOL | LOCATION | SYMBOL | LOCATION |
|--------|----------|--------|----------|--------|----------|--------|----------|--------|----------|
| DDUM1  | 0        | TIME   | 100      | DELT   | 104      | DELMIN | 108      | FINTI  | 10C      |
| PRDFL  | 110      | OUTDFL | 114      | ZZ0001 | 118      | ZZ0003 | 11C      | ZZ0005 | 120      |
| ZZ0007 | 124      | ZZ0009 | 128      | ZZ0011 | 12C      | M1DDT  | 130      | M2DDT  | 134      |
| M3DDT  | 138      | M4DDT  | 13C      | M5DDT  | 140      | M6DDT  | 144      | ZZ0002 | 148      |
| ZZ0004 | 14C      | ZZ0006 | 150      | ZZ0008 | 154      | ZZ0010 | 158      | VP     | 15C      |
| ZZ0000 | 160      | K      | 164      | F      | 168      | KS     | 16C      | S3     | 170      |
| S0     | 174      | V      | 178      | S1     | 17C      | S2     | 180      | MALAR* | 184      |
| S4     | 188      | S5     | 18C      | S6     | 190      | DDUM2  | 194      |        |          |
| DDUM3  | 7E04     | KEEP   | 848C     |        |          |        |          |        |          |

```
TOTAL MEMORY REQUIREMENTS 000268 BYTES

FORTRAN IV G LEVEL 1, MOD 1                    MAIN

    0001        CALL MAINEX
    0002        RETURN
    0003        END

FORTRAN IV G LEVEL 1, MOD 1                    MAIN

                        SUBPROGRAMS CALLED
SYMBOL  LOCATION   SYMBOL  LOCATION   SYMBOL  LOCATION   SYMBOL  LOCATION
MAINEX  90         IRCOM#  94

TOTAL MEMORY REQUIREMENTS 0000F4 BYTES

E-LEVEL LINKAGE EDITOR OPTIONS SPECIFIED OVLY
*****DEJEXE  DOES NOT EXIST BUT HAS BEEN ADDED TO DATA SET
```

```
                    *** CSMP/360 SIMULATION DATA ***

TITLE    FIXED FILM REACTORS

PARAM    K= 1432., F= 48., KS= 121., VP=1.

PARAM    SO= 400.

METHOD   MILNE

TIMER    DELT=0.10, FINTIM=1.5

PRINT    S1,S2,S3,S4,S5,S6

PREPAR   S1,S2,S3,S4,S5,S6

TIMER VARIABLES
DELT   = 1.5000E-02
DELMIN = 1.5000E-07
FINTIM = 1.5000E 00
PRDEL  = 1.5000E-02
OUTDEL = 1.5000E-02
```

FIXED FILM REACTORS

MILNE INTEGRATION

| TIME (hrs) | S1 | S2 | S3 | S4 | S5 | S6 |
|---|---|---|---|---|---|---|
| 0.0        | 5.0097E 02 | 4.7717E 02 | 4.5361E 02 | 4.3033E 02 | 4.0733E 02 | 3.8463E 02 |
| 1.5000E-02 | 4.3702E 02 | 4.5691E 02 | 4.4906E 02 | 4.2954E 02 | 4.0722E 02 | 3.8462E 02 |
| 3.0000E-02 | 4.0614E 02 | 4.2490E 02 | 4.3181E 02 | 4.2312E 02 | 4.0536E 02 | 3.8418E 02 |
| 4.5000E-02 | 3.9126E 02 | 3.9873E 02 | 4.0851E 02 | 4.0910E 02 | 3.9896E 02 | 3.8181E 02 |
| 6.0000E-02 | 3.8408E 02 | 3.8094E 02 | 3.8632E 02 | 3.9053E 02 | 3.8722E 02 | 3.7584E 02 |
| 7.5000E-02 | 3.8062E 02 | 3.6988E 02 | 3.6856E 02 | 3.7144E 02 | 3.7178E 02 | 3.6581E 02 |
| 9.0000E-02 | 3.7996E 02 | 3.6336E 02 | 3.5575E 02 | 3.5461E 02 | 3.5516E 02 | 3.5265E 02 |
| 1.0500E-01 | 3.7815E 02 | 3.5964E 02 | 3.4711E 02 | 3.4121E 02 | 3.3953E 02 | 3.3804E 02 |
| 1.2000E-01 | 3.7777E 02 | 3.5757E 02 | 3.4157E 02 | 3.3129E 02 | 3.2618E 02 | 3.2366E 02 |
| 1.3500E-01 | 3.7758E 02 | 3.5644E 02 | 3.3814E 02 | 3.2433E 02 | 3.1558E 02 | 3.1072E 02 |
| 1.5000E-01 | 3.7749E 02 | 3.5583E 02 | 3.3608E 02 | 3.1966E 02 | 3.0763E 02 | 2.9987E 02 |
| 1.6500E-01 | 3.7745E 02 | 3.5551E 02 | 3.3486E 02 | 3.1662E 02 | 3.0191E 02 | 2.9127E 02 |
| 1.8000E-01 | 3.7743E 02 | 3.5534E 02 | 3.3416E 02 | 3.1470E 02 | 2.9796E 02 | 2.8476E 02 |
| 1.9500E-01 | 3.7742E 02 | 3.5525E 02 | 3.3377E 02 | 3.1351E 02 | 2.9531E 02 | 2.8002E 02 |
| 2.1000E-01 | 3.7741E 02 | 3.5520E 02 | 3.3354E 02 | 3.1279E 02 | 2.9358E 02 | 2.7667E 02 |
| 2.2500E-01 | 3.7741E 02 | 3.5518E 02 | 3.3342E 02 | 3.1236E 02 | 2.9248E 02 | 2.7437E 02 |
| 2.4000E-01 | 3.7741E 02 | 3.5517E 02 | 3.3335E 02 | 3.1211E 02 | 2.9178E 02 | 2.7283E 02 |
| 2.5500E-01 | 3.7741E 02 | 3.5516E 02 | 3.3331E 02 | 3.1197E 02 | 2.9135E 02 | 2.7183E 02 |

# CHAPTER 9

# DYNAMIC MODELS AND CONTROL STRATEGIES FOR BIOLOGICAL WASTEWATER TREATMENT PROCESSES

John F. Andrews

Department of Civil Engineering
University of Houston
Houston, Texas

## INTRODUCTION

The need for consideration of the dynamic behavior and operational characteristics of wastewater treatment processes is frequently greater than that for industrial processes because of the large temporal variations which occur in wastewater composition, concentration, and flow rate. However, most wastewater treatment processes are in a primitive state with respect to process operation when compared with industrial processes. Gross failures, such as the bulking of activated sludge and "sour" anaerobic digesters, are all too frequent. In addition, there are significant variations in treatment plant efficiency, not only from one plant to another, but also from day-to-day and hour-to-hour in the same plant. Daily variations from 60-95% efficiency in BOD removal are not uncommon and Thomann,[1] by a statistical analysis of the variation in effluent quality from eight treatment plants, has shown that these variations can have a significant effect on the water quality of the receiving stream.

Improved performance of wastewater treatment plants can be obtained in several ways. Included among these are increasing the quantity and quality of operating personnel, more adequate consideration of dynamic behavior and operational characteristics during the design phase, and improved control strategies for plant operations. The need for an increase in the quantity and quality of operating personnel is well known and has been summarized in an Environmental Protection Agency (EPA) report.[2] The need for more consideration of dynamic behavior and operational character-istics during the design phase is receiving increasing attention[3] and some specific items of importance will be presented here. The potential for im-provement of plant performance through the development and use of improved control strategies is not as well recognized and will be one of the major items discussed in this chapter.

Dynamic mathematical models are usually necessary for the description of time variant phenomena, as is commonly encountered in wastewater treat-ment processes, and increasing efforts are being devoted to their develop-ment. Models for different types of reactors can be developed by applying material and energy balances using the fundamental transport, stoichiometric, thermochemical, and kinetic relationships. The models will usually consist of sets of nonlinear differential equations for which analytical solutions are not available. However, solutions to these equations or prediction of pro-cess performance with respect to time, can be obtained by computer simulation. Dynamic modeling and computer simulation are useful tools in developing better procedures for process start-up, prediction and prevention of process failures, and improvement of process performance by considera-tion of dynamic behavior during both the design of a process and its associated control system.

Control strategies are primarily involved with the handling and use of information. This may be done manually, as is frequently the case in wastewater treatment plant operations, or by automatic control systems. Automatic control systems can range from simple on-off control, similar to that used for temperature control in homes, or on-line computer control for an entire plant. Regardless of the type of control, some of the same basic questions must be answered and included among these are: (a) What information should be collected? (b) How should the information be transmitted? (c) How should the information be processed? (d) What control actions should be taken?

The design of a control systems is intimately related to the dynamic behavior of a process since the need for control, whether it be manual or automatic, is brought about by dynamic or transient behavior. Control systems frequently represent an economical means for improving the per-formance of wastewater treatment processes and should receive more con-sideration for both existing plants and in the design of new plants. A trade off is possible between a control system and the controlled process. Sizes of processes, and therefore capital costs, can frequently be decreased by increasing operational efforts either manually or by automatic control systems. There is, of course, an optimum since the value of a control strategy should always be balanced against its cost. However, current plant designs may be far from this optimum since little attention is devoted to dynamic behavior and operational characteristics during the design phase.

There are many potential benefits from the development and use of dynamic models for wastewater treatment processes and the incorporation of modern control systems into wastewater treatment plant design. Included among these benefits are:

1. *Performance.* Maintenance of plant efficiency nearer the maximum by improved operation could result in significant decreases in the pollution load placed on our water resources. For example, consider a plant with a maximum BOD removal efficiency of 95%, a variation in efficiency of 80-95%, and an average of 87.5%. If through better operation, plant efficiency could be maintained between 90 and 95% with an average of 92.5%, the BOD load to the receiving waters would be reduced by 40% without increasing the maximum capacity of the plant. Actual reductions may even be greater than this as demonstrated in the work reported by West.[4]

2. *Productivity.* Another significant advantage of good operation can be improved productivity through an increase in the amount of waste that can be treated per unit of process capacity. There have been several instances where operators have been able to significantly increase process capacity above the design capacity through changes in operational procedure.[5,6]

3. *Reliability.* It is well recognized that bypassing of raw wastewater and gross process failures occur all too frequently.[7] The frequency of these occurrences could be greatly reduced by improved operational procedures whether they be manual or automatic.

4. *Process stability.* Some processes used for wastewater treatment are more stable than others and, therefore, require less attention to operation. Although this is of considerable importance, quantitative comparisons of the stability of different processes are not available. Dynamic models are needed for making these comparisons of process stability and modern control systems can be used to improve process stability.

5. *Operating personnel.* Dependable and rugged control systems can be used to minimize the need for operation when it is anticipated that plants will be operated by unskilled personnel, as is often the case for smaller plants. In larger plants, where labor problems might be encountered, automation would permit plant management personnel to operate plants for limited periods of time.

6. *Operational costs.* Operational costs, such as power and chemical costs, can frequently be reduced by considering the dynamic behavior of the process and designing a control system to regulate power and chemical additions in accordance with the need for such additions. This could be of special importance for the new physiochemical processes where operational costs are relatively high as compared to conventional processes.

7. *Start-up procedures.* Dynamic models can be used to develop more reliable and faster start-up procedures. This would be of considerable value for processes such as anaerobic digestion and activated sludge which have long start-up times.

8. *Manual operational guides.* Control strategies developed through the use of dynamic models may be implemented either manually or by automatic control systems. Manual implementation may be accomplished by the use

of written operational guides or flow charts which present a sequence of operations to be performed upon occurrence of a certain event. These guides or charts are called procedural models and would be of value for those plants that are predominately manually operated.

9. *Dynamic operation.* It is frequently assumed that steady-state operation is best for maximizing performance. However, this is not always so, and dynamic models could be used to explore the possibility of operating a process in a dynamic fashion so as to maximize performance. As an example, Hawkes has shown that there is an optimum dosing frequency for the trickling filter.[8] Alternating double filtration is also an example of a process purposefully operated in a dynamic fashion.

10. *Variable efficiency operation.* It is conceivable that a treatment plant could be operated at variable efficiency to match the assimilative capacity of the receiving body of water which usually varies with time. This would require a good knowledge of the dynamic behavior of both the treatment plant and receiving waters and could be accomplished through the use of modern control systems.

As illustrated above, dynamic models and control systems do offer many potential benefits; however, it should be emphasized that the development of dynamic models for wastewater treatment processes and the use of these models for the improvement of control strategies is a difficult task and is in its infancy. On the other hand, many of the conventional control systems discussed in this chapter can now be applied.

## PROCESS DYNAMICS

Changes are always taking place in the inputs, outputs, or environment of a process as well as in the characteristics of the process itself. It is important to identify the nature of these changes and the rates at which they occur. Some may be so slow that they need not be considered for a particular process while others may be so rapid or of such short duration that they also have little appreciable effect on the process. However, there are many transients that do affect the behavior of a process and these should be considered in analysis, design, and operation.

In Figure 9.1a, the process is depicted as a "black box" with time varying inputs and outputs. The outputs can be predicted from the inputs if a dynamic mathematical model, sometimes called a transfer function, is available for the process and this might be called the transient response analysis problem. A knowledge of the inputs and outputs will permit the development of a model, and this can be called the experimental problem. The design problem would be where the inputs and required outputs are known and the type and characteristics of the process must be determined. This is usually an iterative procedure since it may be necessary to examine several different processes to select the process best suited for transforming the inputs into the required outputs.

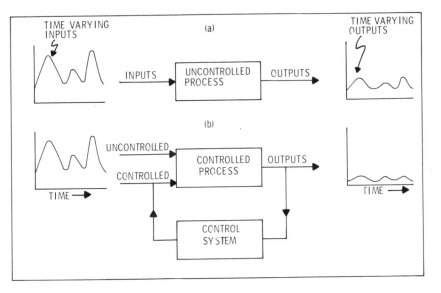

**Figure 9.1.** Time varying inputs and outputs for controlled and uncontrolled process.

It is also important to make a distinction between controllable and uncontrollable inputs to a process and this leads to consideration of a control system for the process as illustrated in Figure 9.1b. The characteristics of the process to be controlled may be fixed; *i.e.,* an existing wastewater treatment plant, in which case the design of the control system represents a problem similar to that of process design. The inputs, outputs, and dynamic characteristics of the control system must be established in much the same iterative fashion as for the process. However, in the design of a new plant, it is possible to strike a better balance between the effort and expense devoted to the process and that devoted to the control system. The major item of significance is that both the control system and the controlled process affect the process outputs and the two should therefore be designed as an integrated system.

Mathematical modeling, computer simulation, transient response analysis, and stability analysis are some of the basic tools used in the study of process dynamics. These tools come from many different disciplines and can only be briefly discussed here. (For a more detailed discussion, see the books by Chestnut,[9] Denbigh and Turner,[10] Himmelblau and Bischoff,[11] Franks,[12] Perlmutter,[13] Shilling,[14] McLeod,[15] and Cooper and McGillem.[16]

## Dynamic Models

Mathematical modeling is a technique frequently used (and sometimes abused!) in today's scientific and engineering investigations. However, modeling is not new since scaled-down physical models have long been used in such diverse areas as astronomy (planetariums), hydraulic engineering (river models), architecture (building models), and chemical engineering (pilot plants). Even the hypothesis formulated in applying the scientific model can be thought of as a verbal model of a system.

Mathematical models are commonly used for more quantitative description of process performance and consist of one or more equations relating the important inputs, outputs, and characteristics of the process. To those not familiar with systems engineering terminology, the term "mathematical model" is sometimes frightening since it may bring to mind large sets of complex equations. However, this need not be the case since most of the common engineering design formulae may also be called mathematical models. In fact, as simple an expression as $y = mx + b$ can be considered as a mathematical model where the system output (y) is related to the system input (x) by the system parameters (m,b). For more complex systems or more adequate description of system performance, it may be necessary to use larger numbers of equations and more powerful mathematical tools such as differential and partial differential equations, difference equations, and probability theory. However, Occam's razor applies here as well as to verbal hypotheses in that the simplest possible expressions should be used. From an engineering point of view, mathematical elegance is secondary; a model that is too complex might be subject to either misuse or disuse.

In developing models, it must also be realized that they are evolutionary in nature and subject to change as more knowledge is gained about the process. A model that is quite adequate as a first approximation might be replaced at a later date by a more exact model with better estimates of the coefficients, fewer empirical relationships, and inclusion of more variables. This evolutionary nature of models is not always recognized and can lead to reluctance on the part of an investigator to either modify or discard a model in the same fashion that investigators in past years have sometimes been reluctant to modify or discard verbal hypotheses.

Mathematical models may be classified in many different ways, and one of the most important for wastewater treatment processes is the distinction between dynamic and steady-state models. Most models currently in use are based on the assumption of steady-state. Steady-state models have proved their value on a qualitative basis by indicating needed changes in process design and also have the advantage of experimental and computational simplicity. However, in most instances they are not adequate to describe process operation since the inputs to the processes are far from constant and there is considerable variation in effluent quality with respect to time. Wastewater treatment processes should be modeled as dynamic systems and the models will usually consist of sets of nonlinear differential equations.

The development or review of dynamic mathematical models for all processes used in wastewater treatment is beyond the scope of this chapter and the ability of the author. However, the basic principles involved can be illustrated by the development of a model for the continuous addition of an inert substance or tracer to a continuous flow stirred tank reactor (CFSTR) as illustrated in Figure 9.2. The model is developed by making a material

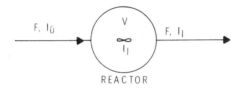

$$F, I_0 \qquad \begin{matrix} V \\ \infty \\ I_1 \end{matrix} \qquad F, I_1$$

REACTOR

V = REACTOR VOLUME
F = FLOW RATE
$I_0$ = INFLUENT TRACER CONCENTRATION
$I_1$ = EFFLUENT TRACER CONCENTRATION

**Figure 9.2.** CFSTR with inert tracer input.

balance on the inert tracer. The general form for a material balance is presented in Equation 1 and the material balance on the tracer is given in Equation 2.

| Rate of Material Flow into Reactor | + | Rate of Appearance of Disappearance of Material due to Reaction | = | Rate of Material Flow Out of Reactor | + | Rate of Accumulation of Material in Reactor | (1) |
|---|---|---|---|---|---|---|---|
| (1) | | (2) | | (3) | | (4) | |

$$F I_0 \quad + \quad 0 \quad = \quad F I_1 \quad + \quad V \frac{dI_1}{dt} \qquad (2)$$

where

F = flow rate
V = reactor volume
I = tracer concentration
$_0$ = subscript denoting reactor influent
$_1$ = subscript denoting reactor effluent

Since the reactor is stirred and the contents are homogeneous, the concentration of tracer in the reactor and in the reactor effluent are identical. The reaction term is zero since the tracer is inert and does not participate in any reactors. When the flow rate and reactor volume are constant, the process can be classified as a first-order, linear system with constant coefficients. The order is determined by the highest order derivative of the output and the system is linear since all derivatives and variables are raised only to the first power and there are no products of derivatives and/or variables.

The development of a model for the growth of microorganisms on an inhibitory substrate in a CFSTR (Figure 9.3), illustrates a model consisting

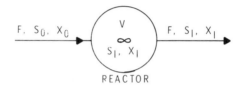

$V$ = REACTOR VOLUME
$F$ = FLOW RATE
$X_0$ = INFLUENT ORGANISM CONCENTRATION
$X_1$ = EFFLUENT ORGANISM CONCENTRATION
$S_0$ = INFLUENT SUBSTRATE CONCENTRATION
$S_1$ = EFFLUENT SUBSTRATE CONCENTRATION

**Figure 9.3.** Biological CFSTR with inhibitory substrate input.

of two interacting differential equations and the inclusion of some basic stoichiometric and kinetic relationships. These relationships, given in Equations 3, 4, and 5, are used in organism and substrate balances on the reactor (Equations 6 and 7) to develop the model for the process. The assumption has been made that there are no organisms in the influent to the reactor and the influent term in the organism balance is therefore zero. For a more detailed explanation of this model, and its application to the anaerobic digestion process, the reader is referred to Andrews[17,18] and Andrews and Graef.[19]

The two models presented illustrate that for a specific process, one or more of the terms given in the general material balance (Equation 1) may be zero thus simplifying the resultant model. Another example is a batch reactor in which the first and third terms are zero since there is no flow of

$$\frac{dX}{dt} = -Y \frac{dS}{dt} \tag{3}$$

$$\frac{dX}{dt} = \mu X \tag{4}$$

$$\mu = \frac{\hat{\mu}}{1 + \frac{K_S}{S} + \frac{S}{K_I}} \tag{5}$$

where

$X$ = organism concentration
$S$ = substrate concentration
$Y$ = yield coefficient
$\mu$ = specific growth rate
$\hat{\mu}$ = maximum specific growth rate
$K_S$ = saturation coefficient
$K_I$ = inhibition coefficient

**General Material Balance**

Influent + Reaction = Effluent + Accumulation

**Organism Balance**

$$0 \quad + \quad \mu X_1 V \quad = \quad F X_1 \quad + \quad V \frac{dX_1}{dt} \tag{6}$$

**Substrate Balance**

$$F S_0 \quad - \quad \frac{\mu}{Y} X_1 V \quad = \quad F S_1 \quad + \quad V \frac{dS_1}{dt} \tag{7}$$

material into or out of the reactor. If the inert tracer for the model presented in Equation 2 was added as a pulse instead of continuously, term one would also become zero in this equation immediately after addition of the tracer. Term four, the accumulation term, becomes zero at steady state. The elimination of term four, or the assumption of steady state, is one of the key differences between dynamic and steady-state models and has been widely used since it reduces the mathematical complexity of the resultant models. For the examples presented, the assumption of steady state reduces the models from differential equations to algebraic equations.

Each of the terms presented in the general material balance equation may have more than one component. For example, there may be several streams containing the material of interest that enter or leave the reactor. The materials in the reactor might also engage in more than one reaction and, a term frequently included in the organism balance for a biological reactor is an organism decay term to account for the disappearance of organism mass by autooxidation.

## Computer Simulation

After a dynamic mathematical model has been developed for a process, the equations that comprise the model must be solved to predict the behavior of the process with respect to time. This procedure is known as simulation and can be defined as the use of a model to explore the effects of changing conditions on the real system. Obviously, the model must be a reasonable representation of the real system for the results to be meaningful, since the simulation results can be no better than the mathematical model and data on which they are based.

Several methods are available for obtaining solutions to differential equations and for simple models, such as the CFSTR with continuous addition of an inert tracer (Equation 2), separation of variables can be used as illustrated in Equations 8 through 13. In solving this equation, the integration constant, C, is evaluated from the boundary condition on $I_1$ where at time equal zero, there is no inert tracer in the reactor. Equation 13 therefore represents the response of the reactor to a step change in the influent tracer concentration. At a time slightly less than zero ($t = 0^-$), the influent tracer concentration is zero ($I_0 = 0$), and at a time slightly greater than zero ($t = 0^+$), the influent tracer concentration has some finite value.

$$V \frac{dI_1}{dt} = F I_0 - F I_1 \tag{8}$$

$$\frac{dI_1}{I_0 - I_1} = \frac{F}{V} dt \tag{9}$$

$$-\ln(I_0 - I_1) = \frac{F}{V} t + C \tag{10}$$

$$@ \ T = 0, \ I_1 = 0, \ \text{and} \ C = -\ln I_0$$

$$-\ln (I_0 - I_1) = \frac{F}{V} t - \ln I_0 \tag{11}$$

$$\frac{I_0 - I_1}{I_0} = \exp(-\frac{F}{V} t) \tag{12}$$

$$I_1 = I_0 \ [1 - \exp(-\frac{F}{V} t)] \tag{13}$$

An alternate to the classical methods for solving differential equations is presented by the Laplace transform, and this technique has been widely used in the study of process dynamics. The Laplace transform technique uses an operator notation to transform differential equations into algebraic equations which can be manipulated more easily.  After the algebraic problem is solved, the desired answer is obtained by an inverse transformation. An analogue can be drawn to the use of logarithms for transforming multiplication to addition and exponentiation to multiplication.

Unfortunately, the Laplace transform is limited in its use, since it can only be applied to linear differential equations with constant coefficients. Most engineering systems, including wastewater treatment processes, contain significant nonlinear elements and considerable attention has therefore been devoted to linearizing the equations so that solutions can be obtained. Most nonlinear curves cannot be approximated by straight lines over their entire range; however, by restricting the range of the variables, a portion of the curve can frequently be considered as linear with little error. The resulting equations can then be solved using classical methods or the Laplace transform. More details on the application of the Laplace transform and linearization techniques to process dynamics can be found in the textbooks of Perlmutter[13] and Shilling.[14]

The Laplace transform and linearization techniques have been of considerable value in predicting the time dependent behavior of processes. However, considerable study is required to become proficient in the use of these techniques and for more complex models, they are cumbersome, time consuming, and frequently not even applicable. This is especially true for dynamic models of wastewater treatment processes, which are largely made up of highly nonlinear equations and have, therefore, presented a sizeable computational bottleneck. In the past, efforts at developing dynamic models for these processes were frequently of no practical value since the equations could not be solved. However, computers have largely eliminated this bottleneck with analog, digital, and hybrid computers being used for process simulation.

The early use of computers was largely restricted to specialists, since a considerable amount of time was required to learn to use a computer. This problem has been overcome by the development of continuous-system simulation languages for the digital computer.[12,20] These languages are heavily user-oriented, thus permitting the process engineer to concentrate on model development and simulation results rather than on the details of the computations. They are especially delightful for those not experienced in using the computer since they can be learned in a relatively short period of time. For example, students with no prior computer experience are usually able to write their own programs for significant simulations after less than four hours instruction. The practicing engineer can obtain access to these simulation languages through computer time-sharing services using a remote terminal in his own office.

The simulations presented in this chapter were performed using CSMP/360 on the IBM 370 digital computer.[20] This program or language may be thought of as being one level above such languages as FORTRAN since CSMP statements much more closely resemble ordinary mathematics and are automatically translated into FORTRAN by the computer. Unlike most other computer languages, CSMP statements, with few exceptions, can be written in any order and are automatically sorted by the computer to establish the correct order of information flow. CSMP provides a number of standard functions such as integrators, comparators, limiters, etc., which are used in building a mathematical model. These standard functions are augmented by the usual FORTRAN functions such as square root, sine, and logarithms. The basic arithmetic operator symbols such as (*) for multiplication and (/) for division are the same as those used in FORTRAN. An easily used and fixed format is provided for tabular and graphic output of selected variables at selected increments of time.

Writing a CSMP program is best illustrated by example and a program for the dynamic model of a CFSTR with continuous addition of an inert tracer (Equation 8) is given below. The asterisk in column one of the

```
  *           CFSTR WITH INERT TRACER INPUT

  *           V=REACTOR VOLUME(L), F=FLOW RATE(L/HR)
  *           I0,I1=INFLUENT,EFFLUENT TRACER CONCENTRATION(MG/L)

PARAM         V=10.0, F=2.0, I0=100.0
INCON         I1IT=0.0

DYNAM

              I1DOT=F*I0/V-F*I1/V
              I1=INTGRL(I1IT,I1DOT)

METHOD        RKSFX
TIMER         DELT=0.01, FINTIM=20.0, PRDEL=0.5, OUTDEL=0.5
PRINT         I1
PRTPLT        I1
TITLE         CFSTR WITH INERT TRACER INPUT
LABEL         CFSTR WITH INERT TRACER INPUT
END
STOP
ENDJOB
```

punched card denotes a comment card which can be used to insert appropriate comments or explanations in the program. The PARAM and INCON cards are used to insert numerical values, or data, for parameters, constants, and initial conditions. The DYNAM section contains the mathematical model and the CSMP functions needed to solve the model. In this instance, INTGRL is the only CSMP function used with I1DOT representing the differential equation to be integrated and I1IT representing the initial condition for this equation. The other cards are control statements and include METHOD, which specifies the numerical integration technique to be used (six are available), the integration interval or step size (DELT), the length of time over which tabular or graphical outputs are to be given (PRDEL,

OUTDEL), and the variables to be printed or plotted (PRINT, PRPLT). An example of the tabular output is presented below. Standard E notation is used to indicate the power of 10 to which the numbers in the columns must be raised.

```
          CFSTR WITH INERT TRACER INPUT

     TIME                   I1
     0.0                    0.0
     5.0000E-01             9.5120E 00
     1.0000E 00             1.8211E 01
     1.5000E 00             2.5923E 01
```

The example presented is very simple and uses only a small fraction of the power of CSMP. A total of 34 functions are available and only one of these (INTGRL) has been used. Regular FORTRAN statements can be used in the program and a TERMIN section can be added after the DYNAM section to control subsequent runs as might be needed in optimization problems. The language can handle sizable numbers of equations since 300 INTGRL blocks are provided. The use of some of the additional functions available will be illustrated later.

In developing mathematical models, it is desirable to iterate between model development, computer simulation, physical experimentation, and field observations since these complement one another. Knowledge gained in simulation is useful for modifying the model, guiding physical experimentation, and establishing the type and frequency of field observations needed. This iterative technique also points out another important aspect of modeling and simulation, this being the need for model verification. The ease and speed with which computer simulations can frequently be made may lead to a neglect of this very important aspect of model development and in the extreme, can result in one becoming so enamoured with the techniques that the purpose for using them is almost forgotten. Computer simulation can lead to the generation of large quantities of worthless results if the model is not a reasonable representation of the real process.

### Transient Response Analysis

After a dynamic mathematical model has been developed for a process and a simulation technique selected, the time varying behavior of the outputs can be predicted from the inputs. This procedure is called transient response analysis. The techniques used in transient response analysis are also of value in developing models. This is accomplished by subjecting the process to standardized inputs or signals, known as forcing functions, and comparing the output response with those for standardized models. These forcing functions may be periodic or nonperiodic and deterministic or random. A periodic signal repeats itself at regular intervals in time with the signal frequency being the number of times the signal is repeated in a given unit of time. Deterministic signals can be predicted exactly in time while random signals have a probability function associated with them. Random signals will not be further considered in this paper.

Transient response analysis can be illustrated using the dynamic model for a CFSTR with inert tracer input and the CSMP program for this model. The model (Equation 2) is usually recast into the form shown in Equation 14 where V/F is called the time constant ($\tau$) of the reactor and is a measure of the time required for the process output to respond to a disturbance in

$$\tau \frac{dI_1}{dt} + I_1 = I_0 \tag{14}$$

the input. $I_0$ is the forcing function. Four nonperiodic forcing functions commonly used are the step, impulse, pulse, and ramp functions. These functions, and the response of the model to these functions, are presented in Figure 9.4. The CSMP program gives the response of the reactor at time $\geqslant 0$, to a step increase of $I_0$ from 0.0 to 100.0 mg/l. This was accomplished by setting I1IT on the INCON card equal to 0.0 and I0 on the PARAM card equal to 100.0. The response to an impulse is obtained by the reverse, *i.e.*, setting I1IT equal to 100.0 and I0 equal to 0.0. The response to a pulse is obtained by removing I0 from the PARAM card and placing a card in the DYNAM section using the COMPAR function. This card, for a pulse height of 100.0 mg/l and a width of 5.0 hr, would read as follows:

$$I0=100.0*COMPAR(5.0,TIME)$$

The value of the COMPAR function is equal to one as long as TIME is less than five hours and is equal to zero when time is greater than five hours. In obtaining the response to a ramp function, I0 is also removed from the PARAM card and a LIMIT function, as listed below, is placed in the DYNAM section. This gives a ramp increase in I0 of 20.0 mg/l-hr up to a maximum of 100.0 mg/l.

$$I0=LIMIT(0.0,100.0,20.0*TIME)$$

The time constant ($\tau$) is an important measure of the dynamic characteristics of a reactor and is numerically equal to the time required for the output concentration ($I_1$) to equal 63.2% of the input or steady-state concentration ($I_0$) for a step forcing of the reactor. The effect of different time constants on the rate of response is illustrated in Figure 9.5. The time constant can be varied by changing either F or V. In interpreting the

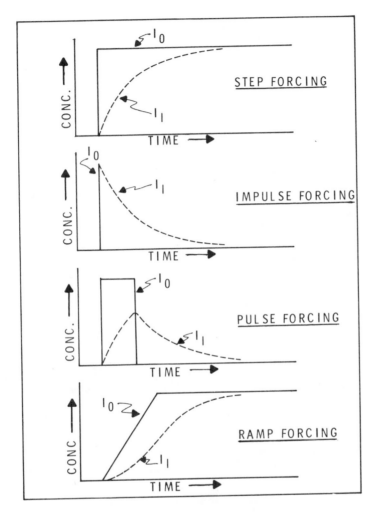

**Figure 9.4.** Response of a CFSTR to nonperiodic inputs of an inert tracer.

effect of the time constant, it should be remembered that this illustration is for the addition of an inert tracer to a CFSTR. The addition of a reacting substance and inclusion of a reaction term in the model could substantially change the value of the time constant as demonstrated by Perlmutter.[13]

Two common periodic inputs are the sinusoid and the pulse train. The response to a pulse train would be the same as illustrated in Figure 9.4, except repeated and perhaps without time for the process to come to

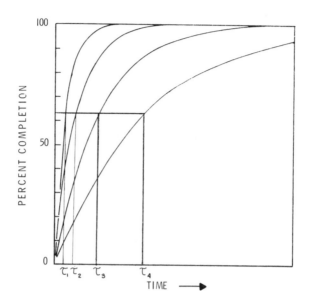

**Figure 9.5.** Effect of time constant on the response of a CFSTR to the step input of an inert tracer.

steady state between pulses. A pulse train can be generated in CSMP using a combination of the IMPULS and PULSE functions. The periodic feeding of an anaerobic digester or the dosing of a trickling filter can be represented by pulse trains and sinusoids are frequently used as first approximations to represent the influent flow rate variations to a wastewater treatment plant. After a transient period, a linear system forced sinusoidally will response sinusoidally. The output wave will have the same frequency as the input but will differ in amplitude and exhibit a phase shift as shown in Figure 9.6. To obtain this response for the CFSTR with inert tracer input, I0 is removed from the PARAM card and a SINE function is placed in the DYNAM section. A typical sinusoidal forcing function, for an average value of I0 equal to 150.0 mg/l and a frequency of one cycle per hour, would read as follows:

$$I0=150.0+100.0*SINE(0.0,6.2832,0.0)$$

One cycle per hour is equivalent to 360° or 6.2832 radians per hour. The maximum and minimum values of I0 would be 250.0 and 50.0 mg/l, respectively.

If the CFSTR is forced sinusoidally at several different frequencies, a diagram known as a Bode plot can be prepared as shown in Figure 9.7.

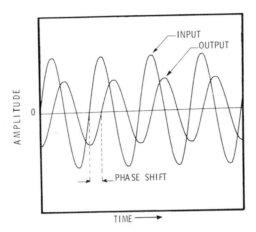

**Figure 9.6.** Response of a CFSTR to the sinusoidal input of an inert tracer.

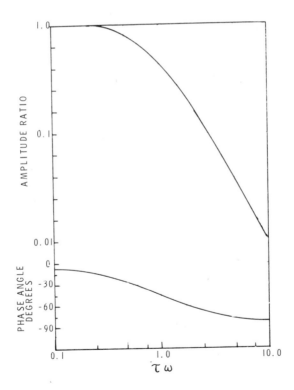

**Figure 9.7.** Bode plot for a CFSTR with inert tracer input.

The abscissa is plotted on a logarithmic scale and has been normalized by multiplying the time constant ($\tau$) of the process by the frequency ($\omega$) of the input. The upper curve is a plot of the ratio of output amplitude to the input amplitude, also on a logarithmic scale, vs $\omega\tau$ and illustrates how the amplitude of the output is attenuated as it passes through the process. At low frequencies, there is very little attenuation with the amplitude of the output being approximately equal to the amplitude of the input. At high frequencies, the input signal is almost completely attenuated and the sinusoidal variation might not even be detected. The lower portion of the Bode plot illustrates the phase angle or lag between the input and output signals. The phase angle increases as the frequency increases. The amplitude portion of the Bode plot illustrates that only a certain band or range of frequencies is important in dynamic process analysis. The phase angle portion would be of importance, for example, when attempts are made to calculate process efficiency based on grab samples of the influent and effluent.

The transient responses which have been presented for the standard forcing functions are for a linear, first-order system, and it would be expected that these responses would be different for other systems. A plug flow reactor is an example of a zero-order system, and the response of such a reactor to a pulse input of an inert tracer is shown in Figure 9.8. No mixing is assumed to occur in a plug flow reactor and the pulse input appears unchanged in shape in the reactor effluent but delayed by a time equal to the time constant or residence time (V/F) of the reactor. This time delay is called dead time or pure time delay and is frequently encountered in wastewater treatment plants and their associated control systems.

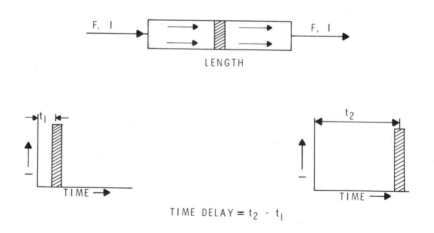

**Figure 9.8.** Response of a plug flow reactor to a pulse input.

Second-order systems are also encountered and typical responses of a second-order system to a step forcing are illustrated in Figure 9.9. Two coefficients, a time constant ($\tau$) and a damping coefficient ($\psi$), are required to describe a second-order system. A second-order system can exhibit oscillatory behavior, which is greatly influenced by the numerical value of the damping coefficient. If $0 < \psi < 1.0$, the oscillations will eventually die out or be damped as shown in Figures 9.9a and 9.9b. Sustained oscillations will occur

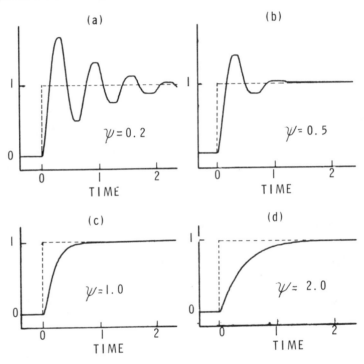

**Figure 9.9.** Response of second-order systems to a step input.

if $\psi = 0$, and if $\psi < 0$ the system would be unstable and the oscillations would increase with respect to time. For $\psi = 1.0$, the system is said to be critically damped and will reach the new steady state in a minimum time and without oscillations. The system is said to be overdamped and longer times will be required to reach the new steady state if $\psi > 1.0$ as shown in Figure 9.9d.

The application of transient response analysis to wastewater treatment plants is illustrated by Bryant's analysis of the dynamic hydraulic behavior of a plant.[21] The plant analyzed was a conventional activated sludge plant with a design flow rate of 2.0 mgd. The response to an input pulse with a

design flow rate of 2.0 mgd. The response to an input pulse with a magnitude of 1.0 mgd and a duration of 1.2 hr is given in Figure 9.10. The effluent flow rate is essentially undamped in magnitude and has a time

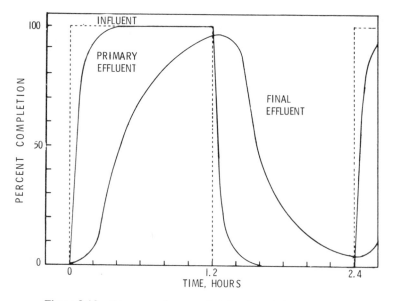

**Figure 9.10.** Response of conventional activated sludge plant to pulse forcing of flow rate.

constant of about 0.5 hr. The amplitude portion of the Bode plot (Figure 9.11) shows that the plant behaves as a low pass filter in that no appreciable damping of input flow rate variations occurs at frequencies less than 10 cycles per day. Frequencies greater than 100 cycles per day are essentially completely damped. These are logical answers when it is considered that wastewater treatment plants are almost constant volume systems and therefore have relatively small time constants with respect to flow rate variations. Only a small change in water level is necessary for a considerable increase in effluent flow rate since V-notch weirs are commonly used as outlet devices. Figure 9.11 illustrates that a conventional wastewater treatment plant does not effectively damp the normal daily cycle of flow rate variations. However, this is not true for damping of fluctuations in concentrations since both mixing and reaction occur in many of the plant processes and the effects of these are not included in the hydraulic model.

The application of standard forcing functions can give considerable insight to the dynamic behavior of a process. However, for a more quantitative evaluation, it is necessary to use inputs corresponding to those

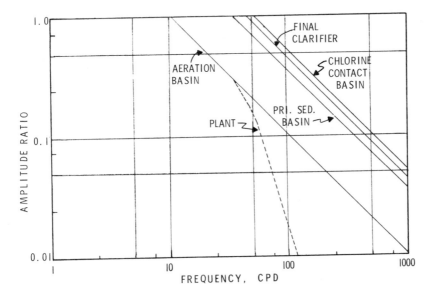

**Figure 9.11.** Frequency response of conventional activated sludge to hydraulic forcing.

actually observed in the field. This can be accomplished in CSMP by using the function generator elements AFGEN or NLFGEN which allow the input of actual field data such as influent flow rate vs. time. The effect of random variations can be superimposed on these deterministic inputs by using the noise or random number generators (GAUSS, RNDGEN) provided in CSMP.

**Process Stability**

Wastewater treatment processes can exhibit instability as evidenced by the occurrence of gross failures such as bulking activated sludge, "sour" anaerobic digesters, and sludge blanket loss in upflow clarifiers. It is recognized that some processes are more stable than others and that this should be considered in selecting a process for a specific application. However, quantitative measures of stability are not available for these processes with an exception being those given by Graef and Andrews for the anaerobic digestion process.[22]

Dynamic modeling and simulation can be used to evaluate the effects of various inputs and process characteristics with respect to stability. A distinction must be made between the stability of the process and that of the control system, since either or both can exhibit instability and there are obviously strong interactions between the two systems. (For a more detailed explanation of process and control system stability, the reader is referred to any of the textbooks on control theory or process dynamics.[23,24]) Biological process stability will be illustrated here by computer simulation, using the dynamic model (Equations 5, 6, and 7) developed for the growth of microorganisms on an inhibitory substrate. The CSMP program for this model is given below:

```
*           BIOLOGICAL CFSTR WITH INHIBITORY SUBSTRATE INPUT

*           V=REACTOR VOLUME(L), F=FLOW RATE(L/HR)
*           SO,S1=INFLUENT,EFFLUENT SUBSTRATE CONCENTRATION(GM/L)
*           X1=EFFLUENT ORGANISM CONCENTRATION(GM/L)
*           Y=YIELD COEFFICIENT(GM X/G S)
*           MU=SPECIFIC GROWTH RATE(1/HR)
*           MUHAT=MAXIMUM SPECIFIC GROWTH RATE(1/HR)
*           KS,KI=SATURATION,INHIBITION COEFFICIENTS(GM/L)

PARAM       F=4.0, SO=5.0
CONST       V=12.0, Y=0.5, MUHAT=1.0, KS=0.03, KI=2.0
INCON       X1IT=2.485, S1IT=0.030

DYNAM

            MU=MUHAT/(1.0+(KS/S1)+(S1/KI))
            S1DOT=F*SO/V-S1*F/V=MU*X1/Y
            X1DOT=MU*X1-F*X1/V
            S1=INTGRL(S1IT,S1DOT)
            X1-INTGRL(X1IT,X1DOT)

METHOD      RKSFX
TIMER       DELT=0.001, FINTIM=16.0, PRDEL=0.1, OUTDEL=0.1
PRINT       S1,X1
PRTPLT      S1,X1
TITLE       BIOLOGICAL CFSTR WITH INHIBITORY SUBSTRATE INPUT
LABEL       BIOLOGICAL CFSTR WITH INHIBITORY SUBSTRATE INPUT
END
STOP
ENDJOB
```

If a simulation was made using the numerical values given in the above program, there would be no appreciable change in X1 or S1 with respect to time since the values given are the steady-state values.

The model presented can exhibit instability for either of two reasons; these being too high a hydraulic flow rate or too sudden an increase in the influent substrate concentration. Too high a hydraulic flow rate decreases the hydraulic residence time (V/F) to below that at which the microorganisms can grow and the organisms will "wash out" of the reactor. The washout residence time for the example presented is 1.24 hr, and the results of step forcing the residence time for its given value of 3.0 hr (V/F = 12.0/4.0) to a value of one hour (F=12.0), are presented in Figure 9.12. The organism concentration decreases from the original steady-state value of 2.485 g/l and approaches zero as a limit. The substrate concentration increases and approaches the influent substrate concentration of 5.0 g/l as a limit.

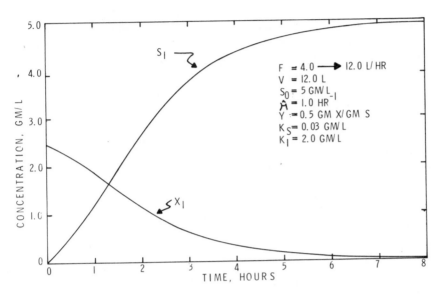

**Figure 9.12.** Step forcing of flow rate for a biological CFSTR.

Instability due to a step change in the influent substrate concentration is illustrated by the dashed curves in Figure 9.13 which give the response of $S_1$ and $X_1$ to a step forcing in $S_0$ from 5.0 to 20.0 g/l. The behavior

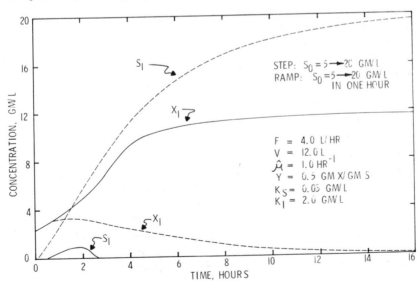

**Figure 9.13.** Step and ramp forcing of influent substrate for a biological CFSTR.

of the process is much the same as that for the step change in flow rate. However, an important distinction between the two modes of failure is that the increase in flow rate will always cause failure regardless of the rate at which the flow rate is changed whereas failure due to the increase in influent substrate concentration can be avoided if the rate of increase is low enough. This is shown in Figure 9.13 by the solid curves which give the response when $S_0$ is increased from 5.0 to 20.0 g/l over a one-hour period (ramp-forcing). This forcing does not cause process failure.

Process design changes which could be used to improve stability would be to increase the reactor volume or to use organism separation and concentration with recycle to the reactor. Incorporation of an adequate control strategy can also increase process stability. The effects of these parameters, as well as others, have been investigated for the anaerobic digester by Graef and Andrews[22] and will be briefly discussed later in this chapter.

## CONTROL STRATEGIES

Control strategies are primarily involved with the handling of information. This can be done manually or by automatic control systems. Environmental engineers are familiar with the theory and technology involved in the collection, transportation, processing, and distribution of materials and energy. However, they are not as accustomed to thinking of information in the same terms, although this can be of equal or greater importance.

The technology involved in information handling is of more recent vintage than that for materials and energy; however, many of the same concepts are applicable. The handling of materials, energy, and information all involve collection, transportation, storing, processing, and distribution. Flow diagrams are used in the handling of materials, energy, and information, and examples of information flow diagrams are given in Figures 9.14a and 9.14b where the temperature of a process is to be controlled either manually or automatically. The temperature is changed from its desired or reference value by some input disturbance such as a change in environmental temperature or heat input to the process. This change is measured by a sensor such as a thermometer. In a manual control system (Figure 9.14a), the measured temperature is transmitted to the man in the control loop by visual observation. The man processes this information by mentally comparing the observed temperature with the desired temperature, and adjusts the heat input to the process in an attempt to bring the temperature back to its desired value. Several iterations of this procedure might be needed before the desired temperature is attained. The man has "closed the loop" by "feedback" of information from the process output to the process input.

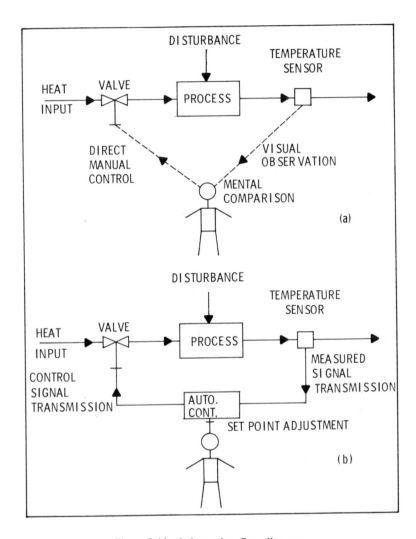

**Figure 9.14.** Information flow diagrams.

In the automatic system, the man is replaced in the feedback loop by a controller. The temperature sensor transmits a signal to the controller. An additional device, a transducer, might be needed between the sensor and controller for amplification or changing the form of the signal so that it can be transmitted to or understood by the controller. The controller compares the signal with a stored reference signal or set point to determine if an error exists. If an error does exist, the controller computes, by means of a control algorithm, the amount of control action needed. It then transmits a signal to a final control element, the valve in this instance, to adjust the heat input to the process. A transducer might also be needed between the controller and the final control element. The automatic system is also iterative since the computed adjustment of the control valve may not give the desired temperature. It should be noted that the man continues to participate in the feedback loop on an intermittent basis, since he must select the controller set point on the basis of his judgment and experience.

Regardless of whether control is to be manual, automatic, or a mixture of the two, some of the same basic questions must be answered in development of a control strategy. Included are:

1. What measurements should be made for initiation of the control strategy? Measurements might be made on the process influent or effluent, the process environment, or could be internal to the process. Associated items of importance are the required accuracy and frequency, time required for making the measurements, and the availability of instruments. The dynamic behavior of the process and its associated control system must be considered in establishing information needs.

2. What control actions should be taken? This is of particular importance in wastewater treatment since many plants have been designed without adequate consideration of dynamic behavior or operational characteristics. In most plants, possible control actions are therefore very limited.

3. How should the information be transmitted from the sensor to the controller and from the controller to the final control element? This is an elementary question; however, in many wastewater treatment plants, information which has been collected is not used for initiation of control but is simply filed for the record. The control loop is therefore broken.

4. How should the information be processed to determine the type and amount of control action needed? A variety of control algorithms involving such calculations as conversion of units, comparison with desired values, averaging, multiplication, integration and statistical analysis might be required.

Dynamic modeling and simulation can be of considerable value in answering questions one and two. However, of equal or greater importance is an intimate knowledge of the process since the answers to these questions are highly dependent upon the wastewater to be processed, process characteristics, and environment in which the process will be operated. Much additional research will be needed to quantify the answers to questions one and two.

A sizable amount of literature relative to questions three and four is available in the field of control engineering. Some of the more pertinent material will be briefly discussed here; for more details, the reader should consult

Perlmutter,[13] Shilling,[14] Hougen,[25] Tucker and Wills,[26] and Lloyd and Anderson.[27]

**Classical Control Algorithms**

The information flow for a simplified feedback control loop is illustrated in the block diagram given in Figure 9.15. The output signal from the block representing the process goes to a comparator where it is subtracted from the set point or desired value of the signal. The resultant signal from

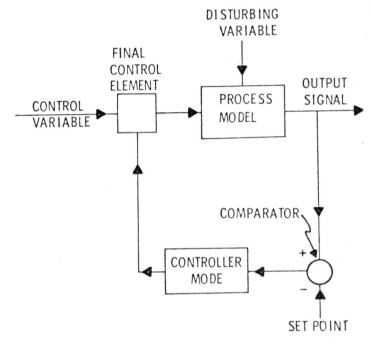

**Figure 9.15.** Block diagram of a simplified feedback control loop.

the comparator is called the error signal and goes to the controller where various mathematical operations called control algorithms or control modes are performed on the error signal to calculate the required value of the signal to be transmitted to the final control element. The final control element then modifies one of the input variables to the process so as to bring the process output signal closer to the desired value.

In simulating the process and its associated control system, it is necessary to consider the dynamic characteristics of all of the components of the total system. The blocks given in Figure 9.15 can be considered to represent dynamic mathematical models or transfer functions of the process and controller. It might also be necessary to include blocks representing the dynamic characteristics of such elements as the sensor, transducer, transmission lines, and final control element. Examples are the time delay involved in pumping a sample for analysis to a residual chlorine analyzer and the relationship between valve stem position and flow rate through a control valve. For simplicity, characteristics such as these have not been included in the control algorithms presented herein; however, it is essential that they be included for real systems since they can substantially modify the dynamic behavior of the system.

The simplest form of a control algorithm is on-off control which means that the final control element is either in the completely open or completely closed position depending on whether the error signal is positive or negative. The CSMP program for a biological CFSTR, with an input of inhibitory substrate, can be modified to incorporate on-off control as shown below.

```
*       ON-OFF CONTROL OF A BIOLOGICAL CFSTR

*
*       V=REACTOR VOLUME(L), F=FLOW RATE(L/HR)
*       S0,S1=INFLUENT,EFFLUENT SUBSTRATE CONCENTRATIONS(GM/L)
*       X1=EFFLUENT ORGANISM CONCENTRATION(GM/L)
*       Y=YIELD COEFFICIENT(GM X/GM S)
*       MU=SPECIFIC GROWTH RATE(1/HR)
*       MUHAT=MAXIMUM SPECIFIC GROWTH RATE(1/HR)
*       KS,KI=SATURATION,INHIBITION COEFFICIENTS(GM/L)
*       SP=SET POINT(GM/L)
*       FC=CONTROLLED FLOW RATE TO REACTOR(L/HR)

PARAM   SP=0.03
CONST   V=12.0, Y=0.5, MUHAT=1.0, KS=0.03, KI=2.0
INCON   X1IT=2.485, S1IT=0.030

DYNAM
        F=4.0+4.0*SINE(0.0,1.5708,0.0)
        S0=5.0+5.0*SINE(0.0,1.5708,0.0)
        MU=MUHAT/(1.0+(KS/S1)+(S1/KI))
        S1DOT=FC*S0/V-S1*FC/V-MU*X1/Y
        X1DOT=MU*X1-FC*X1/V
        S1=INTGRL(S1IT,S1DOT)
        X1=INTGRL(X1IT,X1DOT)

        FC=F*COMPAR(SP,S1)

METHOD  RKSFX
TIMER   DELT=0.001, FINTIM=16.0, PRDEL=0.1, OUTDEL=0.1
PRINT   S1,X1,FC
PRTPLT  S1,X1,FC
TITLE   ON-OFF CONTROL OF A BIOLOGICAL CFSTR
LABEL   ON-OFF CONTROL OF A BIOLOGICAL CFSTR
END
```

The measured signal is the effluent substrate concentration and the controlled variable is the flow rate to the reactor. Flow rate and substrate concentration inputs to the reactor are sinusoidal and in phase with a period of four hours and minimum and maximum values of 0.0 and 8.0

liters/hr and 0.0 and 10.0 g/l, respectively. Control action is provided by
the COMPAR function in that whenever the effluent substrate concentration
exceeds the set point concentration, the flow into the reactor drops to zero.
Whenever the effluent substrate concentration is less than the set point,
the flow into the reactor is sinusoidal as given by the SINE function. The
results of the simulation after the first four hours of operation are presented
in Figure 9.16, which shows the controlled and uncontrolled effluent substrate concentration as well as the times during which the flow to the
reactor is turned off.

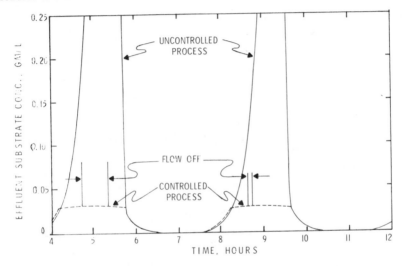

**Figure 9.16.** On-off control of flow for a biological CFSTR.

Although the simulation presented is adequate to illustrate the principles
of on-off control, the problem presented may not be very practical from a
wastewater treatment point of view. Serious time delays may be encountered in measuring the effluent substrate concentration and there may be
no way to dispose of the wastewater when the flow is turned off. Other
signals, such as the wastewater flow rate, and other control actions, such
as recycle of concentrated microorganisms, might be more effective. It
would also be possible, through design, to damp out the fluctuations in
influent flow rate and substrate concentration by constructing a variable
volume mixing tank upstream from the reactor. Dynamic modeling and
simulation could be used to evaluate the effectiveness of these alternate
strategies.

A very commonly used control algorithm is that provided by the three-
mode controller equation which is also known as PID control. This

equation, given below, allows calculation of the amount of control action as a function of four terms. The first term is called the bias and is provided to allow the amount of control action to be set at approximately 50% of the full value whenever the measured variable equals the set point. The

$$C_A = K_B + K_p e + K_{IN} \int_0^t e \, dt + K_{DI} \frac{de}{dt}$$ (15)

$$(1) \quad (2) \quad (3) \quad (4)$$

where:

$C_A$ = amount of control action
$K_B$ = bias control coefficient
$K_P$ = proportional control coefficient
$K_{IN}$ = integral control coefficient
$K_{DI}$ = derivative control coefficient

second term provides proportional control in which the amount of control action is proportional to the error signal. The third term gives control action proportional to the integral of the error signal and therefore reflects the length of time that the error has persisted. The fourth term is proportional to the derivative of the error signal and thus reflects the rate of change of the error.

The number of terms to be used and the values selected for their coefficients depends on the dynamic behavior of the process. Determination of the values for the coefficients is known as "controller tuning" and is primarily a trial and error process although there are optimization programs available. Simulation of proportional control for the biological CFSTR is accomplished in the CSMP program by replacing the COMPAR function with the cards shown below:

```
E=SP-S1
FP=LIMIT(0.0,8.0,FB+KP*E)
FC=AMIN1(F,FP)
```

Values for FB, the bias flow, and KP, the proportional control coefficient which is also known as the gain, should be inserted into the program on a PARAM or CONST card. The LIMIT card places upper and lower limits on the flow rate calculated from the control algorithm which is FB+KP*E. This card is necessary to avoid negative control flow rates into the reactor, which are a physical impossibility. The AMIN1 card compares the flow rate as specified by the sinusoidal input with that calculated from the LIMIT function and uses the smaller of the two as the controlled flow rate to the reactor. This card is necessary since the controlled flow rate should not exceed the flow rate specified by the SINE function.

A simulation is not presented for proportional control of the biological CFSTR since, as far as effluent substrate concentration is concerned, there is little difference between the effectiveness of the two control algorithms. However, proportional control as well as other control modes, could be of value for different forcing functions. The effect of the control model on the amount of control action needed should also be examined since for the example used, it would be desirable to minimize the amount of wastewater diverted from the reactor.

## Advanced Control Algorithms

On-off and PID controllers are widely used and have proven their value in a variety of process control applications. However, there are many situations in which process performance could be further improved by the use of more advanced control modes. The simpler of these could be the addition of circuits to take into account such items as dead time in sensors and sampling apparatus, and filters for removing high frequency components of the signal which one might not want to consider in initiation of a control action. More sophisticated control algorithms are being increasingly used and included among these are multiple-variable control, feedforward control, and various optimal control strategies.

Two common types of multiple-variable control are ratio and cascade control. Ratio control simply means maintaining one variable, the controlled variable, in a preset ratio to a second variable, the disturbing or uncontrolled variable. A good example of this is found in the activated sludge process (Figure 9.17a) where the recycled sludge flow rate is maintained as a set fraction of the wastewater flow rate to the aeration basin. This type of control is usually initiated in an attempt to maintain a more constant concentration of mixed-liquor suspended solids (MLSS) in the aeration basin. It would be classified as an open loop control system since there is no feedback from the key variable of interest, the MLSS. Closed loop control, using MLSS as the measured variable and recycle flow rate as the controlled variable, is also illustrated in Figure 9.17a.

Cascade control involves two controllers where a primary or master controller is used to adjust the set point of a secondary or slave controller. The secondary controller is usually, although not always, placed in a loop internal to the loop in which the master controller is placed and has a faster response than the primary controller. An example of cascade control for the activated sludge process is illustrated in Figure 9.17b. This is a combination of the two control systems shown in Figure 9.17a. The

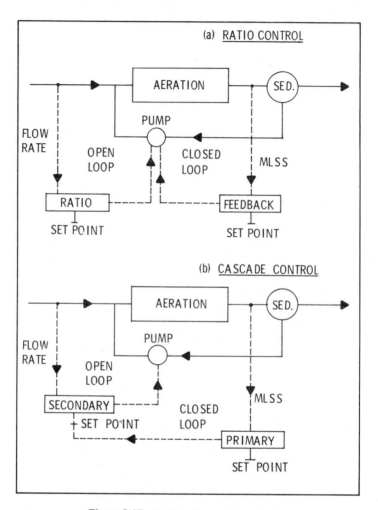

**Figure 9.17.** Ratio and cascade control.

measured variable for the primary controller is the MLSS in the effluent from the aeration basin, whereas the measured variable for the secondary controller is the wastewater flow rate into the aeration basin. The primary controller would establish the ratio of recycle sludge flow to wastewater flow to be maintained by the secondary controller.

Feedforward control (Figure 9.18) is similar to ratio control in that it is open loop control. Information for feedforward control is obtained by measuring the inputs to the process instead of the outputs as in feedback control. The amount and type of control action needed is then predicted

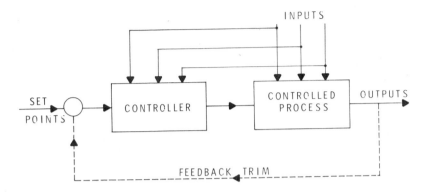

**Figure 9.18.** Feedforward control with feedback trim.

by a high-speed simulation of the process using a dynamic mathematical model. Feedforward control is theoretically capable of perfect control since no error need exist, as for feedback control, before the control action is initiated. This can be of special importance for processes with large time constants as commonly encountered in wastewater treatment.

Little knowledge of the process is required for feedback control; however, considerable time might elapse after the control action is initiated before the process is restored to the proper operating condition. Although this disadvantage can be overcome by using feedforward control, considerable knowledge of the process is required to develop the necessary mathematical models: Feedforward control strategies for wastewater treatment processes will have to incorporate a considerable amount of feedback control until their dynamic behavior is more clearly defined.

Considerable attention has been devoted in the control engineering literature to optimal control strategies. As a first step, it is necessary to establish an objective function, with associated constraints on the variables, for quantitative definition of the word "optimum." For example, in a treatment plant "optimum" might mean to control the plant at minimum operating cost but with the constraint that the effluent BOD always be equal to or less than 20 mg/l. The optimization program would then take into account all significant process variables and calculate the process conditions and control changes necessary to obtain optimum performance. The control system would then initiate these changes.

Many different tools are available for optimization and included among these are the calculus of variations, linear programming, dynamic programming, and various statistical techniques. These are primarily the same techniques as those used in optimal design of plants or development of optimal management strategies. The reader is referred to the books of Lee, *et al.*[28] and Savas[29] for an introduction to the application of these techniques to control systems.

### Computer Control Systems

Control systems involving many control loops, or using advanced control algorithms, require considerable computing power and it is only logical that digital computers are being increasingly used for process control. By mid-1968, 1700 process control computers were installed or ordered for a variety of applications in industries throughout the United States and 5900 installations are forecast by 1975.[30] A report on the feasibility of computer control for wastewater treatment plants was prepared by the American Public Works Association in 1970[31] and computers are now being installed in several plants. The author is aware of at least 28 plants throughout the world that have either installed or ordered process control computers. Included among these are plants located in Stockholm, Paris, Tokyo, Melbourne, Philadelphia, Los Angeles, New York, Chicago, Detroit, and Atlanta. Only a brief discussion of computer control can be presented here; the books of Lee, *et al.*,[28] Savas,[29] and Lowe and Hidden[32] should be consulted for more information.

Most readers are familiar with the ordinary business or scientific computer installation in which the user carries a deck of cards to the computer center and receives, after some period of time, a printed output containing the results of his requested computations. A process control computer installation differs considerably from this in that provision is made for very rapid or "real time" two-way communication between the computer and the process, and the computer and the user who, in this case, is the plant operator. A simplified block diagram of a process control computer installation is presented in Figure 9.19. Considerable attention must be devoted

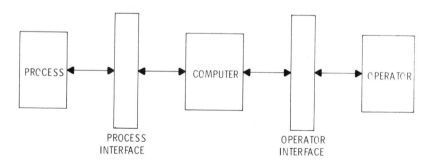

**Figure 9.19.** Computer process control.

to the two blocks labeled "interface" which lie between the computer and the process, and the computer and the operator. The process interface consists of such units as multiplexers, analog to digital (A/D) and digital to analog (D/A) convertors, and amplifiers which are provided to facilitate two-way communication between the computer and the process. The multiplexer enables the computer to sequentially examine signals from a large number of points throughout the plant and, after processing of these signals to determine the appropriate control action, to send signals to the appropriate final control elements. A/D and D/A converters are necessary since many of the inputs and required output signals are analog whereas the computer operates only on digital signals.

The interface between the computer and the operator is also of key importance, and considerable attention must be devoted to facilitating rapid and easy communication between the computer and the operator. Special keyboards are usually provided for the operating console so that the operator needs little knowledge of the internal details of the computer and can concentrate on control of the process. Protection against "button pushers" is provided by keyed locks which restrict access to certain portions of the computer program. Other communication devices provided include cathode ray tube displays of tabular and graphical data, electric typewriters and teletypes, audible and visible alarms, process flow diagrams and control loop diagram displays by computer controlled slide projectors, and trend plotting recorders.

A block diagram of a fully developed computer controlled plant is presented in Figure 9.20. However, computer control can be installed in stages

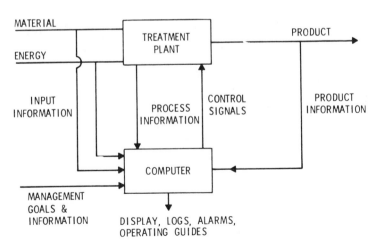

**Figure 9.20.** A computer controlled treatment plant (after Savas[27]).

and it is important to realize that a digital computer can be of substantial value for information handling even when a human operator must be used to completely or partially replace any of the information links shown between the treatment plant and the computer. Several possible configurations are illustrated in Figure 9.21. The solid lines indicate direct transmission without human delay or handling while the dashed lines indicate that human delay or handling is involved. The use of the off-line mode (Figures 9.21a,b), with a combination of manual and automatic data collection, has been discussed by Guarino and Radizul[33] for wastewater plant operations in Philadelphia. Guarino, *et al.*[34] have more recently reported on an in-line mode (Figure 9.21c) for which they have coined the word "near real time." In this mode, data is entered by the operator directly into a time-sharing computer service via a remote terminal located at the treatment plant. The data is processed by the computer to determine the appropriate control actions which are transmitted back to the operator via the remote terminal.

The term "on-line" refers to the direct transmission of data from the process to the computer as shown in Figures 9.21d and 9.21e. An objection frequently raised to the incorporation of control systems in wastewater treatment plants, is that reliable on-line sensors are not available for many important measurements. However, this problem is not as serious as might be expected since many changes in treatment processes occur relatively slowly and there is frequently adequate time for a man to participate in the data collection pathway by performing analyses on-site or in the plant laboratory with data input to the computer via remote terminals such as simple card readers or teletypes. An example might be the performance of wet chemical analyses using an automated wet chemical analyzer such as the Technicon Autoanalyzer. The important thing is that the information be entered into the control strategy instead of being simply filed for the record.

The configuration illustrated in Figure 9.21d is commonly known as data processing and monitoring and is usually the first step taken toward computer control. A more detailed diagram of this configuration is given in Figure 9.22. The computer collects data, processes it into a more meaningful form, and displays it to the operator. There may also be substantial manual entry of data. The operator then adjusts the set points of his automatic controllers. In this mode of operation, the operator is still in the feedback loop and the computer only assists him in running the plant. Typical functions which may be performed by the computer are:

1. Scan process sensing instruments, check for instrument malfunction, and convert raw data into engineering units.
2. Process data into a more meaningful form for the operator by such operations as smoothing, curve fitting, integration, differentiation, and statistical analysis.

------ TRANSMISSION WITH HUMAN INTERVENTION
———— DIRECT TRANSMISSION

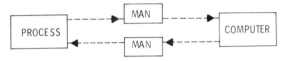

(a) OFF-LINE MODE, MANUAL DATA COLLECTION

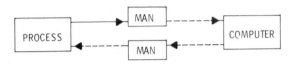

(b) OFF-LINE MODE, AUTOMATIC DATA COLLECTION

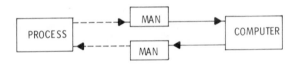

(c) IN-LINE MODE, DIRECT DATA INPUT & OUTPUT
    (NEAR REAL TIME)

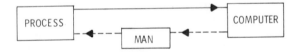

(d) ON-LINE, OPEN-LOOP MODE

(e) ON-LINE, CLOSED-LOOP MODE

**Figure 9.21.** Operational modes for computer-process systems.

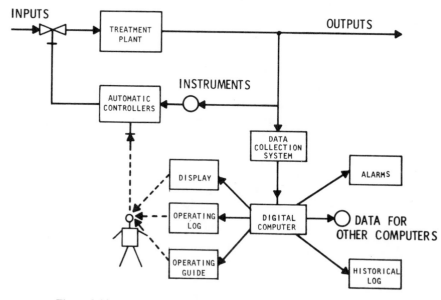

**Figure 9.22.** Data processing and monitoring by digital computer (after Lee, *et al.*[26]).

3. Monitor and report on the status of process equipment. For example, the computer can monitor the on-off condition of pumps, valves, motors, and compressors. It can check for overheating of bearings and excessive vibration of motors and compressors.

4. Compare process variables against high-low limits and sound alarms.

5. Determine normally unmeasured variables by computation from measured variables. An example would be calculation of oxygen utilization rate in the activated sludge process from a mass balance on oxygen.

6. Prepare operating logs and display information to the operator. This can be in tabular form, plotted trend charts of particular variables, or graphical displays on a cathode ray tube.

7. Store maintenance schedules and notify operator of needed maintenance.

8. Furnish the operator with an operating guide upon request. For example, procedures to be followed in case of a process upset.

9. Furnish data for other computers or provide reduced operating data to higher management. Examples might be the furnishing of data on costs to the business computer in an accounting section or monthly reports on operation to a state pollution control authority.

Control algorithms used in computer control systems range from simple on-off control to some form of optimal control. However, only the simpler algorithms are being used in wastewater treatment, since these processes are poorly defined from a dynamic point of view, and control strategies are still very empirical. This points out one of the major advantages of using a digital computer for control; the control strategy can be easily changed by reprogramming (software changes) the computer whereas conventional controllers normally require replacement or rewiring (hardware changes). Still another advantage is that the computing power and storage capabilities of the digital computer enable it to handle complex interactions between variables and implement the more advanced control strategies such as feedforward and optimal control.

Prospective users should recognize that although computer control offers many potential benefits, the adaptation of a process to computer control can be a difficult and time-consuming task. A substantial amount of technical manpower is required to implement computer process control and it has been reported that from 2-21 man-years have been required to complete computer control projects in industrial plants.[30] Bailey has discussed some of the problems which have been encountered in adapting industrial processes to computer control.[35] It is expected that considerable development work will be required to take full advantage of computer control for wastewater treatment plants since the processes involved are complex with poorly understood behavior.

## CONTROL STRATEGIES FOR THE STEP FEED ACTIVATED SLUDGE PROCESS

The activated sludge process may be classified as a continuous flow, enrichment culture of aerobic microorganisms with the predominating species being selected by the characteristics of the wastewater and the environmental conditions created through process design and operation. There are several versions of the process in current use with the primary differences consisting of variations in such factors as rate of waste application, type of reactor, contacting pattern between the biological and liquid phases, and amount and type of recycle. The conventional process consists of two units; an aerobic biological reactor followed by a solids-liquid separator. The inputs to the reactor consist of the wastewater, a concentrated suspension of organisms recycled from the separator, and an air supply to provide both oxygen and mixing. Reactor dimensions and the mixing provided are usually such that the hydraulic regime is somewhere between plug flow and complete mixing. The organics in the wastewater serve as substrate for the microorganisms and are converted in the reactor to more activated sludge, carbon dioxide, and water.

The step feed process is a multi-stage, multi-stream system in which the influent wastewater is admitted to the reactor at discrete points along its length. For reasons of economy of construction, most activated sludge reactors are arranged in a "folded" fashion, as shown in Figure 9.23, and therefore have the characteristics of a multi-stage system. For the reactor

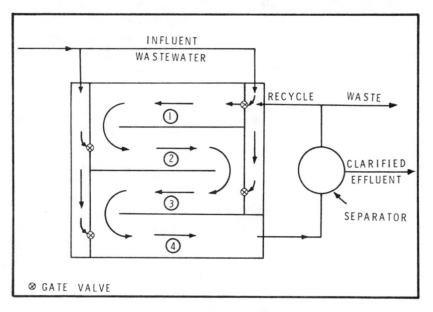

**Figure 9.23.** Step feed process.

shown, wastewater can be admitted at four discrete points thus making it also a multi-stream system. A more uniform rate of oxygen utilization can be obtained along the length of the reactor and a larger mass of sludge per unit of reactor volume can also be carried in the system since the concentrated recycled sludge is progressively diluted with wastewater as it passes through the reactor. For a given process loading intensity (lb BOD/lb MLSS in reactor-day), considerably smaller reactors can, therefore, be used for the step feed process than for the conventional process. Unfortunately, this emphasis on the capital cost advantage of the step feed process has detracted attention from its operational advantage, the ability to change the points at which wastewater is added along the length of the reactor, which might be of equal or more importance. The additional control action provided by step feed will be the subject of this section. For more detail than is given here, the reader is referred to the papers by Andrews[36] and Andrews and Lee.[37]

## Process Model

A dynamic mathematical model for the step feed activated sludge process can be developed using the same basic material balances, with only minor modifications, as have been presented for the CFSTR reactor for growth of microorganisms on an inhibitory substrate (Equations 6 and 7). For a substrate that is not inhibitory, the specific growth rate would be expressed as given in Equation 16 instead of as in Equation 5. Organism decay is also incorporated by using Equation 17 instead of Equation 4.

$$\mu = \frac{\hat{\mu}}{K_S + S} \tag{16}$$

$$\frac{dX}{dt} = (\mu - K_D) X \tag{17}$$

where

$K_D$ = specific organism decay rate.

These simple relationships do have deficiencies, especially when they are applied to the mixed-culture, mixed-substrate processes utilized in most wastewater treatment plants. Andrews should be consulted for a more detailed discussion of some of these deficiencies.[36,38] However, even simple relationships such as these can be used for process improvement and also serve as a valuable framework for modifications.

Organism and substrate balances for a single CFSTR with organism concentration and recycle (Figure 9.24) are given in Equations 18 and 19. It is assumed that there are no organisms in the influent wastewater and that the substrate concentration in the recycle flow is equal to that in the reactor effluent. Only the portion of the system shown in heavy lines is included in the material balances.

**General Material Balance**

Influent + Reaction = Effluent + Accumulation

**Organism Balance**

$$R F X_R + \mu X_1 V - K_D X_1 V = (F + R F) X_1 + V \frac{dX_1}{dt} \tag{18}$$

**Substrate Balance**

$$F S_0 + R F S_1 - \frac{\mu}{Y} X_1 V = (F + RF) X_1 + V \frac{dS_1}{dt} \tag{19}$$

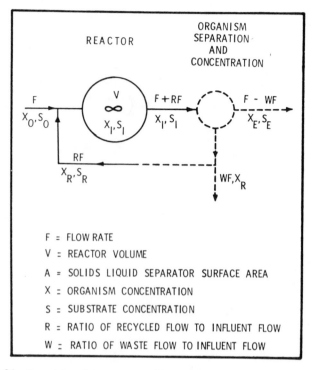

F = FLOW RATE
V = REACTOR VOLUME
A = SOLIDS LIQUID SEPARATOR SURFACE AREA
X = ORGANISM CONCENTRATION
S = SUBSTRATE CONCENTRATION
R = RATIO OF RECYCLED FLOW TO INFLUENT FLOW
W = RATIO OF WASTE FLOW TO INFLUENT FLOW

**Figure 9.24.** Complete mixing reactor with organism concentration and recycle.

where

R = ratio of recycled flow to influent flow and subscript denoting concentration in recycled flow.

The step feed process (Figure 9.23) can be approximated as four CFSTRs in series with each stage, except for minor modifications in the influent and effluent terms, being represented by Equations 18 and 19. The digital computer is ideally suited to handle calculations involving reactors in series, since the basic equations are almost identical and it can be easily instructed to consider the output from one reactor as one of the inputs to the following reactor. The MACRO function in CSMP is used for this purpose. The dynamic model for a single CFSTR is given in a MACRO function and the computer is then instructed by statements in the DYNAM section to let the output from one reactor as well as additional wastewater, be the inputs to the following reactor.

The model which has been presented is incomplete in that it does not consider the interactions between the biological reactor and the solids-liquid separator. As shown in Figure 9.24, not all of the organisms produced are recycled since a portion, $F X_E$, escapes in the overflow from the separator and another portion, $W X_R$, is wasted. The organisms contained in the overflow from the separator represent an additional burden of pollutants on the receiving waters and must be taken into account in determining the total BOD (soluble plus insoluble) in the process effluent. Pflanz, in a study of the performance of several activated sludge plants, has shown that for a sludge with fixed settling characteristics, the concentration of suspended solids in the clarified effluent is approximately proportional to the solids flux applied to the solids-liquid separator.[39] This empirical relationship may be expressed as

$$X_E = \frac{K_E (F + R F) X_1}{A} \qquad (20)$$

where:

$X_E$ = concentration of suspended solids in clarified effluent.

$K_E$ = proportionality coefficient

$A$ = surface area of solids-liquid separator

$\dfrac{(F + R F) X_1}{A}$ = solids flux for solids-liquid separator

The total BOD in the process effluent can be calculated by establishing the BOD equivalent of the suspended solids and adding this to the substrate BOD calculated from Equation 19.

Another factor not adequately considered in the model is the thickening function of the solids-liquid separator. The concentration of solids in the recycle flow is not a constant, as is frequently assumed, but is instead dependent upon, among other factors, the settling characteristics of the sludge, physical characteristics of the separator, temperature, recycle rate, and the solids flux to the separator. The concentration of solids in the recycle flow, in turn, influences the concentration of solids, which can be maintained in the reactor, the solids flux to the separator, and the sludge gas. The settling characteristics of the sludge will also be affected since this is a function of the sludge age. Dick and Javaheri have presented some excellent quantitative examples, which illustrate the importance of considering the interactions between the biological reactor and the separator in the design of the activated sludge process.[40]

## Process Control

In the conventional activated sludge process, the operator has a relatively limited choice of control actions, these being: (a) air flow rate, (b) sludge recycle flow rate, and (c) sludge wasting rate. Of these, air flow rate is primarily of importance in controlling process economics and does not appear to have much effect on process efficiency as long as the dissolved oxygen in the reactor remains above some minimum level. Exceptions to this may be encountered in the control of filamentous organism growth[41] and the use of high purity oxygen instead of air.[42] Common strategies for the control of air flow rate include the use of ratio control to regulate the air flow rate in proportion to the wastewater flow rate or conventional feedback control, in which the air flow rate is regulated to maintain a constant dissolved oxygen concentration in the reactor.

Ratio control to regulate the sludge recycle flow rate in proportion to the wastewater flow rate (Figure 9.17a) is also a common control strategy.[43] However, this is not always successful, especially when poorly settling sludges are encountered, since it does not take into account possible changes in the concentration of solids in the recycle flow. An increase in the recycle sludge flow rate can, within limits, increase the mass of sludge in the reactor. However, it also increases the solids loading to the solids-liquid separator as well as creating additional turbulence in the separator, and can sometimes result in an increased carryover of solids in the process effluent.

There is a net growth of sludge in the activated sludge process and sludge must therefore be intermittently or continuously wasted from the system. One control strategy is to waste that amount of sludge each day which will maintain a constant mass of sludge in the reactor; another is to waste sludge whenever the sludge blanket level in the solids-liquid separator exceeds a certain depth. Sludge wasting may also be used when poorly settling or bulking sludge is encountered in order to prevent the sludge blanket level from rising in the separator until sludge is discharged in the effluent. However, on a long-term basis this may be the wrong control action since if the bulking is due to an overload of organic materials, as is frequently the case, sludge wasting will decrease the amount of sludge in the reactor thus resulting in still further overloading with possible process failure.

Two or more of the above mentioned control actions may be combined and an example of this has been given by Brouzes, in which he controlled sludge wasting to maintain a constant specific growth rate of the sludge.[44] Using a special purpose analog computer, he regulated and measured the air flow rate required to maintain a constant dissolved oxygen concentration in the reactor. Assuming a constant oxygen transfer efficiency, the computer then calculated the oxygen uptake rate and related this, using empirical coefficients, to the sludge production rate. The computer then calculated and controlled the rate of sludge wasting to maintain a constant specific growth rate of the sludge. A safety override control was provided to actuate sludge wasting whenever the sludge blanket level exceeded a certain depth in the separator.

A multi-stage, multi-stream reactor system, such as the step feed activated sludge process, permits a fourth control action to be taken, this being the ability to regulate the points at which wastewater is added along the length of the reactor.

## Process Control by Step Feed

Steady-state solutions for three possible operational modes of the step feed process are shown in Figure 9.25. The first version (Figure 9.25a) may be considered equivalent to the conventional activated sludge process, since this process is frequently approximated as several complete mixing reactors in series with the entire wastewater flow going to the first stage. For these examples, it has been assumed that the organism concentration that can be maintained in the recycle flow is independent of the behavior of the solids-liquid separator. Although this is not correct, as has been previously discussed, the examples can still serve to illustrate several key points. $M_{XR}$ is the total mass of organisms contained in all four reactors in a series and it has been assumed that the mass of organisms stored in the separator, $M_{XC}$, is equal to 25% of that contained in the series of reactors for the conventional process (Figure 9.25a). $R_{XP}$ is the net rate of organism production and includes both organisms in the overflow from the separator and those drawn off as waste sludge. $\theta_X$ is the organism residence time or "age" and is calculated as $M_{XR}/R_{XP}$. SF is the solids flux to the separator. The highest quality effluent, as measured by soluble substrate only, is produced by the conventional process ($S_4 = 1$ mg/l) whereas the poorest quality effluent ($S_4 = 14$ mg/l) is produced by the process where a separate flow of substrate is supplied to all four reactors. In practice, the actual effluent qualities may be more nearly alike as has been indicated by Gould.[5] For all three operational modes, the organism concentration in the last reactor, and therefore the solids flux to the separator, are approximately equal. The major difference in the three systems is the total mass of organisms contained in the reactors and, therefore, the organism age which exerts a significant effect on the flocculation and settling characteristics of the sludge. Changes in the total mass of organisms also influences the net rate of organism production, which is important in designing processes for treatment of the waste sludge produced.

The ability to change the points at which wastewater is added along the length of a reactor may be one of the most effective types of control for

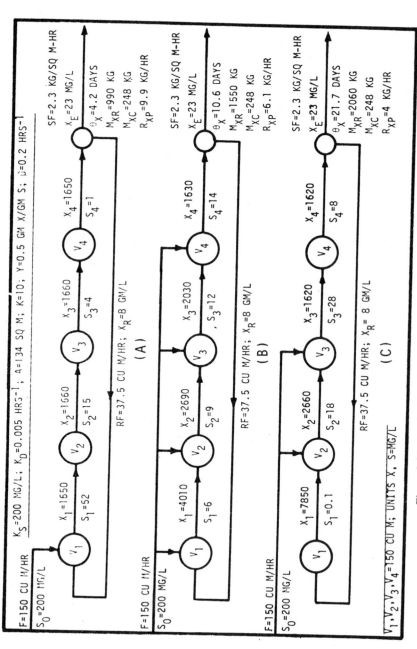

**Figure 9.25.** Example operational modes for the step feed process.

poorly settling or bulking sludge. Figure 9.26 shows the transient effects of suddenly shifting from the conventional process (Figure 9.25a) to the operational mode shown in Figure 9.25c. The organisms in the solids-liquid separator are rapidly transferred to the reactors and there is also a rapid decrease in the solids flux to the separator. Both of these responses would have the short-term effect (hours) of decreasing the mass of solids carried over in the effluent from the separator. A long-term effect (days) would also be obtained in the event that the poor settling was due to organic overloading, a larger mass of organisms would ultimately be carried in the reactors thus decreasing the overloading and improving the flocculation and settling characteristics of the sludge. This would require days to take effect since only a limited mass of solids is available for rapid transfer from the separator to the reactors; the additional organisms needed to reach steady-state for the new operational mode must be obtained through organism growth and this is slow because of the low substrate concentration in most wastewaters.

These predictions are qualitatively verified through the field studies reported by Torpey in his work on the step feed activated sludge process at the Bowery Bay plant in New York City.[45] This plant was normally operated with the influent wastewater flow divided equally between reactors two and three (Figure 9.25c). When poorly settling sludge was encountered, as evidenced by an increase in the Sludge Volume Index (SVI) and a rising sludge blanket level in the separator, the influent wastewater flow **was** shifted to reactors three and four. Following this change, there was a rapid decrease in the solids flux to the separator, a drop in the sludge blanket level, and an increase in the mass of sludge contained in the reactors. These observed changes qualitatively agree with those predicted by the model (Figure 9.26). Long-term effects (within four days) were a further increase in the mass of sludge contained in the reactors, improved settling characteristics of the sludge as evidenced by a decrease in the SVI, and a considerable reduction in the total BOD of the process effluent. After these long-term improvements were obtained, the influent flow was shifted back to reactors two and three, to be prepared for future process upsets.

The simulation results presented should be considered only qualitative in nature since the model on which they are based does not include several significant factors. Included among these are the dynamic behavior of the separator, the effects of time lags on the dynamic response of the microorganisms, and the relationship between sludge age and sludge setting characteristics. Research on these factors includes that of Bryant, *et al.* on a dynamic model of the separator,[46] Blackwell and Bunday on time lags for microorganisms,[47] and Bisogni and Lawrence on sludge settling characteristics.[48]

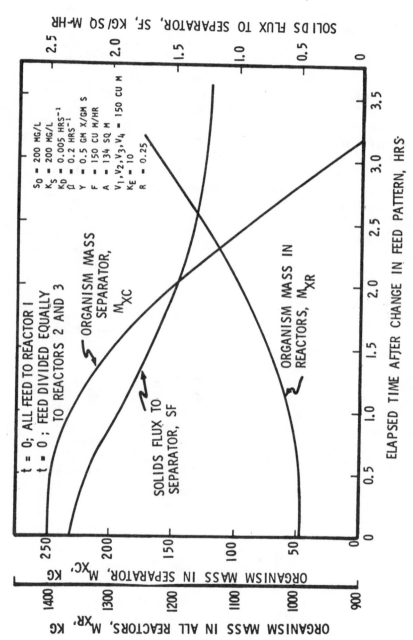

Figure 9.26. Process control by step feed.

The author and co-workers have developed an improved dynamic model of the step feed activated sludge process, which incorporates the effects of the above-mentioned factors, and are using this model to simulate control strategies based on automatically controlling the points of addition of wastewater according to daily variations in flow rate.

Even though the simulations presented are only qualitative, when considered along with Torpey's results, they clearly indicated that changing the points at which wastewater is added along the length of the reactor is an effective control strategy for the activated sludge process. Many activated sludge processes are now designed to permit operation in the step feed mode. However, the operational advantage of the step feed process is not well recognized and changes in the wastewater distribution pattern are seldom used for process control.

## PROCESS STABILITY AND CONTROL STRATEGIES FOR THE ANAEROBIC DIGESTER

The anaerobic digestion process is widely used for the treatment of organic solids found in municipal wastewaters and has several significant advantages over other processes used for this purpose. Among these are a low production of waste sludge, low power requirements, and formation of a useful product, methane gas. The digested sludge is a good soil conditioner, has some fertilizer value, and is being increasingly used for the reclamation of land. Unfortunately, even with all of these advantages the process has in general not enjoyed a good reputation because of its poor record with respect to process stability as indicated through the years by the many reports of "sour" or failing digesters. Many of the problems with the process appear to lie in the area of process operation as evidenced by its successful performance in larger cities where skilled operation is more prevalent. There is a definite need for the development of quantitative measures of stability and investigation of those design and operational factors which can be used to improve stability. In addition, current operating practice consists only of empirical rules and there is a need for the development of quantitative control strategies.

### Dynamic Model

The dynamic model for the anaerobic digestion process is based upon the model which has been presented for the biological CFSTR with an inhibitory substrate input. However, through necessity, the model for the digester must be considerably more complex to predict the dynamic response of the five variables most commonly used for process operation. These variables are: (a) volatile acid concentration, (b) pH, (c) bicarbonate alkalinity, (d) gas flow rate, and (e) gas composition. A summary of the model is presented in Figure 9.27.

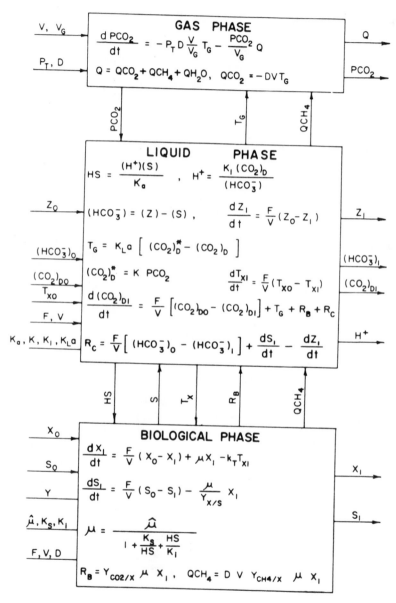

**Figure 9.27.** Summary of mathematical model and information flow.

The substrate and organism balances on which the model is based are the same as those given in Equations 6 and 7 except that the substrate in the inhibition function (Equation 5) is taken as the unionized volatile acid, HS, which is a function of both the total volatile acid concentration and the pH. This is an important modification of the model since it assists in resolving the conflict that has existed in the literature as to whether inhibition is caused by high volatile acid concentration or low pH. Since the concentration of unionized acid is a function of pH *and* total volatile acid concentration, both are important. Another important modification is incorporation of an expression for organism death by a conservative toxic agent, $T_{XI}$. Inclusion of this term allows the model to predict failure due to toxic materials as well as failure by hydraulic and organic overloading (Figures 9.12 and 9.13).

The remainder of the model consists of material balances, in either the liquid or gas phase, on such materials as carbon dioxide, bicarbonate, and cations (Z). Relationships used to express the interactions between variables include yield factors ($Y_{X/S}$, $Y_{CO2/X}$, $Y_{CH4/X}$), ionic equilibrium expressions for the volatile acid and bicarbonate systems, a charge balance on the ionic species in solution, Henry's law, and a mass transfer expression for the transfer of carbon dioxide across the gas-liquid interface. The qualitative validity of the model has been established through experimental results and by computer simulations which predict results commonly observed in the field. For a more detailed development of the model, the reader is referred to the papers of Andrews,[17,18] Andrews and Graef,[19] and Graef and Andrews.[22]

## Process Stability

As previously mentioned, the anaerobic digester has a poor reputation with respect to process stability. The model that has been presented can be used to predict instability and ultimate process failure due to hydraulic, organic, and toxic material overloading. More importantly, it can also be used to indicate those design and operational factors which can be used to improve process stability. Included among these factors are increases in residence time, alkalinity, influent substrate concentration, digested sludge recycle, loading frequency, and various control strategies.

Quantification of the effects of design and operational factors on digester stability is a tedious and time-consuming task even with CSMP. Fortunately, these problems can be overcome by using a hybrid computer for the simulations and the results reported herein were obtained by using an EAI 680 analog computer interfaced with a PDP 15 digital computer. Graef should be consulted for the details of the programming involved.[49] The high speed of the analog computer permitted days of digester operation to be compressed into seconds and the "hands on" aspects of the hybrid computer permitted the rapid evaluation of the effect of a wide range of the factors of interest.

The stability analysis procedure utilized involved simulating a change in a factor of interest, such as digester residence time, by changing a potentiometer (pot) setting on the analog computer. A step forcing in influent substrate concentration was then simulated with the results being rapidly displayed on a cathode ray tube (CRT). If this forcing was insufficient to cause process failure, the influent substrate concentration was increased by means of a second pot and the step-forcing again simulated. This trial-and-error procedure was repeated until the step increase in substrate concentration, which just caused process failure, was determined. The pot representing digester residence time was then changed to a new value and the trial-and-error procedure repeated at this new residence time. By plotting the locus of points of critical substrate loading rates vs. residence time, or other factor of interest, it was possible to obtain a semiquantitative measure of digester stability. This is illustrated in Figure 9.28, which shows the increased stability that is obtained by increasing either digester residence time or alkalinity. As an example, increasing the residence time from 10 to 15 days at an alkalinity of 2000 mg/l, permits an increase in the maximum step loading from 56-61 mmol/l/day.

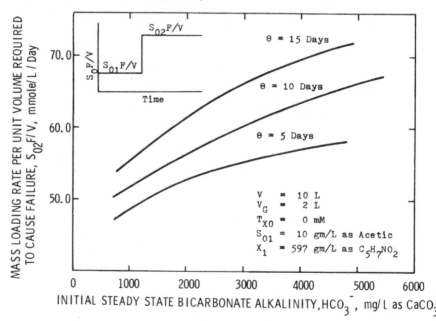

**Figure 9.28.** Process stability as function of $HCO_3$ and $\theta$—organic overload.

For an evaluation of other factors influencing digester stability, the reader is referred to Graef and Andrews.[22] Their results suggest that sludge thickening is of special value for improving digester stability since it results in increased residence times, higher bicarbonate alkalinity concentrations, and increased influent substrate concentrations. Simulation studies indicated that increases in any of these three factors would result in improved process stability.

**Process Control**

In selecting a control strategy for the anaerobic digester, a wide variety of output variables are available for initiation of the control action. The variables most commonly used are volatile acid concentration, pH, bicarbonate alkalinity, gas-flow rate, and gas composition. Combinations of these variables are also used and included among these are the volatile acid/alkalinity ratio, unionized acid concentration, and rate of methane production. There has been considerable speculation as to which variable, or combination of variables, is the best indicator of impending digester difficulty. Graef and Andrews[22] have shown that this is dependent upon the type of overloading to which the digester has been subjected and their work should be consulted for a more detailed evaluation of possible process condition indicators. In selecting a variable for initiation of control action, the available analytical techniques must also be considered and, on this basis, preference would be given to gas phase measurements over liquid or solid phase measurements.

After selecting a variable for measurement, an appropriate control action must be initiated. Control actions available in anaerobic digestion include a temporary halt in organic loading, a decrease in the rate of organic loading, addition of a base such as lime or soda ash, dilution of the digester contents, or the addition of well-digesting sludge from another digester. A control action proposed by Graef and Andrews includes the scrubbing of carbon dioxide from the digester gas with subsequent recycle of the gas.[22] This provides process control by adjusting the digester pH upward through removal of a weak acid, carbon dioxide. The most effective type of control action is dependent upon the type of overloading to which the digester has been subjected.

A simulation of digester control using pH as the feedback signal and base addition as the control action is presented in Figure 9.29. Proportional control is used with the rate of base addition being proportional to the deviation of pH from a set point of 6.95. Figure 9.29 indicates the response of four variables, volatile acid concentration, rate of methane production, percentage of carbon dioxide in the gas phase, and pH, to an organic

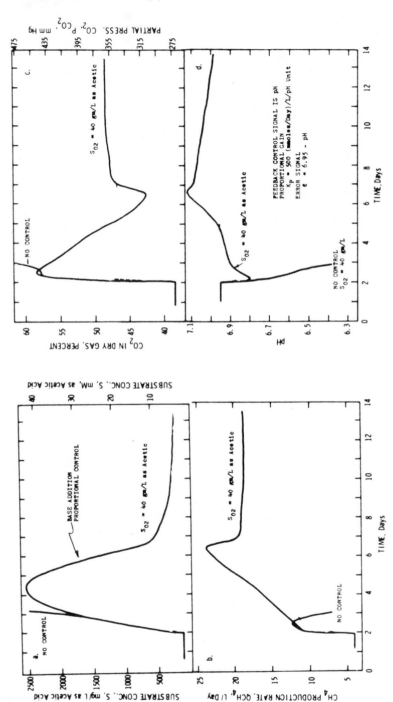

Figure 9.29.  Controlled response of $S_1$, $QCH_4$, % $CO_2$ and pH for proportional base addition control of organic overload.

overloading which in the absence of control would cause process failure. With pH control, failure is avoided and the process recovers in a relatively short period of time. However, it should be noted that other simulations indicated that this control strategy offered little protection against toxic and hydraulic overloading.

Recirculation of digesting sludge from a second-stage digester was first advocated as a control action by Buswell in the 1930s[50] and, in the author's opinion, is one of the strongest arguments for the two-stage digestion process. Figure 9.30 presents a simulation of digester control using the rate of methane production as the feedback signal and recirculation of digesting sludge as the control action. An on-off controller mode is used for the sludge recirculation pump and the simulations are for the pulse addition of a conservative toxic substance which, in the absence of control, would cause process failure. This control strategy appeared to be one of the best of those investigated with respect to prevention of failure by toxic overloading.

The examples presented have illustrated the use of control strategies for the prevention of gross process failure. However, control can also be used to increase the maximum organic and hydraulic loading rates which a digester can handle or to improve the quality of the digester effluent. For example, pilot plant studies by Torpey have demonstrated that stable operation can be attained at residence times as low as three days.[51] However, current digester design criteria usually call for residence time in the range of 15-25 days. Improved operational strategies could result in substantial reductions in the required volumes of new digesters and permit existing digesters to meet increased demands without requiring plant expansion.

## SUMMARY

The performance of wastewater treatment plants could be substantially improved by more consideration of dynamic behavior during both plant design and operation. Modern control systems for modification of dynamic behavior can be incorporated during the design phase and manual operation can be improved through the use of written operational guides based on transient response analysis.

Several powerful tools are available for the study of dynamic behavior and included among these are dynamic mathematical models, computer simulation, transient response analysis, and techniques for evaluation of process stability. Dynamic models can be developed by applying material and energy balances to a reactor using fundamental transport, stoichiometric, thermochemical, and kinetic relationships. The equations which comprise the model are usually best solved by computer simulation using one of the

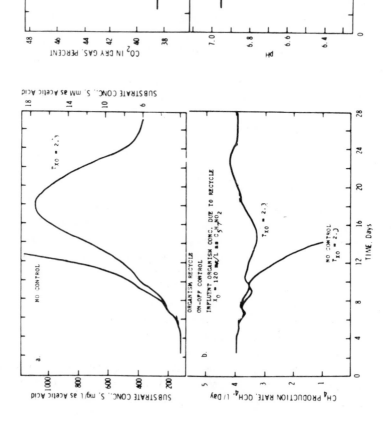

**Figure 9.30.** Controlled response of $S_1$, $QCH_4$, $\% CO_2$ and pH for on-off organism recycle control of toxic overload.

continuous system simulation languages such as CSMP/360. Solution of the model, or prediction of the time varying behavior of the outputs of the model, is known as transient response analysis. Wastewater treatment processes can exhibit instability and dynamic modeling and computer simulation can be used to evaluate the effects of various inputs and process characteristics with respect to stability. Two simple reactors, a CFSTR with an inert tracer input and a CFSTR for the growth of microorganisms on an inhibitory substrate, are used to illustrate the basic principles of dynamic modeling, simulation programming, transient response analysis, and process stability.

Control systems are primarily involved with the handling of information. This can be done manually or by automatic control systems. However, in either case, some of the same basic questions must be answered, including (a) What information should be collected? (b) How should the information be transmitted? (c) How should the information be processed? and (d) What control actions should be taken? Once these questions are answered, it is usually found that control systems frequently represent an economical means for improving plant performance.

A wide variety of control algorithms are available and included among these are on-off, PID, ratio, cascade, and feedforward control. An example is presented for the application of on-off control to a biological CFSTR. Many of these algorithms can be implemented using conventional control hardware; however, the more sophisticated controller modes are best implemented by the use of a digital computer for process control. Computer control systems also offer many other advantages and are being used increasingly in wastewater treatment plant operations.

Examples of dynamic models and control strategies for two wastewater treatment processes, the step feed activated sludge process and the anaerobic digester, are presented. In addition, dynamic modeling and computer simulation are used to indicate those design and operational factors for the anaerobic digester which can be changed to improve process stability. Included among these are increases in residence time, alkalinity, influent substrate concentration, digested sludge recycle, and loading frequency.

The step feed activated sludge process permits the operator to change the points at which wastewater is added along the length of the reactor. Both computer simulation and the field work of Torpey clearly indicate that this is an effective control strategy for poorly settling sludge and may be of even wider applicability. However, this operational advantage of the step feed process is not well recognized in either plant design or operations and changes in the wastewater distribution pattern are seldom used for process control.

The anaerobic digester is an example of a process where a wide variety of output variables and control actions are available for development of a control strategy. Two example strategies are presented, these being base addition as the control action with pH as the feedback signal, and recirculation of digested sludge with the rate of methane production as the feedback signal. The strategy to be selected is dependent upon the type of forcing to which the digester is subjected. Improved operational strategies can not only prevent process failure, they can also result in substantial reductions in the required volume of the digester.

## REFERENCES

1. Thomann, R. V. "Variability of Waste Treatment Plant Performance," *J. San. Eng. Div., ASCE* **96** (SA3), 819 (1970).
2. Environmental Protection Agency. *A Report to Congress on Water Pollution Control Manpower Development and Training Activities.* (Washington, D.C.: U.S. Government Printing Office, 1972).
3. Andrews, J. F. "Design-Operation Interactions for Wastewater Treatment Plants," *Water Res.* **6**, 319 (1972).
4. West, A. W. "Case Histories: Improved Activated Sludge Plant Performance by Operations Control," *Proceedings 8th Annual Environmental and Water Resources Engineering Conference*, Vanderbilt University, Nashville, Tennessee (1969).
5. Gould, R. H. "Sewage Aeration Practice in New York City," *Proc. ASCE* **79**, 307 (1953).
6. Torpey, W. N. "High-Rate Digestion of Concentrated Primary and Activated Sludge," *Sewage Ind. Wastes* **26**, 479 (1954).
7. Michel, R. L., A. L. Pelmoter, and R. C. Palange. "Operation and Maintenance of Municipal Waste Treatment Plants," *J. Water Poll. Control Fed.* **41**, 335 (1969).
8. Hawkes, H. A. *The Ecology of Waste Water Treatment.* (London: Pergammon Press, 1963).
9. Chestnut, H. *Systems Engineering Tools.* (New York: John Wiley and Sons, 1965).
10. Denbigh, K. G. and J. C. R. Turner. *Chemical Reactor Theory*, 2nd ed. (London: Cambridge University Press, 1971).
11. Himmelblau, D. M. and K. B. Bischoff. *Process Analysis and Simulation: Deterministic Systems.* (New York: John Wiley and Sons, 1968).
12. Franks, R. G. E. *Mathematical Modeling in Chemical Engineering.* (New York: John Wiley and Sons, 1967).
13. Perlmutter, D.D. *Introduction to Chemical Process Control.* (New York: John Wiley and Sons, 1965).
14. Shilling, G. D. *Process Dynamics and Control.* (New York: Holt, Rinehart and Winston, 1963).
15. McLeon, J. *Simulation.* (New York: McGraw-Hill Book Co., 1968).
16. Cooper, G. R. and C. D. McGillem. *Methods of Signal and System Analysis.* (New York: Holt, Rinehart and Winston, 1967).

17. Andrews, J. F. "A Mathematical Model for the Continuous Culture of Microorganisms Utilizing Inhibitory Substrates," *Biotech. Bioeng.* **10**, 707 (1968).

18. Andrews, J. F. "Dynamic Model of the Anaerobic Digestion Process," *J. San. Eng. Div., ASCE* **95**(SA1), 95 (1969).

19. Andrews, J. F. and S. P. Graef. "Dynamic Modeling and Simulation of the Anaerobic Digestion Process," *Anaerobic Biological Treatment Processes*, Advances in Chemistry Series No. 105 (Washington, D.C.: American Chemical Society, 1971).

20. International Business Machines Corp. *System 360 Continuous System Modeling Program (360A-CX-16X) User's Manual*. (New York: International Business Machines Corp., 1968).

21. Bryant, J. O. "Transient and Frequency Response Analysis of Water and Wastewater Treatment Systems," *Proceedings 4th Annual Workshop*, American Association of Professors in Sanitary Engineering, Clemson University, Clemson, South Carolina (1969).

22. Graef, S. P. and J. F. Andrews. "Process Stability and Control Strategies for the Anaerobic Digester," Presented at the 45th Annual Conference, Water Pollution Control Federation, Atlanta, Georgia (October 10, 1972).

23. Coughanowr, D. R. and L. B. Koppel. *Process Systems Analysis and Control*. (New York: McGraw-Hill Book Co., 1965).

24. Harrott, P. *Process Control*. (New York: McGraw-Hill Book Co., 1964).

25. Hougen, J. O. *Measurements and Control Applications for Practicing Engineers*. (Boston: Cahners Books, 1972).

26. Tucker, G. K. and D. M. Wills. *A Simplified Technique of Control System Engineering*. (Fort Washington, Pa.: Honeywell, Inc., 1962).

27. Floyd, S. G. and G. D. Anderson. *Industrial Process Control*. (Marshalltown, Iowa: Fisher Controls Co., 1971).

28. Lee, T. H., G. E. Adams, and W. M. Gaines. *Computer Process Control: Modeling and Optimization*. (New York: John Wiley and Sons, 1968).

29. Savas, E. S. *Computer Control of Industrial Processes*. (New York: McGraw-Hill Book Co., 1965).

30. U. S. Department of Labor. *Outlook for Computer Process Control*, Bulletin 1658 (Washington, D.C.: U.S. Government Printing Office, 1970).

31. American Public Works Association. *Feasibility of Computer Control of Wastewater Treatment*. (Chicago: American Public Works Association, 1970).

32. Lowe, E. I. and A. E. Hidden. *Computer Control in Process Industries*. (London: Peter Peregrinus, Ltd., 1971).

33. Guarino, C. F. and J. V. Radizul. "Data Processing in Philadelphia," *J. Water Poll. Control Fed.* **40**, 1385 (1968).

34. Guarino, C. F., H. D. Gilman, M. D. Nelson, and C. M. Koch. "Computer Control of Wastewater Treatment," *J. Water Poll. Control Fed.* **44**, 1718 (1972).

35. Bailey, S. J. "On-Line Computer Users Polled," *Control Engineering*, **16**, 86 (January, 1969).

36. Andrews, J. F. "Kinetics and Mathematical Modeling," *Ecological Aspects of Wastewater Treatment*, H. A. Hawkes and C. R. Curds, Eds. (London: Academic Press, 1974).

37. Andrews, J. F. and C. R. Lee. "Dynamics and Control of a Multi-Stage Biological Process," *Proceedings 4th International Fermentation Symposium*, Kyoto, Japan, March, 1972 (1974).

38. Andrews, J. F. "Kinetic Models of Biological Waste Treatment Processes," *Biological Waste Treatment*, R. P. Canale, Ed. (New York: John Wiley and Sons, 1971).

39. Pflanz, P. "Performance of (Activated Sludge) Secondary Sedimentation Basins," *Advances in Water Pollution Research*, S. H. Jenkins, Ed. (London: Pergamon Press, 1969).

40. Dick, R. I. and A. R. Javaheri. "Discussion of Unified Basis for Biological Treatment Design and Operation," *J. San. Eng. Div., ASCE* 97(SA2), 234 (1971).

41. Ryder, R. A. "Dissolved Oxygen Control in Activated Sludge," *Proceedings 24th Industrial Waste Conference,* Purdue University, Lafayette, Indiana (1969).

42. Albertsson, J. G., J. R. McWhirter, E. K. Robinson, and N. P. Vahldieck. *Investigation of the Use of High Purity Oxygen Aeration in the Conventional Activated Sludge Process.* (Washington, D.C. Federal Water Quality Administration, 1970).

43. Babcock, R. H. *Instrumentation and Control in Water Supply and Wastewater Disposal.* (Chicago: Reuben H. Donnelley Corp., 1968).

44. Brouzes, P. "Automated Activated Sludge Plants with Respiratory Metabolism Control," *Advances in Water Pollution Research,* S. H. Jenkins, Ed. (London: Pergamon Press, 1969).

45. Torpey, W. N. "Practical Results of Step Aeration," *Sewage Works J.* 20, 781 (1948).

46. Bryant, J. O., L. C. Wilcox, and J. F. Andrews. "Continuous Time Simulation of Wastewater Treatment Plants," Presented at the 69th National Amer. Inst. Chem. Eng. Meeting, Cincinnati, Ohio (May, 1971).

47. Blackwell, L. G. and H. R. Bungay. "Dynamics of Biosorption and Substrate Utilization in an Activated Sludge Contact Stabilization Process," Presented at the 69th National Amer. Inst. Chem. Eng. Meeting, Cincinnati, Ohio (May, 1971).

48. Bisogni, J. J. and A. W. Lawrence. "Relationships Between Biological Solids Retention Time and Settling Characteristics of Activated Sludge," *Water Res.* 5, 753 (1971).

49. Graef, S. P. "Dynamics and Control Strategies for the Anaerobic Digester," Ph.D. Thesis, Clemson University, Clemson, South Carolina (1972).

50. Buswell, A. M. and W. D. Hatfield. *Anaerobic Fermentation.* Bulletin No. 32, Illinois State Water Survey (1936).

51. Torpey, W. N. "Loading to Failure of a Pilot High Rate Digester," *Sewage Ind. Wastes* 27, 121 (1955).

# CHAPTER 10

## MODELING CONCEPTS FOR SYSTEMS OF TREATMENT PROCESSES: DESIGN AND OPERATION

Martin P. Wanielista, Christian S. Bauer

Department of Civil Engineering and
Environmental Sciences
Florida Technological University
Orlando, Florida

### INTRODUCTION

Analysis and design of water treatment facilities has more often than not been more of an art than an organized scientific procedure. Factors ranging from the somewhat poorly understood physical-chemical-biological dynamics of large stochastic/hydraulic/nonlinear systems such as those represented in modern treatment plants to the nonavailability of accurate and reliable instrumentation for such systems have contributed to this situation. Fortunately, however, with the advent of the modern high-speed digital computer, and the development of user-oriented software simulation languages, the prospects for significant improvements in the operations and economics of operation of treatment systems are quite good. This chapter will treat the use of one of these languages, the IBM System/360 Continuous System Modeling Program, or CSMP, for the construction of dynamic time-varying models of conventional water treatment plant components. Responses of conventional water treatment facilities to perturbations in raw water characteristics and operational changes will be presented. The operational changes and design alterations will form sets of strategies to be investigated for more efficient and economical operation of facilities.

379

The primary educational discipline providing the understanding of this approach to problem solving can be defined as systems engineering. With this discipline, the design engineer must approach the solution considering each plant configuration as a functioning entity rather than the design of each unit by itself. It is hoped that this chapter will encourage other workers to begin experimentation with this approach and that this type of design approach be included in the educational program of advanced environmental engineering curriculum.

## BASIC MODELING LANGUAGE ELEMENTS

The CSMP/360 language development has been described.[1] IBM manuals presented an introduction to the use of the language,[2] and a detailed user's guide[3] containing all of the program options and structural details. The use of the CSMP software in various application simulations has been described.[4-6] This chapter will present only a few of the language capabilities. (Interested readers are directed to the cited references for additional information.)

By design, CSMP/360 is a "user-oriented" language. In practical terms, this means that the analyst need only to describe the system to be simulated in the form of FORTRAN-type statements or in terms of CSMP/360 building block elements, called FUNCTIONS or MACROS, or some combination of the two. This model description is punched into computer cards or typed into a remote terminal and the program is run. The internal logic of CSMP checks for syntactical errors in the user's model, and assuming no errors have been encountered, sets up a machine code to exercise a time varying simulation run and print or plot (or both) values of user-specified variables. The mechanics of the internal operations of equation sorting, derivative computation, and numerical integration step scheduling are performed automatically without user intervention. In a large model, with many variables and complex interacting equation structures, this feature alone can materially reduce the trials required to produce a running model.

### Integration

One of the most frequently used CSMP/360 language elements is the built-in function to perform integration with respect to time. This element has the form in a program as follows:

$$VOUT = INTGRL\ (IC,\ RATE)$$

where

VOUT     = name of variable representing output value of integration
INTGRL  = name of integrator function
IC          = initial condition argument (output of integrator at time = 0)
RATE     = input variable value

Mathematically, the effect of the block on the variable values defined as above is:

$$VOUT = \int_0^t (RATE)dt + IC$$

with "new" VOUT values becoming available every $\Delta t$ time units. $\Delta t$ is a user-selected time increment for all system integrators to use.

The equivalence between the CSMP "INTGRL" function and the standard electronic analog computer integrator, again with the variables as defined earlier, is

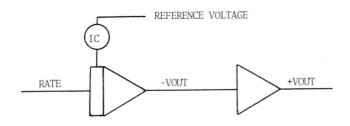

The extra summer is included in the above diagram to denote the fact that CSMP integrators do not generate the negative of the integral of the inputs, but rather the true integrated value. This is a consequence of the use of numerical techniques to perform the integration, and can be a source of confusion when attempting to structure a CSMP program from an existing electronic analog computer program circuit diagram.

One other point should be made with respect to CSMP/Electronic Analog Computer correspondences. The CSMP integrator does not require scaling to keep input and output values within specified limits. Again, since all processing values are determined numerically, the maximum values permissible are a function of the word size and data storage capabilities of the host computer. In the case of the System 360/370 computers, this range is approximately $-10^{75}$ to 0 to $10^{75}$. The need for scaling typically requires additional time and effort in an analog model study; CSMP offers a potential time savings here in the model development phase.

**System Data Representation**

Much of the operational data the environmental engineer is concerned with is in the form of graphs of flow rate vs. time, plant efficiencies vs. concentration levels, and other similar functional relationships. This type of data can be incorporated easily into a CSMP model through several possible approaches. If a functional relationship between several quantities may be expressed as an algebraic formula, this expression may be translated into the corresponding FORTRAN code and directly inserted into the program. For example, suppose one has the mathematical form

$$X = e^{2\beta}/K_1 + 2Z(t)$$

This can be expressed in CSMP with a FORTRAN statement of the form

ALPHA = (EXP (2*BETA)/K1) + (2*Z of t)

It should be noted that all FORTRAN variables used in CSMP are considered to be floating point, including those whose names begin with I, J, K, L, M, and N, unless explicitly declared to be integer-valued. This is a source of some confusion to experienced FORTRAN programmers just starting out with CSMP.

An approach for treating variable relationships not following a standard mathematical form is available in CSMP through the use of the "FUNCTION," "AFGEN," and "NLFGEN" features of the system program. The use of these operators can best be described through presentation of a simple example, one incidentally which will provide the input data for the water treatment plant model described later in this chapter. Consider the data of Figure 10.1. This graph represents the observed flow rates of raw water (influent) at a typical treatment plant over 24 hours. It was required to use this data to test the effectiveness of a CSMP model of the treatment plant operations, and therefore, the curve had to be translated into a machine-recognizable form. The CSMP "FUNCTION" statement allows the user to specify a series of pairs of (X,Y) data points for numerical computer storage. The X values, or independent values, must be in ascending numerical sequence. It is then possible to address the discretized curve with any given value of the independent variable, and obtain the corresponding dependent variable value. The functions "AFGEN" and "NLFGEN" provide linear point interpolation and quadratic Lagrange interpolation, respectively for these generations. Thus, in the case of the flow data curve, the values of flow rates at hourly intervals were "picked-off" the curve and used in a FUNCTION statement of the form

FUNCTION FLOW = (0.0, 1234.), (3600., 4567.), ...

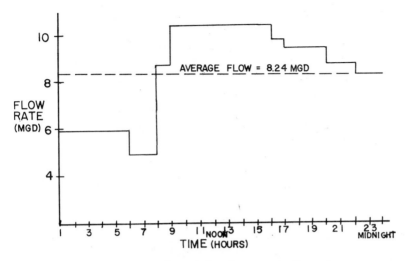

**Figure 10.1.** Daily variations in raw water flow rate.

where the entries on the right-hand side of the equality represented pairs of (TIME, FLOW RATE) values. Time was considered in seconds to permit flexible manipulations with the model. In the body of the model, "TIME" is a variable automatically updated and computed internally by the CSMP logic through a simple relationship of the form:

$$\text{TIME} = \text{TIME} + \Delta t$$

executed once each time step. Thus, when a new value of the system flow rate input variable was needed, a simple CSMP statement of the form:

$$Q = \text{AFGEN (FLOW, TIME)}$$

was all that was required. The arguments of the AFGEN function are respectively, the function to be treated, and the value of the independent variable to be used in the interpolation process. The result of the interpolation procedure is returned in the AFGEN function, and becomes available for use in arithmetic or assignment statements such as the one above.

**MACRO Capabilities**

CSMP is not unlike many other user-oriented languages in its provisions for MACRO generation and use. A MACRO is a set of statements and/or other CSMP functional elements designed to perform some computational task. The user may have need for this particular task at several places in a simulation model. Rather than copying the same code several times, the user may elect to define a building block element, or MACRO, once at the beginning of the program and assign a name to it. Then, whenever the code is needed in the model, the user need only specify the name of the appropriate MACRO along with the desired inputs and parameters, and the CSMP processor will automatically insert the appropriate set of statements into the program. This option can save considerable programming and card punching time in large simulation efforts.

To illustrate the use of the CSMP MACRO capability, a simple example will be presented. The authors had the need for a delay element in their work to represent the time lags generated by various system components in a water treatment plant. This general subject area has been treated in a wide variety of books and papers on process dynamics, but a basic presentation of the concepts involved is described by Forrester[7] and a CSMP implementation of the third-order DYNAMO DELAY MACRO has also been described.[8] The response of this element is exemplified in Figure 10.2.

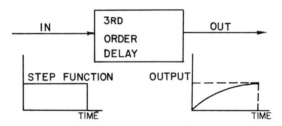

**Figure 10.2.** REALPL delay function.

Some of the work of Bryant used system response characteristics described by fifth-order linear system response characteristics,[9] so a MACRO of the following form was implemented to provide a fifth-order delay function:

```
MACRO    OUT = DELAYS (IC, DELAY, IN)

         T1  = REALPL (IC, DELAY, IN)

         T2  = REALPL (IC, DELAY, T1)

         T3  = REALPL (IC, DELAY, T2)

         T4  = REALPL (IC, DELAY, T3)

         OUT = REALPL (IC, DELAY, T4)

ENDMAC
```

The "REALPL" function is a CSMP component which simulates the first-order transfer function $(1/(KS + 1)$; five of these blocks in a sequence provide the required signal processing.

When the main segment of the simulation program is treating one of the hardware components in the actual system with delay inducing properties, it is possible to invoke the MACRO simply by naming it and supplying the required arguments, *e.g.,*

$$FLOW\ 2 = DELAYS\ (0.0,\ VF,\ COLIFF)$$

**Output Options**

As mentioned earlier, CSMP provides a capability for the printer-plotting of simulation variable values, as well as for the standard numerical value printing typically required in a simulation study. To print variable values from a simulation run, the user need only insert a program statement of the form

$$PRINT\ X,\ Y,\ Z.$$

Up to 49 variables may be printed for each simulation run. The printing interval is controlled by the user and can be set to a multiple of the specified integration interval to gain the dynamic accuracy of small step sizes without the voluminous data output resulting from the printing of all variable values at each integration step.

The value of the system "TIME" variable is automatically printed when any variable value is plotted. Plotting of selected simulation variables is done with a statement of the form

$$PRTPLOT\ X$$

Up to 10 variables may be plotted for each simulation run. Each variable requires a separate "PRTPLOT" card. Automatic scaling is performed on the variable values to assure maximum resolution on the output listing, although the user is free to supply his own scale ranges if desired. The numerical values of the plotted points, as well as the corresponding values of the system "TIME" variable, are automatically printed with the graph, and the user is free to specify up to three additional variables to be printed on each plot for cross-referencing purposes. It should be noted, however, that only one variable at a time is actually plotted, and that this plot must have "TIME" as the independent variable. To allow for the provision of user-written programs to provide multivariable plots, variable cross-plots, or the use of external plotting hardware, such as X-Y plotters, drum plotters, or CRT displays, the CSMP language has a statement of the form

PREPARE X, Y, Z, . . .

This statement will allow up to 49 variables in addition go the system "TIME" variable, supplied automatically, to be written on an external tape or disk storage unit. The data formats for this process are described in an IBM system manual.[10] This option allows a considerable degree of flexibility to the user in past-run data analysis and report generation.

There are many other statements and functions available in CSMP. The ones presented above were done to elucidate the computer program of Appendix A and are the ones most frequently used in simulation.

## AN ILLUSTRATIVE PROBLEM

Using the CSMP language elements, a representative model of conventional water treatment units is developed. The conventional coagulation and rapid gravity filtration units were chosen because of their general use in treating water. This system configuration lends itself to an adequate explanation for the use of a simulation language for design and operation. Chlorination facilities were not added at this time because of the tutorial nature of the model.

The approach used to define a mathematical model of a series of water treatment units was basically empirical.[11] The input data necessary for this type of model construction was available from the literature. Every effort was made to obtain realistic data. As a starting condition for size of unit selection, design of the water treatment units was done following recommended design criteria,[12] such as a filtration rate of 2 gpm/sq ft, and a chemical sedimentation time of 4 hr. These sizes of units provided the basis for evaluating the performance of the two units when raw water quality inputs were variable with time. These random variations are the first factors which make the simulation presented in this chapter different from the conventional method of design.

Conventional methods of design only consider the variability of raw water inputs to determine a maximum level of impurities that must be removed. The level of impurities to be removed was considered to be consistent with meeting drinking water standards.[13] The methods of design presented in this chapter consider the variability of input raw water and flow rate to choose design sizes and operating policy. The characteristic variability of one of the inputs is illustrated in Figure 10.3. The probability distribution function on raw water coliform bacteria indicates that the conventional design assumption of constant input quality is not true, and the conventional designs have a "built-in" safety factor during most of the operation. These raw water cumulative frequency distributions

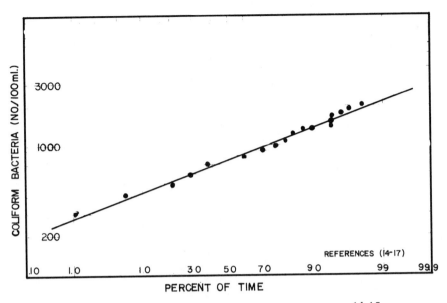

**Figure 10.3.** Frequency diagram for coliform bacteria.[14-17]

are available from surveillance systems data[14-17] or can be obtained from local pollution control agencies. These data were grouped into classes of equal intervals and the frequency distributions were obtained.

Using the input data on turbidity, the quantity of alum required for a certain coagulation efficiency can be related to the input turbidity level. For waters of less than 500 ppm of turbidity, Hudson related alum dosage in mg/l to influent turbidity.[18] In addition, Dostal and Robeck[19] produced frequency distributions on influent turbidity and related alum dosage to the influent turbidity. Figure 10.4 illustrates the relationships of alum dosage to influent turbidity. An equation expressing this relationship is

$$D_A = 10.4 \log T_j + 2.9 \qquad (1)$$

where

$D_A$ = dosage of alum (mg/l)
$T_j$ = influent turbidity (S.U.)

Equation 1 is the line of best fit obtained by a regression analysis of $D_A$ on $\log T_j$. Since turbidity can be measured by remote sensing equipment, it is possible with a computer to tabulate and update equations on alum dosages and removal efficiencies. Computers are gaining acceptance for the automatic

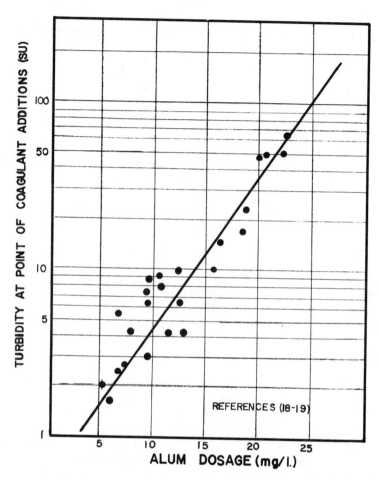

**Figure 10.4.** Alum dosage required.[18,19]

control of treatment plants.[20] By the remote sensing of turbidity, the above calculations are possible.

When one mg/l of alum is added to raw water, the total alkalinity will be reduced by 0.45 mg/l if the theoretical reactions are duplicated in practice.[21] Scott reported decreases in total alkalinity of from 0.38 to 0.405 mg/l per mg/l of alum.[22] Riddick reported a value of 0.415.[23] A value of 0.41 is assumed for the illustrative problem. When lime (CaO) is added to water, the total alkalinity should increase by about 1.79 mg/l per one mg/l of lime.

For effective floc formation minimum concentration of total alkalinity between 10 and 50 mg/l is usually maintained after the addition of chemicals, but 35 mg/l is generally used. Therefore, the equation representing chemical additions to induce coagulation, expressed in terms of feed rates is

$$A_1 = -\frac{1.18}{Q} X_2 + \frac{5.15}{Q} X_3 \geqslant 35 \tag{2}$$

where

$A_1$ = input total alkalinity (mg/l)
$Q$ = flow rate (mgd)
$X_2$ = alum feed rate (lb/hr), which is a function of the dosage, $D_A$
$X_3$ = lime feed rate (lb/hr)

From turbidity and coliform bacteria data obtained on the operation of coagulation units, empirical equations were developed relating effluent conditions to influent conditions, flocculation times, and sedimentation times.[11] The empirical relationships were good (correlation coefficient = 0.95). Once flocculation occurs, terminal velocity increases and removals have been shown to be a function of detention times.[24,25] If flocculation time were always greater than 30 minutes, it was assumed the following equations could be used for removal efficiencies by coagulation:

$$\log (COLIFS) = \log (COLIF) - 0.25 (SEDTIM)$$
$$\log (TURBS) = \log (TURBIN) - 0.33 (SEDIM) \tag{3}$$

where

COLIFS = effluent coliform bacteria (#/100 ml)
COLIF = influent coliform bacteria (#/100 ml)
TURBS = effluent turbidity (S.U.)
TURBIN = influent turbidity (S.U.)
SEDTIM = sedimentation time (hr)

Better empirical equations for the efficiencies of removal can be found.[11] The model was simplified for tutorial purposes.

Rapid gravity filtration is used mainly for the removal of coliform bacteria and turbidity. The removal efficiencies are controlled primarily by the medium head loss, influent impurities, and the rate of filtration. The rate of filtration can be related to the surface area of a filter by the following equation:

$$S = \frac{(Q_D + Q_B) \times 10^4}{1440 \times r} \tag{4}$$

where

$S$ = surface area of the filter (100 sq ft)
$r$ = filtration rate (gpm/sq ft)

$Q_D$    = quantity of water required (mgd)
$Q_B$    = quantity of backwash water (mgd)
1440  = conversion factor (min/day)

The quantity of water for backwashing and frequency of backwashing are not considered in this model development. Additional equations would be necessary relating head loss to filtration rate and influent impurities. It was assumed for this model that 40 minutes each day would be required for backwashing one filter and 3%/day of the plant flow rate would be used for backwashing.

A mixed media filter is used in the example problem because it is characterized by longer filter runs and as good or better removal efficiencies relative to a single media sand or a single media anthracite filter.[26]   A mixed media of 18 in. of anthracite on 6 in. sand is used in the simulation.

Using the empirical equations that relate effluent conditions to influent conditions and filtration rate,[11] the following equations are used to represent the removal efficiencies of filtration.

$$\log (TURFIN) = 0.12 - \frac{0.148 \ (SA)}{DAYQ} + 0.81 \log (TURBSF)$$

(5)

$$\log (BACFIN) = 0.83 - \frac{0.328 \ (SA)}{DAYQ} + 0.38 \log (COLIFF)$$

where

TURFIN = effluent turbidity from the rapid gravity filter (S.U.)
TURBSF = influent turbidity to the rapid gravity filter and effluent from coagulation (S.U.)
BACFIN = effluent coliform bacteria from the rapid gravity filter (#/100 ml)
COLIFF = influent coliform bacteria to the rapid gravity filter (#/100 ml)
SA      = filter surface area (100 sq ft)
DAYQ   = flow rate at any time (mgd)

The data used to derive the above equations came from actual plant operations.[27,28]

In the simulation, the above equations (1-5) were used to represent the plant operation given input data on water requirements, raw water quality distribution, and effluent (clear well) quality. To simulate actual time delays for the detention of water in the coagulation units, a fifth-order delay function can be used by dividing the basin into five equal volume compartments as shown in Figure 10.5. Turbidity delay is only shown in Figure 10.5; however, the model also delays flow and coliform bacteria. Mass balance equations can now be constructed around each compartment, or

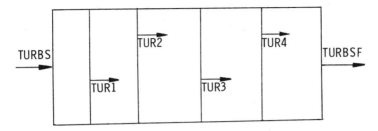

**Figure 10.5** Schematic of a delay function for turbidity in the coagulation process.

TIME CHANGE = INPUT - OUTPUT ± ADDITIONS

for compartment 1:

$$TUR1 = TURBS\ (Q)/V - TUR1\ (Q)/V \pm 0$$

where:

TUR1    = d(TUR1)/dt = rate of change with time of turbidity from compartment 1

TURBS    = turbidity to be delayed (S.U.)

Q    = flow rate (m³/hr)

V    = compartment volume (m³)

TUR1    = turbidity from compartment 1 (S.U.)

for the remaining compartments:

$$TUR2 = TUR1\ (Q)/V - TUR2\ (Q)/V$$

$$TUR3 = TUR2\ (Q)/V - TUR3\ (Q)/V$$

$$TUR4 = TUR3\ (Q)/V - TUR4\ (Q)/V$$

$$TURBSF = TUR4\ (Q)/V - TURBSF\ (Q)/V$$

These differential equations can be simulated using various programming features of CSMP. A first-order lag for each compartment using the laplace function was used. The results of using the laplace transform were very good as judged by the computer output shown in Appendix A. The flow rate was lagged after the coagulation process. This lag is a function of the flow rate at a specific time and the sizes of the flocculation and sedimentation basins. This lag is evident by reviewing the output sheets.

The first two pages of the output represent the computer program. All constants used are identified in the program. The third output page presents the upper and lower values of important variables in the simulation and the time of the day they occur. The next pages illustrate some of the variable inputs (flow rate, coliform, bacteria) and outputs (effluent turbidity and coliform bacteria) in print-plot format.

The responses of effluent conditions to perturbations in raw water characteristics can be clearly examined from the output. For a fixed operating condition of a conventional water treatment facility design, the output can be predicted for various input and examinations of output are made very easily without expensive changes in the field (physically changing the size of a unit or the feed rates of chemicals).

## EXPECTED STUDENT RESPONSES

After presenting the above problem in a classroom situation, it is possible to test the students' understanding of the general concepts of designing and operating a water treatment plant given random inputs and their understanding of the computer simulation language.

The student is asked to obtain a computer output of the above model and evaluate the results. The results of the above model on coliform bacteria do not meet with USPHS drinking water standards of a coliform bacteria density of one/100 ml.[13] A design change or operational change will have to be implemented. An operational change to reduce flow rate during the afternoon hours is one possibility. The question of storage capacity after treatment will have to be evaluated if the flow rate in the plant is to be changed. A probability function on demand and storage capacity can be introduced. Longer simulation runs with provisions for backwashing the filter defined by filtration head loss data and incorporation of sludge disposal is also possible. Another possible operational change would be the addition of more chemical coagulants to obtain better removal efficiencies in coagulation or chemical aids for filtration. The availability of data on removal efficiencies when more chemicals are added must be defined by the student. The most frequent student response is to add a chlorination facility to aid in reducing the coliform bacteria.

A more realistic model to include chlorination and the associated changes in total alkalinity and reduction of bacteria is introduced (this could be presented in a test). The student then has to add a chlorination tank with a retention time. The lagging of flow and quality changes will have to be incorporated into the computer program. This will test the student's ability to use the CSMP language and his knowledge of the unit processes in a series configuration. Students are referred to the work of Bryant and Wanielista to help them identify the quality changes and design parameters.[10,11]

Every time a program is submitted for a computer run and assuming the program works, only one design and operating configuration can be tested with the previous model. However, with one submission many sets of designs and operation configurations can be tested. To accomplish this recycling through the program, a segment of the program called "TERMINAL" is written to perform computations at the end of each run. In the "TERMINAL" section, FORTRAN statements are written to test the results of the simulation run. The user can automatically examine the output and make a design or operational change without resubmitting the computer program for another run. Cost data can be entered in the "TERMINAL" section to perform a cost minimization study. This procedure will test the student's knowledge of the CSMP language and FORTRAN programming.

Another function available in CSMP is "GAUSS," which can be used to generate a normal distribution. However, any type of distribution can be generated with another function called "RNDGEN" which randomly generates numbers that can be used to simulate any type of distribution on quality or flow.

There is much flexibility in this approach for examining the design and operation of water treatment plants. Examination of student responses and understanding appears almost limitless.

## SUMMARY

A mathematical model of the conventional coagulation and rapid gravity filtration units of a water treatment plant has been developed to examine a new design procedure and operational procedures given random inputs. Primary attention in the chapter was given to the development of the model and the use of CSMP to simulate the model. The model is realistic; however, it can be improved by the collection of additional data to define other variables of operation and design so that with a higher degree of accuracy plant performance can be predicted. The model has been used before to train and educate graduate engineers in the use of the CSMP language and to provide a better understanding of coagulation and filtration. The understanding of the student can easily be measured.

## REFERENCES

1. Brennan, R. D. and M. Y. Silberberg. "The System/360 Continuous System Modeling Program," *Simulation*, 301 (December 1968).
2. System/360 Continuous System Modeling Program: Application Description, IBM Corporiation Form GH20-0240.
3. System/360 Continuous System Modeling Program User's Manual, IBM Corporation Form GH20-0367.
4. Gordon, Geoffrey. *System Simulation*. (Englewood Cliffs, New Jersey: Prentice-Hall, Inc., 1969).

5.  Patten, B. C. (Ed.). *Systems Analysis and Simulation in Ecology*, Vol. 1. (New York: Academic Press, 1971).
6.  Chubb, B. A. "Application of a Continuous System Modeling Program to Control System Design," *Proc. of the Joint Automatic Cont. Conf.*, 350 (1970).
7.  Forrester, J. W. *Industrial Dynamics*. (Cambridge: MIT Press, 1961).
8.  International Business Machines Corporation. *Continuous System Simulation for the Management Scientist.*
9.  Bryant, James O., Jr. "Continuous Time Simulation of the Conventional Activated Sludge Wastewater Renovation System," Ph.D. Thesis, Clemson University, Clemson, South Carolina (August 1972).
10. System/360 Continuous System Modeling Program: System Manual, IBM Corporation Form Y20-0111-0.
11. Wanielista, M. P. "An Evaluation of the Design and Operation of Water Treatment Facilities," Ph.D. Thesis, Cornell University, Ithaca, New York (January 1971).
12. New York State Department of Health. *Recommended Standards for Water Works* (May 25, 1962).
13. United States Public Health Service. *Drinking Water Standards* Public Health Service Publication 956 (1962).
14. Public Health Service. *National Water Quality Network.* (Washington, D.C.: U.S. Department of Health, Education and Welfare, 1959-1962) Four Volumes.
15. Public Health Service. *Water Pollution Surveillance System,* Vol. 5: *Ohio and Tennessee River Basins.* (Washington, D.C.: U.S. Department of Health, Education and Welfare, 1963).
16. Public Health Service. *Water Pollution Surveillance System,* Vol. 2: *North Atlantic Basin.* (Washington, D.C.: U.S. Department of Health, Education and Welfare, 1963).
17. Public Health Service. *Water Pollution Surveillance System,* Vol. 7: *Missouri River Basin.* (Washington, D.C.: U.S. Department of Health, Education and Welfare, 1963).
18. Hudson, H. E. "Physical Aspects of Flocculation," *J. Amer. Water Works Assoc.* **57**, 885 (July 1965).
19. Dostal, K. A. and G. G. Robeck. "Studies of Modifications in Treatment of Lake Erie Water," *J. Amer. Water Works Assoc.* **58**, 1489 (November 1966).
20. Hatada, Hideji and Keiji Nishijima. "Computer Control of Moniwa Water Treatment Plant," *Toshiba Rev.,* 5 (April 1971).
21. Hardenberg, W. A. and E. B. Rodie. *Water Supply and Waste Disposal* Scranton, Pa.: International Textbook Co., 1963).
22. Scott, Kimberly, Ey, and Waring. "Fluoride in Ohio River Water Supplies—Its Effect, Occurrence and Reduction," *J. Amer. Water Works Assoc.* **29**, 9 (January 1937).
23. Riddick, T. M. "Role of the Zeta Potential in Coagulation Involving Hydrous Oxides," *TAPPI*, **47**, 171A (January 1964).
24. Camp, T. R. "Sedimentation and the Design of Settling Tanks," *Trans. ASCE*, 895 (1946).
25. Fitch, E. B. "The Significance of Detention in Sedimentation," *Sewage Ind. Wastes* **29**, 1123 (1957).

26. Robeck, G., K. A. Dostal, and R. L. Woodward. "Studies of Modifications in Water Filtration," *J. Amer. Water Works Assoc.* **56**, 198 (February 1964).
27. "High-Rate and Dual-Media Filtration Study in a Northwestern Ohio Water Plant," *USPHS—Taft Center* (January 1966).
28. "High-Rate Filtration Study at Easley South Carolina Water Plant," *USPHS—Taft Center* (August 1965).

## APPENDIX A

**Computer Program and Results**

```
****CONTINUOUS SYSTEM MODELING PROGRAM****

***PROBLEM INPUT STATEMENTS***

*
*
*       COAGULATION, FILTRATION, AND CHLORINATION PROCESSES
*       USING RANDOM WATER QUALITY INPUTS
*       AND VARIABLE FLOW RATES.  THE
*       FLOW RATES AND DESIGN CRITERIA
*       ARE CHANGED AT THE BEGINNING OF EACH
*       RUN TO EVALUATE THEIR EFFECTS.
*
*
*       DESIGN CRITERIA ARE--
*           SEDIMENTATION DETENTION TIME = 4.0 HRS.
*           FLOCCULATION DETENTION TIME = 1.0 HRS.
*           FILTRATION RATE = 2 GPM/SQ. FT.
*           CHLORINE DETENTION TIME = 0.5 HRS.
*
*
*       ********** BY MARTY WANIELISTA **********
*       *************** AND CHRIS BAUER ****************
*
```

```
*
*
*
MACRO OUT=DELAY5(IC,DELAY,IN)
       T1=REALPL(IC,DELAY/5.0,IN)
       T2=REALPL(IC,DELAY/5.0,T1)
       T3=REALPL(IC,DELAY/5.0,T2)
       T4=REALPL(IC,DELAY/5.0,T3)
       OUT=REALPL(IC,DELAY/5.0,T4)
ENDMAC
*
*
INITIAL
CONSTANT  K1=10.4,K2=2.9,MALKOT=35.,K3= .41,K4=1.79
CONSTANT  K5=.35,AVEQ=8.24,STH=4.0,FTH=1.0,SA=29.4,CTH=.5
CONSTANT  K6=0.12,K7=.148,K8=.81,K9=.83,K10=.328,K11=.38
*
*          IMPORTANT CONSTANTS ARE
*          MALKOT= TOTAL ALKINITY AFTER CHEMICAL ADDITIONS TO
*                  MAINTAIN REMOVAL EFFICIENCIES
*          AVEQ  = AVERAGE DAILY FLOW (MGD).
*          STH   = SEDIMENTATION DETENTION TIME (HOURS).
*          FTH   = FLOCCULATION DETENTION TIME (HOURS).
*          CTH   = CHLORINE DETENTION TIME (HOURS)
*          SA    = SURFACE AREA OF FILTERS (100 SQ.FT.)
*          K1=CONSTANT IN ALUM DOSAGE TERM.
*          K2=CONSTANT IN ALUM DOSAGE TERM.
*          K3=DECREASE IN TOTAL ALKINITY PER ONE
*             MG/L ADDITION OF ALUM (MG/L ALK/MG/L ALUM).
*          K4=INCREASE IN TOTAL ALKALINITY PER ONE
*             MG/L ADDITION OF LIME (MG/L ALK / MG/L LIME).
*          K5=8.34/24 (LBS/GAL/HR/DAY).
*          K6=CONSTANT IN TURBIDITY REMOVAL EQUATION
*             BY FILTRATION.
*          K7=CONSTANT IN TURBIDITY REMOVAL EQUATION
*             BY FILTRATION.


*          K8=CONSTANT IN TURBIDITY REMOVAL EQUATION
*             BY FILTRATION.
*          K9= CONSTANT IN BACTERIA REMOVAL EQUATION
*              BY FILTRATION.
*          K10=CONSTANT IN BACTERIA REMOVAL EQUATION
*              BY FILTRATION.
*          K11=CONSTANT IN BACTERIA REMOVAL EQUATION
*              BY FILTRATION.
*          NOTE$  CONSTANTS STARTING WITH K (EXCEPT K5)
*                 HAVE BEEN EMPIRICALLY DERIVED.
*
*
FUNCTION FLOW=(.0,157.0),(3600.,113.0),(7200.,113.0),(10800.,113.),...
         (14400.,113.),(18000.,113.),(21600.,113.),(25200.,93.0),...
         (28800.,93.0),(32400.,165.),(36000.,197.),(39600.,197.),...
         (43200.,197.),(46800.,197.),(50400.,197.),(54000.,197.),...
         (57600.,197.0),(61200.,185.),(64800.,177.),(68400.,177.0),...
         (72000.,177.0),(75600.,165.),(79200.,165.),(82800.,157.0),...
         (86400.,157.)
*          NOTE$  UNITS OF FLOW IN THE ABOVE FUNCTION
*                 ARE CUBIC-METERS/HOUR.
*
*          FIND SETTLING BASIN VOLUME
*          VOLCM = STH*AVEQ*157.
*          FIND FLOCCULATION BASIN VOLUME
*          VOLFT = FTH*AVEQ*157.
*          FIND CHLORINE BASIN VOLUME
*          VOLCL=CTH*AVEQ*157.
*          UNITS FOR VOLUME ARE CUBIC METERS
*
*
*
```

```
DYNAMIC
NOSORT
*          FIND INFLUENT FLOW RATE
           IF (TIME .GE.24.) GO TO 1
           TIMIN=3600.*TIME
           GO TO 5
    1 CONTINUE
           TIMIN=3600.*(TIME-24.)
    5 CONTINUE
*          GENERATE FLOW RATE FOR PLANT
*          IN CUBIC METERS/HOUR
*
           Q=AFGEN(FLOW,TIMIN)
           HRLYQ=8.24*Q
           DAYQ=HRLYQ/157.
            IF(TIMIN.EQ.0) GO TO 10
*
*          FIND INFLUENT TURBIDITY
           LTURB=GAUSS(1,1.846,0.267)
        TURBIN=10**LTURB
*          FIND INFLUENT COLIFORM BACTERIA
           LCOLI=GAUSS(3,2.875,0.204)
        COLIF=10**LCOLI
*
*          FIND ALUM FEED RATE
           LTUR=ALOG10(TURBIN)
           X2   =((K1*LTUR)+(K2))*K5*DAYQ
*
*           ALKIN = TOTAL ALKINITY IN RAW WATER.
           ALKIN=GAUSS(5,24.,4.)
*
*          FIND LIME FEED RATE
           ALKOUT=ALKIN-(K3*X2)/(DAYQ*K5)
           IF(ALKOUT .GE. 35.) GO TO 10
           X3=((MALKOT-ALKOUT)*K5*DAYQ)/K4
           GO TO 20
*
   10 CONTINUE
           X3=0.
   20 CONTINUE
*
SORT
*          FIND SETTLING TIME
           SEDTIM=VOLCM/HRLYQ
*          FIND FLOCCULATION TIME
           FLCTIM=VOLFT/HRLYQ
*          FIND EFFLUENT TURBIDITY
           P=-.333*SEDTIM
           TURBS=(10.**P)*TURBIN


*          FIND EFFLUENT COLIFORM BACTERIA
           R=-.25*SEDTIM
           COLIFS=(10.**R)*COLIF
           DE=FLCTIM+SEDTIM
           FPS=DELAY5(0.0,DE,DAYQ)
           TURBSF=DELAY5(0.,DE,TURBS)
           COLIFF=DELAY5(0.,DE,COLIFS)
           RF=(DAYQ*10000)/(1400*SA)
           TURFIN= (10**(K6-(K7*SA)/DAYQ))*(TURBSF**K8)
           BACFIN= (10**(K9-(K10*SA)/DAYQ))*(COLIFF**K11)
*          FIND FINAL BACTERIA AFTER CHLORINATION
        CLEAN=VOLCL/HRLYQ
        BICLO=.1*BACFIN
           FPC=DELAY5(0.,CLEAN,FPS)
           BOUT=DELAY5(0.,CLEAN,BICLO)
*
*          OUTPUT FORMAT IS SPECIFIED TO
*          GIVE GRAPHICAL DISPLAY
*
```

```
METHOD RECT
PRTPLT DAYQ
LABEL FLOW RATE (MG/D)
PRTPLT TURBIN
LABEL INFLUENT TURBIDITY (S.U.)
PRTPLT COLIF
LABEL INFLUENT COLIFORM BACTERIA (=/100ML)
PRTPLT ALKIN
LABEL INFLUENT ALKALINITY
PRTPLT FPS
LABEL FLOW AFTER SEDIMENTATION BASIN (MG/D)
PRTPLT FPC
LABEL FLOW AFTER CHLURINATION (MG/D)
PRTPLT TURFIN
LABEL FILTRATION EFFLUENT TURBIDITY (S.U.)
PRTPLT BACFIN
LABEL FILTRATION EFFLUENT BACTERIA (=/100ML)
PRTPLT BOUT
LABEL BACTERIA OUT (=/100ML)
RANGE RF,X2,X3,SECTIM,FLCTIM
TIMER PRDEL=48.,OUTDEL=1.,FINTIM=48.,DELT=.05
END
STOP
PROBLEM DURATION 0.0        TO  4.8000E 01
```

| VARIABLE | MINIMUM | TIME | MAXIMUM | TIME |
|---|---|---|---|---|
| DAYQ | 4.8810E 00 | 7.0500E 00 | 1.0339E 01 | 1.0050E 01 |
| TURBIN | 0.0 | 0.0 | 7.1710E 02 | 9.9000E 00 |
| COLIF | 0.0 | 0.0 | 2.8659E 03 | 1.2300E 01 |
| ALKIN | 0.0 | 0.0 | 3.5401E 01 | 3.9950E 01 |
| FPS | 0.0 | 0.0 | 1.0255E 01 | 4.2050E 01 |
| FPC | 0.0 | 0.0 | 1.0254E 01 | 4.2450E 01 |
| TURFIN | 0.0 | 0.0 | 2.4388E 00 | 1.6000E 01 |
| BACFIN | 0.0 | 0.0 | 5.1403E 00 | 1.6000E 01 |
| BOUT | 0.0 | 0.0 | 5.1359E-01 | 1.6250E 01 |
| RF | 1.1859E 00 | 7.0000E 00 | 2.5120E 00 | 1.0000E 01 |
| X2 | 0.0 | 0.0 | 1.1605E 02 | 9.9000E 00 |
| X3 | 0.0 | 0.0 | 6.1872E 01 | 1.6100E 01 |
| SECTIM | 3.1878E 00 | 1.0050E 01 | 6.7527E 00 | 7.0500E 00 |
| FLCTIM | 7.9695E-01 | 1.0050E 01 | 1.6882E 00 | 7.0500E 00 |

INFLUENT TURBIDITY (S.U.)

| | | MINIMUM<br>0.0 | TURBIN VERSUS TIME | MAXIMU<br>7.1710E |
|---|---|---|---|---|
| TIME | TURBIN | I | | I |
| 0.0 | 0.0 | + | | |
| 1.0000E 00 | 1.2672E 02 | --------+ | | |
| 2.0000E 00 | 4.3272E 01 | ---+ | | |
| 3.0000E 00 | 9.8938E 01 | ------+ | | |
| 4.0000E 00 | 6.6642E 01 | ----+ | | |
| 5.0000E 00 | 4.9461E 01 | ---+ | | |
| 6.0000E 00 | 3.2068E 01 | --+ | | |
| 7.0000E 00 | 4.1623E 01 | --+ | | |
| 8.0000E 00 | 7.1437E 01 | ----+ | | |
| 9.0000E 00 | 6.6564E 01 | ----+ | | |
| 1.0000E 01 | 3.4269E 01 | --+ | | |
| 1.1000E 01 | 9.8068E 01 | ------+ | | |
| 1.2000E 01 | 6.1163E 01 | ----+ | | |
| 1.3000E 01 | 5.1242E 01 | ---+ | | |
| 1.4000E 01 | 4.0742E 01 | --+ | | |
| 1.5000E 01 | 6.2780E 01 | ----+ | | |
| 1.6000E 01 | 3.2270E 01 | --+ | | |
| 1.7000E 01 | 8.0962E 01 | -----+ | | |
| 1.8000E 01 | 4.8624E 02 | --------------------------------+ | | |
| 1.9000E 01 | 2.4778E 02 | ----------------+ | | |
| 2.0000E 01 | 1.7507E 02 | -----------+ | | |
| 2.1000E 01 | 4.3558E 01 | ---+ | | |
| 2.2000E 01 | 5.1964E 01 | ---+ | | |
| 2.3000E 01 | 2.5010E 01 | -+ | | |
| 2.4000E 01 | 5.5090E 01 | ---+ | | |
| 2.5000E 01 | 5.2948E 01 | ---+ | | |
| 2.6000E 01 | 1.1349E 02 | -------+ | | |
| 2.7000E 01 | 2.0037E 02 | -------------, | | |
| 2.8000E 01 | 1.0611E 02 | --------+ | | |
| 2.9000E 01 | 1.5257E 02 | ----------+ | | |
| 3.0000E 01 | 5.2836E 01 | ---+ | | |
| 3.1000E 01 | 4.5194E 01 | ---+ | | |
| 3.2000E 01 | 2.4135E 01 | -+ | | |
| 3.3000E 01 | 2.7341E 01 | -+ | | |
| 3.4000E 01 | 4.7319E 01 | ---+ | | |
| 3.5000E 01 | 2.6045E 02 | ------------------+ | | |
| 3.6000E 01 | 6.8401E 01 | ----+ | | |
| 3.7000E 01 | 8.0858E 01 | -----+ | | |
| 3.8000E 01 | 1.8391E 01 | -+ | | |
| 3.9000E 01 | 4.6517E 01 | ---+ | | |
| 4.0000E 01 | 8.6175E 01 | ------+ | | |
| 4.1000E 01 | 5.5983E 01 | ---+ | | |
| 4.2000E 01 | 2.7990E 01 | -+ | | |
| 4.3000E 01 | 1.2636E 02 | --------+ | | |
| 4.4000E 01 | 1.2734E 02 | --------+ | | |
| 4.5000E 01 | 6.0211E 01 | ----+ | | |
| 4.6000E 01 | 1.2871E 02 | --------+ | | |
| 4.7000E 01 | 1.3695E 02 | ---------+ | | |
| 4.8000E 01 | 2.6925E 01 | -+ | | |

INFLUENT COLIFORM BACTERIA (#/100ML)

```
                             MINIMUM              CLLIF   VERSUS TIME           MAXIMU
                               0.0                                             2.8E50-
   TIME              COLIF       I                                                  I
0.0                 0.0          +
1.0000E 00          9.3124E 02   ---------------+
2.0000E 00          1.2475E 03   ---------------------+
3.0000E 00          4.7412E 02   --------+
4.0000E 00          5.2583E 02   ---------+
5.0000E 00          8.6681F C2   ----------------+
6.0000E 00          1.6138E 03   ----------------------------+
7.0000E 00          4.1307E 02   -------+
8.0000E 00          5.8566E 02   ----------+
9.0000E 00          2.2416E 03   ------------------------------------------+
1.0000E 01          1.0055E 03   -----------------+
1.1000E 01          4.2210E 02   -------+
1.2000E 01          9.7652E 02   -----------------+
1.3000E 01          1.4188E C3   --------------------------+
1.4000E 01          9.0083E 02   ----------------+
1.5000E 01          1.1940E 03   ---------------------+
1.6000E 01          1.1115E 03   ---------------------+
1.7000E 01          5.1875E 02   ---------+
1.8000E 01          1.3538E C3   --------------------------+
1.9000E 01          8.2357E 02   --------------+
2.0000E 01          1.6115E 03   ----------------------------+
2.1000E 01          2.5857E 02   ----+
2.2000E 01          6.0817E 02   ----------+
2.3000E 01          1.3381E 03   --------------------------+
2.4000E 01          1.3570E 03   --------------------------+
2.5000E 01          1.0611E C3   ------------------+
2.6000E 01          9.9777E 02   ----------------+
2.7000E 01          1.1541E 03   --------------------+
2.8000E 01          7.7438E 02   -------------+
2.9000E 01          7.7157E C2   -------------+
3.0000E 01          1.0653E 03   ------------------+
3.1000E 01          4.8750E 02   --------+
3.2000E 01          1.4297E C3   --------------------------+
3.3000E 01          1.5361E 03   ---------------------------+
3.4000E 01          8.6365E 02   ---------------+
3.5000E 01          9.7403E 02   ----------------+
3.6000E 01          3.7187E 02   ------+
3.7000E 01          1.1798E 03   --------------------+
3.8000E 01          6.2019E 02   -----------+
3.9000E 01          8.1093E 02   --------------+
4.0000E 01          4.2594E 02   -------+
4.1000E 01          8.2504E 02   --------------+
4.2000E 01          4.6023E 02   --------+
4.3000E 01          1.0621E 03   ------------------+
4.4000E 01          6.1240E 02   ----------+
4.5000E 01          4.8459E 02   --------+
4.6000E 01          1.0535E 03   ------------------+
4.7000E 01          7.8887E 02   -------------+
4.8000E 01          8.0611E 02   --------------+
```

INFLUENT ALKALINITY

```
                            MINIMUM          ALKIN  VERSUS TIME          MAXIMU
                              0.0                                        3.5375E
    TIME            ALKIN         I                                         I
    0.0             0.0           +
    1.0000E 00      1.8313E 01    ------------------------+
    2.0000E 00      2.9543E 01    --------------------------------------+
    3.0000E 00      2.3045E 01    -------------------------------+
    4.0000F 00      2.0092F 01    --------------------------+
    5.0000E 00      1.7858E 01    -----------------------+
    6.0000E 00      2.2427E 01    ------------------------------+
    7.0000E 00      2.9789E 01    --------------------------------------+
    8.0000E 00      2.6839E 01    -----------------------------------+
    9.0000E 00      2.8378E 01    -------------------------------------+
    1.0000E 01      2.6115E 01    ----------------------------------+
    1.1000E 01      2.9664E 01    --------------------------------------+
    1.2000E 01      2.3546E 01    -------------------------------+
    1.3000E 01      2.8186E 01    -------------------------------------+
    1.4000E 01      2.0919E 01    ---------------------------+
    1.5000E 01      2.4984E 01    ---------------------------------+
    1.6000E 01      2.6525E 01    -----------------------------------+
    1.7000E 01      1.9595E 01    --------------------------+
    1.8000E 01      2.7151E 01    ------------------------------------+
    1.9000E 01      3.0059E 01    ---------------------------------------+
    2.0000E 01      2.8088E 01    -------------------------------------+
    2.1000E 01      2.0915E 01    ---------------------------+
    2.2000E 01      1.7124E 01    ----------------------+
    2.3000E 01      2.7203E 01    ------------------------------------+
    2.4000E 01      2.4548E 01    --------------------------------+
    2.5000E 01      2.2461E 01    ------------------------------+
    2.6000E 01      1.9150E 01    -------------------------+
    2.7000E 01      2.6729E 01    -----------------------------------+
    2.8000E 01      2.4218E 01    -------------------------------+
    2.9000E 01      2.2545E 01    ------------------------------+ .
    3.0000E 01      1.7542E 01    -----------------------+
    3.1000E 01      3.0948E 01    ----------------------------------------+
    3.2000E 01      3.1410E 01    -----------------------------------------+
    3.3000E 01      2.7478E 01    ------------------------------------+
    3.4000E 01      2.0612E 01    --------------------------+
    3.5000E 01      3.0174E 01    ----------------------------------------+
    3.6000E 01      2.2436E 01    ------------------------------+
    3.7000E 01      2.3574E 01    -------------------------------+
    3.8000E 01      2.8672E 01    -------------------------------------+
    3.9000E 01      2.2719E 01    ------------------------------+
    4.0000E 01      2.5609E 01    ----------------------------------+
    4.1000E 01      2.5146E 01    ----------------------------------+
    4.2000E 01      2.2036E 01    ------------------------------+
    4.3000E 01      2.6895E 01    -----------------------------------+
    4.4000E 01      2.5242E 01    ----------------------------------+
    4.5000E 01      2.2505E 01    ------------------------------+
    4.6000E 01      2.5016E 01    ----------------------------------+
    4.7000E 01      2.5014E 01    ----------------------------------+
    4.8000E 01      1.7646E 01    -----------------------+
```

FILTRATION EFFLUENT TURBIDITY  (S.U.)

```
                              MINIMUM                TURFIN VERSUS TIME              MAXIML
                                0.0                                                 2.6191E
       TIME              TURFIN       I                                                  I
     0.0                 0.0          +
     1.0000E 00          1.4272E-03   +
     2.0000E 00          2.1239E-02   +
     3.0000E 00          6.7014E-02   -+
     4.0000E 00          1.2457E-01   --+
     5.0000E 00          1.7889E-01   ---+
     6.0000E 00          2.2024E-01   ----+
     7.0000E 00          1.7021E-01   ---+
     8.0000E 00          1.7895E-01   ---+
     9.0000E 00          4.4728E-01   --------+
     1.0000E 01          5.2659E-01   ----------+
     1.1000E 01          6.6359E-01   ------------+
     1.2000E 01          1.0599E 00   -------------------+
     1.3000E 01          1.6422E 00   -----------------------------+
     1.4000E 01          2.1563E 00   --------------------------------------+
     1.5000E 01          2.3760E 00   -----------------------------------------+
     1.6000E 01          2.3717E 00   -----------------------------------------+
     1.7000E 01          2.1724E 00   --------------------------------------+
     1.8000E 01          2.0993E 00   -------------------------------------+
     1.9000E 01          2.1252E 00   -------------------------------------+
     2.0000E 01          2.1774E 00   -------------------------------------+
     2.1000E 01          2.0403E 00   ------------------------------------+
     2.2000E 01          2.0709E 00   ------------------------------------+
     2.3000E 01          1.9132E 00   ----------------------------------+
     2.4000E 01          1.7813E 00   --------------------------------+
     2.5000E 01          1.0134E 00   ------------------+
     2.6000E 01          9.3277E-01   ----------------+
     2.7000E 01          8.5454E-01   ---------------+
     2.8000E 01          7.7499E-01   --------------+
     2.9000E 01          6.9316E-01   ------------+
     3.0000E 01          6.1201E-01   -----------+
     3.1000E 01          3.7734E-01   -------+
     3.2000E 01          3.3892E-01   ------+
     3.3000E 01          7.1787E-01   -------------+
     3.4000E 01          7.1962E-01   -------------+
     3.5000E 01          7.9905E-01   --------------+
     3.6000E 01          1.1743E 00   ---------------------+
     3.7000E 01          1.8262E 00   --------------------------------+
     3.8000E 01          2.4013E 00   -------------------------------------------+
     3.9000E 01          2.6134E 00   -------------------------------------------------+
     4.0000E 01          2.6037F 00   ------------------------------------------------+
     4.1000E 01          2.3918E 00   ---------------------------------------------+
     4.2000E 01          2.2144E 00   --------------------------------------------+
     4.3000E 01          2.1198E 00   ------------------------------------------+
     4.4000E 01          1.9996E 00   ----------------------------------------+
     4.5000E 01          1.7706E 00   ------------------------------------+
     4.6000E 01          1.7430E 00   -----------------------------------+
     4.7000E 01          1.6327E 00   ---------------------------------+
     4.8000E 01          1.6009E 00   ---------------------------------+
```

```
BACTERIA OUT  (#/100ML)

                                   MINIMUM          BOUT   VERSUS TIME          MAXIMU'
                                     0.0                                        5.2560E-
    TIME              BOUT          I                                              I
 0.0               0.0             +
 1.0000E 00        0.0             +
 2.0000E 00        6.8976E-03      +
 3.0000E 00        1.7534E-02      -+
 4.0000E 00        2.7607E-02      --+
 5.0000E 00        3.6176E-02      ---+
 6.0000E 00        4.2965E-02      ----+
 7.0000E 00        4.0225E-02      ---+
 8.0000E 00        2.3636E-02      --+
 9.0000E 00        7.7192E-02      -------+
 1.0000E 01        2.3526E-01      -----------------------+
 1.1000E 01        2.9619E-01      -----------------------------+
 1.2000E 01        3.6090E-01      -----------------------------------+
 1.3000E 01        4.2455E-01      -----------------------------------------+
 1.4000E 01        4.6633E-01      --------------------------------------------+
 1.5000E 01        4.9286E-01      -----------------------------------------------+
 1.6000E 01        5.0530E-01      ------------------------------------------------+
 1.7000E 01        4.6747E-01      --------------------------------------------+
 1.8000E 01        4.1359E-01      ----------------------------------------+
 1.9000E 01        3.8971E-01      --------------------------------------+
 2.0000E 01        3.8345E-01      --------------------------------------+
 2.1000E 01        3.4750E-01      ----------------------------------+
 2.2000E 01        3.1899E-01      ------------------------------+
 2.3000E 01        2.9783E-01      ----------------------------+
 2.4000E 01        2.7513E-01      --------------------------+
 2.5000E 01        1.9526E-01      ------------------+
 2.6000E 01        9.4903E-02      ---------+
 2.7000E 01        9.2716E-02      --------+
 2.8000E 01        9.0871E-02      --------+
 2.9000E 01        8.8396E-02      --------+
 3.0000E 01        8.5228E-02      --------+
 3.1000E 01        6.9294E-02      ------+
 3.2000E 01        3.6686E-02      ---+
 3.3000E 01        1.0647E-01      ----------+
 3.4000E 01        2.8760E-01      ---------------------------+
 3.5000E 01        3.2122E-01      -------------------------------+
 3.6000E 01        3.4947E-01      ----------------------------------+
 3.7000E 01        4.0391E-01      ---------------------------------------+
 3.8000E 01        4.6498E-01      --------------------------------------------+
 3.9000E 01        5.0981E-01      -----------------------------------------------+
 4.0000E 01        5.2560E-01      ------------------------------------------------+
 4.1000E 01        4.7736E-01      ---------------------------------------------+
 4.2000E 01        4.1931E-01      ----------------------------------------+
 4.3000E 01        3.9635E-01      --------------------------------------+
 4.4000E 01        3.8892E-01      -------------------------------------+
 4.5000E 01        3.4664E-01      ----------------------------------+
 4.6000E 01        3.1059E-01      -------------------------------+
 4.7000E 01        2.8461E-01      ----------------------------+
 4.8000E 01        2.6083E-01      --------------------------+
```

# CHAPTER 11

# MODELING CONCEPTS CONSIDERING PROCESS PERFORMANCE, VARIABILITY AND UNCERTAINTY

Paul M. Berthouex

    Department of Civil and Environmental Engineering
The University of Wisconsin
Madison, Wisconsin

## INTRODUCTION

Uncertainty is the gap between what is known and what needs to be known to make the correct decisions. Dealing sensibly with uncertainty is not a byway on the road to responsible business and engineering decision. It is central to it. The subject is elusive and omnipresent.

In sanitary engineering design we hedge bets, we make decisions in steps, and we use information feedback. All these are techniques for dealing with uncertainty; they are natural ways of doing it in fact. Engineers thrive on uncertainty. It surrounds us. It is our business.

> . . . uncertainty is truly a good thing. Could anyone face a life of certainty? Success preknown would be tasteless; defeat and grief known to lie ahead would erode the pleasures of today; such knowledge might often be insupportable. Not only the future, but the present would be completely mapped—no need for experiment, no open end to learning, the siren whistle of adventure silenced, the deep breath before the plunge replaced by the deliberate gaze of total calculation. I think one must find as much satisfaction in the **fact that uncertainty** "is there" as did Sir Edmund Hillary in the fact of Mount Everest.[1]

Unfortunately, though uncertainty adds spice to life, it has a price. There is a cost due to the uncertainty discount—the idea that the bird in the hand is worth two in the bush. Uncertainty often causes alternates to be narrowed and innovation to be eschewed because uncertainty and indecisiveness are held to be unbecoming, especially to young ambitious engineers. The viscosity of the political process, into which pollution control decision are often thrust, can strangle programs having clear benefits. When the benefits are unclear, even to the sponsors, difficulties become enormous. A cost results because uncertainties produce a bias toward conservatism, toward routine ways of solving problems, and toward doing nothing. Such bias can limit targets even more than the uncertainty discount. Another type of cost resulting from uncertainty is a tendency to make it easy to overlook unattractive externalities. It is easy to ignore the impact on a region of discharging a particular waste if no one knows what the impact will be. Or, as a corollary, it is too easy to overvalue externalities.

There is one benefit of uncertainty. Sometimes uncertainty allows people to compromise and act because they do not know exactly what they are agreeing upon.

Given an exact description of kinetics and other technical data for various processes which comprise a system, and having some knowledge of economic factors, an engineer can utilize a variety of mathematical methods to determine an optimal system design. In practice, often neither an *exact* description of the process nor *true* values for process parameters are available, and the engineer seeks a design which is optimal given the existing information. Limitations on money may prohibit elaborate pre-design research. Sometimes it is clear that even further experimentation will not enable the design to be tightened because factors that ultimately determine system performance are expected to change during the design period in an unpredictable way.

Successful engineers often have an intuition about how to handle uncertain situations. The safety factors may be identified explicitly or implicitly. Sometimes a designer may feel that following design codes produces a conservative design and he does not intentionally overdesign beyond that point. Sometimes the designer is unsure of a design flow or treatment parameter to the extent that he voluntarily tries to compensate. At all times he should try to balance the penalty for underdesign against the cost of providing a large safety factor.

The underlying motivation is a desire to determine how the deterministic design should be changed to accommodate the more realistic uncertain or variable situation, and there is a need for techniques which permit the consideration of the consequences arising from the use of uncertain technical data in design. The purpose of this chapter is to outline simple methods that allow the designer to compensate for uncertainty.

## UNCERTAINTY AND VARIABILITY

There is a distinction between uncertainty and variability. Certainty is a situation where only one state is possible, and all other states are impossible. Risk is present when two or more states are possible and the assignment of probabilities to each possible state can be assigned with confidence. Betting on the toss of a coin involves a risk, but the probability of the coin falling heads or tails can be stated with confidence. The situation is well understood and decisions can be made in a prudent way even though the outcome of any one toss is not certain.

Uncertainty characterizes the more common situation in engineering—information is scarce and inadequate. The amount of information, and the confidence of the designer, can vary over a wide range and perhaps approach total confusion or a high degree of uncertainty. Sometimes experiments can provide a basis for successful decision making even though understanding of the process is minimal.

Variability normally implies some knowledge about the behavior of the system, such as knowing the mean value and standard deviation of incoming flows and waste strengths (pH, BOD, suspended solids, acidity, toxicity) and, perhaps, even knowledge of the probability distribution of the different waste characteristics. The next levels of knowledge would be to know about the degree of correlation (of flow at time i with the flow at time i+1), for example, to know something about cross-correlation of flows and concentrations, and to have a time series model incorporating this information.[2-5]

## METHODS FOR CONSIDERING
## UNCERTAINTY OR VARIABILITY

There are several techniques one might consider using to analyze a problem involving uncertainty or variability. The methods, which are not really alternates to each othere, are sensitivity analysis, probability and decision theory, Monte Carlo simulation and propagation of variance.

### Sensitivity Analysis

A practical approach is to test the sensitivity of the process performance and system cost to variation in the level of key system parameters. The method can be useful in analyzing either uncertain or variable situations. Optimistic and pessimistic levels used should differ from the expected values by a degree proportionate to the degree of uncertainty prevailing. Examples of the approach to process design have been published by McBeath and Eliassen.[6]

## Probability Theory, Including Decision Theory

This approach is useful in analyzing situations of special interest having uncertainties, either because the consequences of a particular action are not apparent, or because events may intervene that cannot be controlled or forecast with certainty.

Raiffa[7] suggests that, in general terms, the analysis of a decision problem under uncertainty requires that you:

1. List options available to you for gathering information, for experimentation, and for action.
2. List events that may possibly occur.
3. Arrange in chronological order information you may require and choices you may make as time goes on.
4. Decide how well you like the consequences that result from various courses of action open to you.
5. Judge what the chances are that any particular uncertain event will occur.

After these five steps have been taken, it is possible to determine the strategy you should follow in experimentation and in action. The strategy you choose is not "best" in any universal sense. But, since it is your problem, the strategy to which you are guided should be the best you can choose for the situation at hand, assuming you have correctly included and analyzed your preferences as based on your understanding of the problem. There must be clear alternate courses of action, each with a consequence that can be defined in a consistent measure. Examples are found in many statistics and engineering texts.

## Monte Carlo Simulation

The method yields an estimate of system performance from information on the performance of system components. The name Monte Carlo suggests uncertainty, as well as, perhaps, glamour, and the technique is reminiscent of roulette in that numbers are turned up at random. And, if the method does not seem glamorous, it does at least carry an aura of "fun" and have a good deal of common sense to it. A complex system having numerous components is operated numerically by computer as shown in Figure 11.1. The method has also been called *synthetic sampling* since the idea is to sample at random some of the many possible operating situations that might be found by varying components.

Simulation with a Monte Carlo flavor has been used often in water resources planning, harbor planning and, to some extent, in process design.[8,9] Other applications of the Monte Carlo method that are useful are available.[10-12]

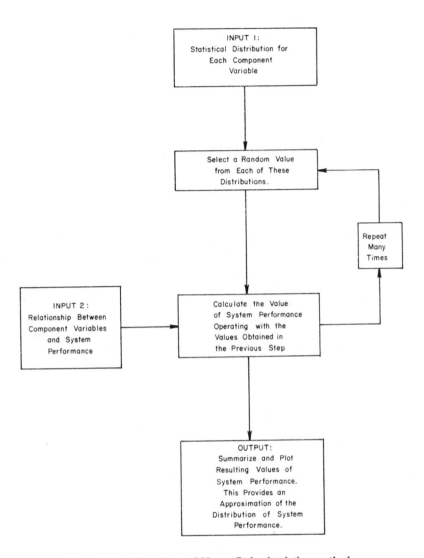

**Figure 11.1.** Flow chart of Monte Carlo simulation method.

## Propagation of Variance

This method permits the designer to formulate subjective probabilistic information on system parameters and components as algebraic expressions. These expressions can be used to estimate uncertainty in system performance. They can also be used with the basic system model in optimization computations.

## FEATURES OF THE PROPAGATION OF VARIANCE MODEL

A processing systems transforms an input Y into an output X according to some transfer function

$$X = f(Y, \underline{D}, \underline{P}) \tag{1}$$

where the $\underline{D}$ are design variables and the $\underline{P}$ are technological parameters. The output is uncertain if either Y or P is uncertain. The case used for illustration treats the input Y as a constant based on the argument that sometime in the future the stated design load will obtain. When this input obtains, the process should transform the design level of Y into the output level of X. The future takes care of furnishing Y, but it cannot guarantee that any particular level of P occurs with the design level; nor do laboratory studies enable prediction of P at some distant future date.

Figure 11.2 represents the functional relationship between some performance characteristic X of a system and some parameter P of that system. The parameters of interest are internal variables or constants as opposed to design variables since the problem is to determine how the design variables should be manipulated to overcome the effect of uncertainty in the parameters. The performance curve is generated by observing the performance of the system as P is varied over a range of values while all other input and parameters are held at their nominal values. The degree and distribution of uncertainty in P is represented by the probability distribution sketched on the P-axis. This uncertainty maps into the distribution of the output X sketched on the X-axis.

If $X_{cr}$ is the critical level of performance and P varies as shown, the process has some expectation of failure. The probability of failure is proportional to the shaded portion under the probability distribution of X. If the probability distribution is constructed so the total area under the curve equals one the probability of failure is the area shaded; otherwise it is the ratio of the shaded area to the total area.

If the mathematical form of the probability density distribution on X is known, the probability of failure can be calculated by integration or by searching statistical tables. For example, when the distribution of X is Gaussian with mean $\mu$ and standard deviation $\sigma$ the probability of X exceeding the target $X_{cr}$ is

$$P_{cr} = \int_{X_{cr}}^{\infty} \frac{1}{2\pi\sigma} \exp\left[-\frac{1}{2}\left(\frac{x-\mu}{\sigma}\right)^2\right] dx \tag{2}$$

If a level of x, say $x_{cr}$, is specified a design performing within the desired limit must have a mean $\mu$ and standard deviation $\sigma$ such that

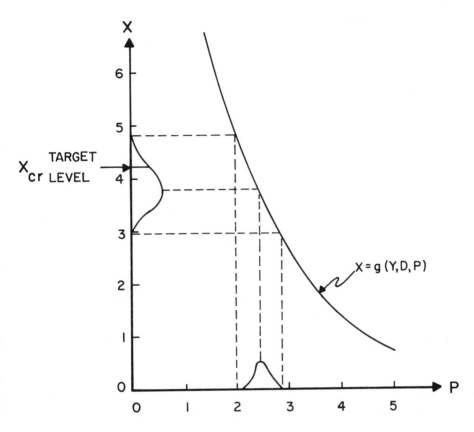

**Figure 11.2.** Given a fixed process input Y, uncertainty in the parameter P generates uncertainty in the process output X.

$$x_{cr} \geq \mu + z_{cr} \cdot \sigma \qquad (3)$$

where $z_{cr}$ is defined for a normal distribution by stating the tolerable probability of failure $P_{cr}$.

## DESIGN ADJUSTMENT TO ACCOMMODATE UNCERTAINTY

The analysis also indicates how the system might be redesigned when the target is not met. The level of $P_1$ might be controlled to raise the mean value or reduce the uncertainty. Figure 11.2 might become Figure 11.3 if temperature were controlled to reduce uncertainty in a reaction rate parameter for example. Another valid alternate is to redesign by using

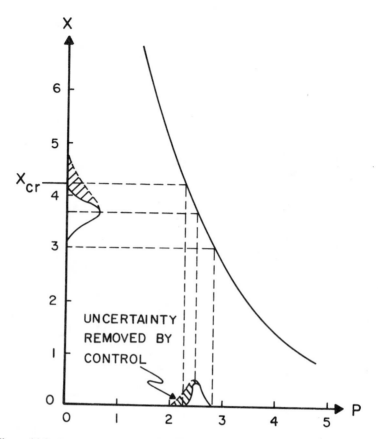

**Figure 11.3.** Process control systems may reduce uncertainty in P to meet target.

different settings for the design variables. The new design values are indicated in Figure 11.4 as D*. The redesign includes a safety factor, or overdesign factor, that accommodates uncertainty in the parameter P.

The method of handling uncertainty in technological design parameters is based on estimating the propagated uncertainty in the output and on adjusting the design variables until the system is designed to perform at the target level with a predetermined level of confidence.

## PROPAGATION OF ERROR

The linear term in Taylor's series expansion of $f(x)$ is

$$\Delta f = f'(x)\Delta x + \frac{1}{2} f''(x)(\Delta x)^2 + \ldots \tag{4}$$

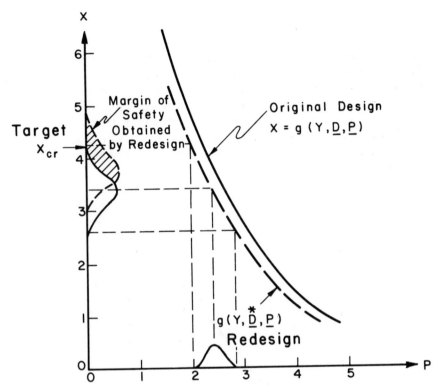

**Figure 11.4.** Redesigned to provide a safety factor or tolerance to uncertainty in P.

can be used to express with sufficient accuracy the effect on f(x) of a small error in x. Figure 11.5 shows this propagation of error relation for a single variable function where f' (x) is the first derivative of f(x) evaluated at x. In practice the true value of x is not known, but it is usually sufficient to evaluate f'(x) at a nearby point.

The above equation says that the error in f(x) will be proportional to the error in x and that the derivative f'(x) is the factor of proportionality. The equation is not exact except when f(x) is a linear function of x. It is close enough for actual use provided the error $\Delta x$ is not too great, or when the higher derivatives of f'(x) are small. Fortunately, in practice these requirements are often met.

If F is a function of several variables, x, y, and z; and if they are in error by the amounts $\Delta x$, $\Delta y$, and $\Delta z$; then F will be in error by some amount $\Delta F$, which can be expressed closely enough by

$$\Delta F = \frac{\Delta F}{\Delta x} \cdot \Delta x + \frac{\Delta F}{\Delta y} \cdot \Delta y + \frac{\Delta F}{\Delta z} \cdot \Delta z \tag{5}$$

where the partial derivatives are evaluated at the point x, y, z, or as near this point as possible. Equation 5 is again based on dropping all higher power and cross product terms from the Taylor expansion. It is not desirable to use Equation 5 to predict errors arising from using wrong values in design because it is unlikely that all the variables would assume their worst values simultaneously. The equation could be used to make a sort of sensitivity analysis, however, it would be better to use the original function F for this purpose.

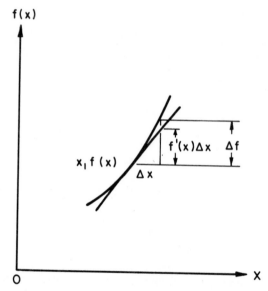

**Figure 11.5.** The relation between the errors in x and f(x), and the approximation $\Delta f = f'(x) \Delta x$.

Equation 5 leads to a relation between the variances of x, y, z, and F. Squaring each side of Equation 5 gives

$$\Delta F^2 = \left( \frac{\partial F}{\partial x} \cdot \Delta x \right)^2 + \left( \frac{\partial F}{\partial y} \cdot \Delta y \right)^2 + \left( \frac{\partial F}{\partial z} \cdot \Delta z \right)^2$$

$$+ 2 \frac{\partial F}{\partial x} \frac{\partial F}{\partial y} \Delta x \Delta y + 2 \frac{\partial F}{\partial x} \frac{\partial F}{\partial z} \Delta x \Delta z + 2 \frac{\partial F}{\partial y} \frac{\partial F}{\partial z} \Delta y \Delta z \tag{6}$$

Now let $\Delta x$, $\Delta y$, and $\Delta z$ take on all possible values within their allowable range of variation. The derivatives evaluated at $(x, y, z)$, are constants while $\Delta x$, $\Delta y$, and $\Delta z$ vary.

Replace each term in Equation 6 by its average value; the result is

$$\sigma_F^2 = \left(\frac{\partial F}{\partial x}\,\sigma_x\right)^2 + \left(\frac{\partial F}{\partial y}\,\sigma_y\right)^2 + \left(\frac{\partial F}{\partial z}\,\sigma_z\right)^2 \qquad (7)$$

$$+ 2\,\frac{\partial F}{\partial x}\,\frac{\partial F}{\partial y}\,\sigma_x\sigma_y r_{xy} + 2\,\frac{\partial F}{\partial x}\,\frac{\partial F}{\partial z}\,\sigma_x\sigma_z r_{xz} + 2\,\frac{\partial F}{\partial y}\,\frac{\partial F}{\partial z}\,\sigma_y\sigma_z r_{yz}$$

where $\sigma_x^2$ is the variance of x, $r_{xy}$ is the correlation between $\Delta x$ and $\Delta y$. This equation is called *propagation of variance*. The variance of parameter estimate is used as a measure of its uncertainty and variance estimates for the parameters may be used to estimate uncertainty in process performance.

The generalized formal mathematical statement of the propagation of variance is

$$\sigma_F^2 = \sum_{j=1}^{M} \left(\frac{\partial F}{\partial P_j}\right)_{\overline{P}_j} \sigma_{\overline{P}_j}^2 + 2 \sum_{r=1}^{N-1} \sum_{s=1}^{N} \rho_{rs}\sigma_{P_r}\sigma_{P_s} \left(\frac{\partial F}{\partial P_r}\right)_{\overline{P}_r} \left(\frac{\partial F}{\partial P_s}\right)_{\overline{P}_s} \qquad (8)$$

where $\sigma_F^2$ is the variance of the measure of system performance, N is the number of contributing parameters, $\sigma_{P_j}^2$ is the variance of the contributing parameter, $P_j$, $\overline{P}_j$ is the mean value of parameter $P_j$, $(\partial F/\partial P_j)\overline{P}_j$ is the first partial derivative of F with respect to $P_j$ evaluated at $\overline{P}_j$, and $\rho_{rs}$ is the correlation coefficient relating parameters $P_r$ and $P_s$. The variance $\sigma^2$ is defined as the second movement of the frequency distribution about its mean which for the Gaussian distribution equals the square of the standard deviation.

The cross product terms are zero if the errors in $P_j$ in Equation 8 (or in x, y, and z of Equation 7) are independent, *i.e.*, uncorrelated. In such a situation the propagation of variance expression, Equation 8, is in general

$$\sigma_F^2 = \sum_{j=1}^{N} \left(\frac{\partial F}{\partial P_j}\,\sigma_{P_j}\right)^2 \qquad (9)$$

The correlation $\rho_{rs}$ will often truly be zero, but even when it is not the terms involving the correlations can usually be neglected. Subjective estimates of the correlations, if they were thought to exist, would be difficult to formulate. Fortunately, they are not necessary. The summation of first derivatives will certainly be sufficient in the design problem when the distributions used are entirely or largely subjective.

Equations 6, 7, 8, and 9 are independent of any assumption concerning the distribution of errors in x, y, z, and F.

Frequency distributions of two parameters P appear on the P axis of Figure 11.6. The distributions represent the designer's available information, including subjective information, of the values that will finally determine the system performance. The range of P indicates the uncertainty considered in the design. The shape of the distribution reflects the confidence associated with each value of P and could be any empirical distribution that reflects the designer's confidence in his design data.

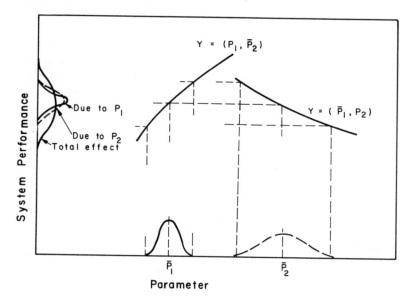

**Figure 11.6.** Uncertainty in the parameters $P_1$ and $P_2$ combine to increase uncertainty in the process output Y.

The resultant variation in Y can be determined by projecting the frequency distribution of P up to the curve Y = f(P) and over to the Y axis. If the curve is linear, or nearly so, the distribution of Y will have the same basic distribution as P. If the curve is highly nonlinear, the distribution will be distorted by the transformation. The resulting distribution of X may become skewed either to the right or left, and it might be either stretched or compressed.

The propagation of these effects through a system involving several processes and described by perhaps fifteen or twenty internal variables is not easily visualized graphically, but the simple concepts described still apply.

The graphical constructions showed that the propagation of variation through a system has the following properties:

1. The variability of the system performance is the cumulative result of the variance contributions from all related internal parameters and the net result is greater than any individual contribution.
2. The contribution of variability from each internal variable depends on its variability, as depicted by the frequency distribution, and the sensitivity of the system as indicated by the slope and the curvature of the $Y = f(P)$ curve.

Some examples involving the propagation of variance equation are given below:

*Example 1* − F is the mean $(\bar{x})$ of a sum of n independent observation $x_i$, each with standard error $\sigma$

$$\bar{x} = \frac{1}{n} (x_1 + x_2 + \ldots + x_n)$$

using Equation 9

$$\sigma_{\bar{x}}^2 = \Sigma \left( \frac{\partial \bar{x}}{\partial x_i} \right)^2 \sigma_x^2$$

Since all the $\dfrac{\partial \bar{x}}{\partial x_i} = \dfrac{1}{n}$ and $\sigma_1 = \sigma_2 = \ldots = \sigma_n$

$$\sigma_{\bar{x}}^2 = \left( \frac{1}{n} \right)^2 \Sigma \sigma_i^2 = \frac{n\sigma^2}{n^2} = \frac{\sigma^2}{n} .$$

This is a familiar result in statistics.

*Example 2*− If $F = a_1 x_1 + a_2 x_2 + \ldots + a_n x_n$ and the variates $x_1, x_2, \ldots, x_n$ are distributed with variances $\sigma_1^2, \sigma_2^2, \ldots, \sigma_n^2$, F is distributed with variance

*Example 3* − If $u = axyz$ then

$$\sigma_u^2 = a^2 \left[ \left( \frac{\sigma_x}{x} \right)^2 + \left( \frac{\sigma_y}{y} \right)^2 + \left( \frac{\sigma_z}{z} \right)^2 \right]$$

*Example 4* − If $u = zxyz$ then

$$\sigma_u^2 = a^2 \left[ \left( \frac{\sigma_x}{x} \right)^2 + \left( \frac{\sigma_y}{y} \right)^2 + \left( \frac{\sigma_z}{z} \right)^2 \right]$$

*Example 5* − If $u = ax^{\alpha} y^{\beta} z^{\gamma}$ then

$$\sigma_u^2 = a^2 \left[ \left(\frac{\alpha\sigma_x}{x}\right)^2 + \left(\frac{\beta\sigma_y}{y}\right)^2 + \left(\frac{\gamma\sigma_z}{z}\right)^2 \right]$$

## VARIANCE OF PROBABILITY DISTRIBUTIONS

To use the propagation of variance equation estimates of the variance of each variable are required. The variance of a random variable is defined as the second moment about the mean of the probability density distribution $q(x)$

$$\sigma = \int_{-\infty}^{+\infty} (x-\mu)^2 \, g(x) \, dx \qquad (10)$$

where $\mu$ is the mean value of $x$.

Clearly then, the probability distribution $g(x)$ must be specified. Most often the shape of the distribution and the range of possible values will be specified subjectively because a body of historical data on which to base calculated distributions does not exist. Often it is impossible or impractical to conduct experiments to elucidate this variability. Even when past data exist there is no reason to ignore personal experience and balanced judgments. Adjustments to allow for uncertainties are in a way guesses. But, even so, they are guesses that must be made, and will be made, either explicitly or implicitly. Reaching a decision without writing down a statement of the range of uncertainty does not avoid the problem; it merely disguises the guess element and distributes it unrationally among different elements in the problem.

*Scientific American* reported an experiment that tested the idea of a relationship between subjective probability and statistical distribution. It was described as follows:

> We show children, ages 10 to 16, a bowl containing blue and yellow beads and inform them that there are equal numbers of blue and yellow beads in the bowl. The experimenter then draws beads from it at random, four at a time, and puts beads in each of 16 cups. The children are asked to tell how many of the cups will contain respectively: (1) four blue beads, (2) three blue and one yellow, (3) two blue and two yellow, (4) one blue and three yellow, (5) four yellow.
>
> On the basis of such experiments we have found the children apparently progress through four stages. The younger children (around 10) merely guess vaguely that the five possible combinations are not equally likely. Those a little more mature realize that the most frequent (or most probable) content of the cups will be two blue and two yellow beads. At the third stage, youngsters advance to the conclusions that one blue and three yellow beads will occur as often as one yellow and

three blue, and that four blue and four yellow also have equal probabilities. Finally, the older children conclude that the combination of one and three is more likely than all four of the same color. These experiments thus show how, *with increasing age and experience, uncertain situations are structured in closer and closer accord with the objectivity of mathematical expectation.*[13]

The uniform (rectangular) distribution can be used when only an estimate of upper and lower bounds are available with no strong preference for any value in the middle of the range.

1. Uniform:

$$g(x) = \frac{1}{b-a} \qquad a \leqslant x \leqslant b \qquad (11)$$

$$g(x) = 0 \qquad \text{elsewhere}$$

$$\sigma^2 = \frac{(b-a)^2}{12} \qquad (12)$$

A step ahead in terms of available knowledge is the triangular distribution where there is a favored value as well as bounds. The triangle described below is isoceles, but of course, skewed triangular distributions are possible.

2. Triangular: (isoceles)

$$g(x) = \frac{1}{a} - \frac{|x|}{a^2} \qquad |x| \leqslant a \qquad (13)$$

$$g(x) = 0 \qquad \text{elsewhere}$$

$$\sigma^2 = \frac{a^2}{6} \qquad (14)$$

The normal distribution is familiar but it is only useful in a truncated form since values of x of infinity are allowed in theory. In reality x is bounded and this may be represented fairly enough by considering values of x within two or three standard deviations of the expected value; which bounds x with about 95% and 99% probability. To use a normal distribution, one must specify a mean and a standard deviation. This is normally not difficult to do.

## APPLICATION TO ACTIVATED SLUDGE DESIGN

The propagation of variance method has been applied to the conventional sludge system shown in Figure 11.7.[14] The treatment system model and the optimization include three important features. The final clarifier has been modeled as an integral part of the activated sludge system. This imposes a realistic constraint on the system design. Uncertainty in the performance of system components is considered. This provides an extra dimension of realism and permits the relative reliability of each process to

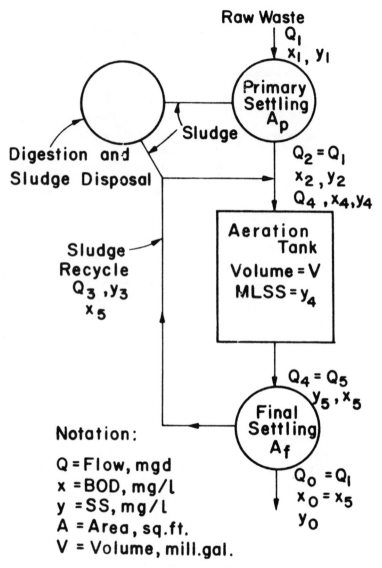

**Figure 11.7.** Conventional activated sludge system.

be considered along with cost and treatment capabilities in determining the optimal design. Operating costs are included.

The system model consists of the process design equations, derived estimates of variances, and cost estimates. Four variables, specifically $Q_1$, $x_1$, $y_1$, and $x_{5\,cr}$, are fixed by considerations outside the system; this leaves four design degrees of freedom. By prudently selecting the four variables to be manipulated in searching for the optimum the calculations may be efficiently ordered. That is, by selecting certain variables as design variables in preference to others it may be possible to avoid trial and error solutions, simultaneous solutions and similar computational complications. Methods for systematically selecting the design variables and ordering the design calculations are available.[8]

The design variables chosen were:

1. The surface area of the primary clarifier, $A_p$,
2. The surface area of the final clarifier, $A_f$,
3. The final clarifier underflow solids concentration, $y_3$, and
4. The effluent BOD concentration, $x_5$.

Use of the effluent BOD concentration as a design variable result from the probabilistic effluent constraint

$$x_{5\,cr} \leqslant x_5 + 2.0 \sqrt{\mathrm{Var}(x_5)} \tag{15}$$

where the critical concentration $x_{5\,cr}$ is specified exogeneously. There are infinite combinations of $x_5$ and $\mathrm{Var}(x_5)$ which could meet this constraint. To specify both would unnecessarily reduce the number of options available in design. Specifying $x_5$ and one measure of the final clarifier underflow stream eliminates trial and error solutions of the activated sludge recycle problem. The variance of $x_5$ can be calculated once process sizes and waste characteristics are known.

### Primary Settling Tanks

There have been different models used for primary settling tank performance;[15-17] analogous expressions for BOD and suspended solids removal can usually be used. It is sufficient for illustration to simply use, in the notation of Figure 11.7,

$$y_2 = f(y_1, Q_1, A_p, a_y, b_y) \tag{16}$$

$$x_2 = g(x_1, Q_1, A_p, a_x, b_x) \tag{17}$$

Estimates of the variance of the effluent, $\mathrm{Var}(x_2)$ and $\mathrm{Var}(y_2)$ are easily derived as functions of the inputs $x_1$, $y_1$, and $Q_1$, the design variable $A_p$, and the model parameters a and b.

The parameters (a and b) are uncertain. Whether or not $Q_1$, $x_1$, and $y_1$ are treated as uncertain depends on the exact statement of the problem. In this work these input values have been treated as constants for the following reason. It is assumed that a design period has been selected and that forecasts of flow and waste strength have been previously made. These forecasts are highly uncertain, but it has been proposed that this uncertainty can be handled when the design period is selected. The problem is not so much whether the design load will be reached but when it will be reached. Therefore, the analysis of uncertainty used for illustration is to design a plant that will perform at the desired level when the design load prevails.

If $Q_1$, $x_1$, and $y_1$ were to be treated as uncertain allowance would have to be made for cross-correlation and simulation might be a better approach.

**Aeration Basin**

The removal of BOD in the aeration basin depends on the concentration of active aerobic bacteria brought into contact with the waste and on the time allowed for reaction. The aeration basin BOD removal model is

$$x_5 = x_4 (1 + ky_4V/Q_4) \tag{18}$$

where k is the BOD removal rate constant, $y_4$ is the suspended solids concentration carried in the aeration basin, V is the aeration basin volume and $Q_4$ is the flow through the basin.

The propagation of variance formula was applied to Equation (18) to estimate the variance of the plant effluent BOD concentration, $Var(x_5)$, which is a measure of the uncertainty in plant performance.

$$Var(x_5) = \beta (Q_4+ky_4V)Var(x_4)+(x_4V)^2 [y_4^2Var(k)+K^2Var(y_4)$$

$$+K^2V^2Var(Q_4)] \tag{19}$$

where $\beta = \dfrac{Q_4^2}{(Q_4+ky_4V)^4}$

$Var(x_4)$, $Var(y_4)$ and $Var(Q_4)$ may be estimated from other design relations.

The input to the aeration basin is determined by a mass balance on the mixed flows of primary effluent and return sludge. The mass balance on solids gives the expected value of the aeration tank suspended solids.

$$Q_4 = Q_2 + Q_3 \tag{20}$$

$$y_4 = (Q_2y_2 + Q_3y_3)/Q_4 \tag{21}$$

Then $Var(Q_4) = Var(Q_3)$ since $Q_2 = Q_1$; $Var(Q_2) = Var(Q_1) = 0$. The variance in $y_4$, derived from equations 20 and 21

$$Var(y_4) = [Q_2^2Var(y_2)+(y_3^2+y_4^2)Var(Q_3)]/Q_2^2 \tag{22}$$

The mass balance on BOD, with $x_5 = x_3$, gives

$$x_4 = (Q_2 x_2 + Q_3 x_5)/Q_4 \tag{23}$$

and

$$\text{Var}(x_4) = [Q_2{}^2 \text{Var}(x_2) + Q_3{}^2 \text{Var}(x_5) + (x_5{}^2 + x_4{}^2) \text{Var}(Q_3)]/Q_4{}^2 \tag{24}$$

The term containing $\text{Var}(x_5)$ could be neglected in Equation 24.

**Final Clarifier**

Final clarifier performance is modeled using a flux-concentration curve, shown in Figure 11.8. The solids flux is G, lb/day/sq ft, or

$$G = 8.34(v,\text{mdg/sq ft}) \ (C,\text{mg/l}) \tag{25}$$

where C is suspended solids concentration.

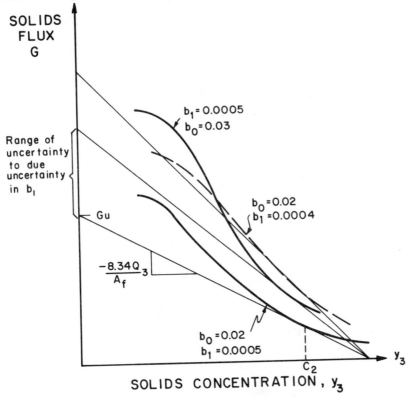

**Figure 11.8.** Graphical analysis of uncertainty propagation due to final clarifier performance. (Vesilind's model)

Vesilind[18] suggested that the settling velocity, v, was hindered by increasing solids concentrations according to

$$v = b_0 \exp(-b_1 C) \tag{26}$$

Dick[19] has presented data supporting a model of the form

$$v = b_0 \, c^{-b_1}. \tag{27}$$

Substituting Equation 26 (or Equation 27) in Equation 25 gives an expression for the flux concentration curve. Three curves generated using different values of $b_0$ and $b_1$ and Equation 26 are shown in Figure 11.8. Clearly small errors in the estimates of either of the two parameters cause great uncertainty in the predicted solids flux. Uncertainty in parameter $b_1$ appears most critical, as shown further by Figure 11.8.

The solids concentration that can be carried in the aeration basin depends upon the performance of the final clarifier, which depends in turn upon the settling properties of the sludge itself. This concentration is uncertain since the settling properties, and thus the exact shape of the solids flux-concentration curve is unknown. We have here information recycle and, because the information is uncertain, the proper adjustment is not obvious.

To apply the propagation of variance formula and equation for underflow concentration as a function of sludge withdrawal rate is needed. By definition

$$G_u = \frac{8.34 y_3 Q_3}{A_f} \tag{28}$$

where y

$y_3$ = underflow suspended solids concentration, mg/l
$Q_3$ = flow, mgd
$A_f$ = surface area, ft$^2$.

By geometry (Figure 11.8) it is also true that

$$G_u = G(C_L) - \frac{dG}{dC}\bigg|_{C=C_L} C_L \tag{29}$$

$$= b_0 b_1 \, C_L^{\,2} \exp(-b_1 C_L)$$

where $G(C_L)$ is Equation 28 (or 27 evaluated at $C_L$, the critical concentration at the point of tangency. The slope of the tangent operating line tangent at $C_L$ is

$$\frac{dG}{dC}\bigg|_{C_L} = 8.34 \, Q_3/A_f \tag{30}$$

From Equation 28, $G_u = (8.34Q_3/A_f)/y_3$, and working with equations 29 and 30 we have

$$G_u = G(C_L) - \frac{dG}{dC}\bigg|_{C=C_L} \cdot C_L = \frac{dG}{dC}\bigg|_{C=C_L} \cdot y_3 \qquad (31)$$

which can be evaluated using an appropriate expression for G. If Equation 26 is used:

$$C_L = \frac{y_3 + (y_3{}^2 - 4y_3/b_1)^{1/2}}{2} \qquad (32)$$

An underflow flux $G_u$ can be estimated using equations 31, then 29; this value $G_u$ can be used in Equation 28.

Uncertainty in the nature of the flux-concentration curve and the final clarifier performance propagates into the aeration basin design as uncertainty in $Q_3$. The value $y_3$ is treated as a constant in the design calculations. Using the basic solids flux expression gives

$$Var(Q_3) = [A_f/8.34y_3]^2 \ Var(G_u) \qquad (33)$$

The variance of $G_u$ as a function of $Var(b_0)$ and $Var(b_1)$ is

$$Var(G_u) = [b_1 C_L{}^2 \exp(-b_1 C_L)]^2 \ Var(b_0)$$

$$+ [C_L{}^2 b_0 \exp(-b_1 C_L) \ (1-b_1 C_L)]^2 \ Var(b_1) \qquad (34)$$

$$+ [C_L b_0 b_1 \exp(-b_1 C_L) \ (2-b_1 C_L)]^2 \ Var(C_L)$$

$Var(C_L)$ must be included since $C_L$ is a function of the uncertain parameter $b_1$.

$$Var(C_L) = [y_3{}^2/(b_1{}^4(y_3{}^2 - 4y_3/b_1))]^2 \ Var(b_1) \qquad (35)$$

## Aeration System

The aeration efficiency $\eta_t$ of many devices has been measured using de-aerated water as the test medium. These experimental efficiencies are used to obtain estimates of efficiencies for wastewater aeration $\eta_w$ as follows:

$$\eta_w = \eta_t \left(\frac{C_{sw} - C_w}{C_{st}}\right) 1.02^{T-20} (\alpha) \qquad (36)$$

where $C_{sw}$ is the saturation value of oxygen in the waste in mg/l, $C_w$ is the oxygen concentration of the waste during operation, $C_{st}$ is the oxygen saturation concentration in tap water at T degrees Centigrade, and $\alpha$ is the ratio of oxygen transfer in waste water to that in tap water. The value of $\eta_t$ can be obtained from equipment manufacturers or from an abundance of information in the literature. The value of $\eta_t$ for mechanical aerators may be uncertain because of scale-up and tank geometry effects. The most uncertain variables are $C_{sw}$ and $\alpha$, both of which are dependent on the nature of the waste being treated.

Equation 36 is linear in both $C_{sw}$ and $\alpha$ so propagation of variance is in proportion to the slope of the functions. Uncertainty in $\alpha$ is most critical because the uncertainty due to $C_{sw}$ is effectively damped out. Furthermore, the designer will usually be able to make a more accurate statement of the saturation concentration than he will of the alpha factor.

The variance in transfer efficiency is calculated using

$$\text{Var}(\eta_w) = [\eta_t 1.02^{T-20}/C_{st}]^2 [\text{Var}(\alpha)(C_{sw}\text{-}C_w)^2 + \alpha^2 \text{Var}(C_{sw})] \tag{37}$$

The last term of Equation 39 is negligible since $\alpha$ is less than one and the variance of $C_{sw}$ will be smaller than the variance of alpha. Therefore, the variance of $\eta$ is essentially independent of the value of alpha, but directly proportional to the variance of alpha.

As an approximation one could take $\eta_t/C_t \sim 1.0$ and $(C_{sw}\text{-}C_w) = 8.0$ mg/l which gives, for $T = 20$,

$$\text{Var}(\eta) \sim 64\, \text{Var}(\alpha) \tag{38}$$

The required air flow to meet an oxygen demand OD is

$$Q_{air} = \frac{OD}{25\eta_w} \tag{39}$$

The 25 factor is approximate and arises from a conversion of 1b oxygen to cu ft. The variance in the required air supply due to uncertainty in several physical and biological factors is

$$\text{Var}(Q_{air}) = \frac{0.0016}{\eta_w^2} \left[ \text{Var}(OD) + \frac{1}{\eta_w^2} \text{Var}(\eta_w) \right] \tag{40}$$

The total oxygen demand (lb/day) neglecting any increase in DO in the flow $Q_4$, is

$$OD = 8.34[a(Q_2x_2\text{-}Q_5x_5)+bf_vy_4V] \tag{41}$$

where $f_v$ is the mixed liquor volatile solids fraction.

$$Var(OD) = 69.2 \left[ (Q_2 x_2 - Q_5 x_5)Var(a) + (aQ_2)^2 Var(x_2) \right.$$

$$\left. + (f_v V)^2 (y_4 Var(b) + b^2 Var(y_4) + Var(f_v)) \right] \tag{42}$$

The variance in oxygen demand increases with increased mixed liquor solids concentration. The importance of $y_4$ is greater than any other effect. This becomes more important when the concentration $y_4$ that can be maintained is uncertain or variable. Increasing the suspended solids level has a beneficial effect in reducing uncertainty in aeration basin effluent BOD concentration. Thus, it is not immediately obvious how the suspended solids concentration should be adjusted to provide an economical safety factor in activated sludge design. These tradeoffs are considered in optimization calculations.

If the aeration system is to have a 95% probability of meeting the expected demand the capacity provided should be

$$Q_{d,air} = Q_{air} + 2 \sqrt{Var(Q_{air})} \tag{43}$$

The possibility of equipment failure at the time of peak load is not included in the overdesign provided; it is suggested that this is a problem which should be considered separately. Reliability theory provides the method of attack.

## DESIGN LOADING CONSTRAINT

The actual loading for a proposed design is $F_t$ and can be calculated once the aeration basin volume and operating conditions are known. For the activated sludge and final settling models to be valid the loading must be less than the tolerable loading, called $F_{max}$. Use of the penalty function avoids an iterative adjustment of the recycle stream to meet all the design equations and constraints. A penalty function provides an effective way of simplifying the computations. Charge against a proposed design the fictitious cost

$$Pen_F = [100(F_t - F_{max})]^2 C_p \quad \text{if } F_t > F_{max} \tag{44}$$

and let

$$Pen_F = 0 \quad \text{when } F_t \leq F_{max}.$$

Again, it is possible to conceive of situations where exceeding $F_{max}$ was partially tolerable and where a realistic penalty for noncompliance could be specified. In this case a penalty function could serve as a completely realistic constraint on the design.

## COST ESTIMATES

Cost estimates for construction have been published by Smith,[15] Chow et al.,[20] Barnard and Eckenfelder,[21] and Patterson and Banker.[23]

### Construction Costs

Only process costs that depend on the values of the design variables need to be included in the cost function. Grit removal and raw waste pumping are examples of components that may be omitted because their design is not influenced by the design of the clarifiers or the activated sludge process. The cost relations presented by Smith are:[15]

| | | |
|---|---|---|
| Primary clarifier | $\$ = 13.4 A_p + 2600 A_p^{0.1}$ | (45) |
| Final clarifier | $\$ = 12.6 A_f + 2190 A_f^{0.127}$ | (46) |
| Aeration basin | $\$ = 175000 v + 36533 v^{0.18}$ | (47) |
| Blowers | $\$ = 10570 + 5.875 Q_{d, air}$ | (48) |
| Return sludge pumps | $\$ = 3650 + 1125 Q_3$ | (49) |

The clarifier areas have units of sq ft, the aeration basin volume is million gallons, the digester volume to cubic feet, $Q_{air}$ is cfm and $Q_3$ is mgd. The cost of sludge drying beds is $2.25 per sq ft.

### Operating Costs

The only operating costs considered are return sludge pumping and aeration power costs. Other costs are assumed fixed, negligible or independent of the activated sludge system design. Operating costs are estimated for each year of the design life using estimated growth curve for average annual flows and waste strengths.

There are no degrees of freedom in the operating cost estimation problem since process sizes have been fixed. The remaining problem is to determine the recycle sludge flow and sludge concentration such that the desired plant effluent is produced. Projected average annual loads are used; $Q_j$ is the flow in year j. A return flow is assumed to start an iterative simultaneous solution of $Q_3$ and $y_3$ to satisfy the final clarifier model and the activated sludge BOD removal model. (One could alternately assume $y_3$ initially and solve for $Q_3$.)

The return sludge flow $Q_{3j}$ and corresponding mixed liquor solid concentration $y_{4j}$ are found that produce the desired effluent $x_{5j}$ in year j. The annual oxygen use (lb/yr) in the jth year is

$$OT_j = 3040[a\bar{Q}_{1j} (\bar{x}_{2j} - x_{5j}) + b\bar{f}_v y_{4j} V]$$ (50)

The bars over x and Q indicate that these are projected annual average values as opposed to the peak flows used to size the units. The j subscripted variables will normally increase each year; either a smooth or arbitrary projected curve can be used.

The annual power requirement for aeration is

$$P_{o_j} = c_p \cdot OT_j \qquad \text{hp hr/yr} \tag{51}$$

where $c_p$ = pounds of oxygen transferred per horsepower-hour.

The annual power for return sludge pumping, which is small compared to that used for aeration is

$$P_{p_j} = 1540 \, H_3 \, \bar{Q}_{3j} \qquad \text{hp hr/yr} \tag{52}$$

where $H_3$ is the operating head (which is reasonably constant over the design period if the sludge piping is oversized as normal).

The total annual power requirement is

$$T_{p_j} = P_{o_j} + P_{p_j} \tag{53}$$

and the cost of power per year is

$$OC_j = 0.7457(c \ \$/\text{kw.hr})T_{p_j} \tag{54}$$

All processes are underloaded during the early years of the design period and the increased detention times available permit the operating solids concentration in the aeration basin to be reduced in the early years. The pattern of this annual rescheduling in operational policy for a 20-year period is shown in Figure 11.9; the resulting power for aeration and return sludge pumping is also shown.

## OPTIMIZATION CALCULATIONS

Numerous methods exist for manipulating the design variables to search for the minimum cost design. Himmelblau has carefully compared several methods; computer programs are also provided.[24] Another source of programs is Kuester and Mize.[25]

In selecting an algorithm to use there are advantages (in practice as well as in teaching) in selecting a simple, rather obvious method, even if in doing so some computational efficiency is lost. The problems encountered in process design optimization normally involve relatively few degrees of freedom so a loss in efficiency is not intolerable.

Process design problems are nearly always nonlinear. Because nonlinear optimization is not so well developed as linear programming troubles with

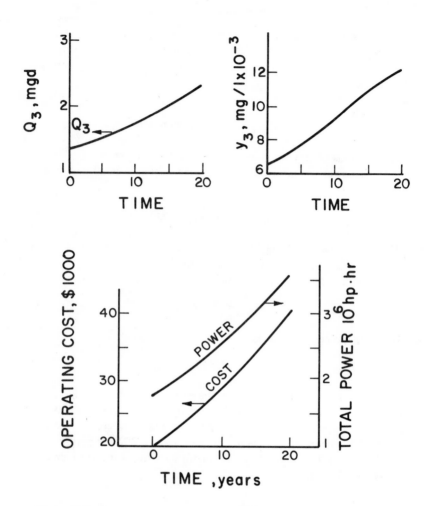

**Figure 11.9.** Typical annual operational adjustments influence power use and annual operating costs.

convergence and local optima can arise. From a practical standpoint it seems prudent to favor algorithms that use logic rather than elaborate mathematics, and that are the more dependable rather than the fastest. Such algorithms seem particularly well suited for use in classes devoted primarily to design rather than to optimization theory.

Hooke and Jeeves (1961) developed a method that is easily implemented and has some advantages over more sophisticated methods. The search method is able to find and follow a valley (or a ridge) on the surface defined by the system cost contours. Since a ridge is a one-dimensional feature, the computational time increases approximately as the first power of the number of variables. The method has been used to solve problems with ten variables; it may be possible to apply it to larger problems. Further, there is no need to directly calculate slopes or directions. The method is based on the conjecture that any adjustments in the design variables which have been successful during early explorations will be worth trying again.

The search begins with an initial guess of the best design. A search increment is established and variables are perturbed one by one to explore the local behavior of the objective function near the starting point. This is the *Exploratory Move.* Changes which produce reductions in the cost are remembered and used to determine a *Pattern Move* since it is assumed that further moves in a direction which once led to lower costs will again be rewarded. If the pattern move is rewarded, the search continues in that direction. If not, the search increment should be reduced and new explorations are made to determine if a change in direction is required also. More conservative explorations are made as the optimum is approached to avoid overlooking a region of promise. Figure 11.10 outlines the logic of the Hooke-Jeeves algorithm.

The first moves require some special steps, but in general the exploratory moves are conducted as shown in Figure 11.11. When this is repeated for all n variables a point $x^B$, called a base point, is reached. The pattern move is a simple step from the current base point $x^B$ to the point

$$x = x^B + (x^B - x^{-B})$$ (55a)

or

$$x = 2x^B - x^{-B}$$ (55b)

where $x^{-B}$ is the previous base point (previous best design). We *do not* test the function value at this new point $x$, but immediately start the exploration again. If the point obtained by the exploratory move is better than the last base point, it becomes a new base point and the exploration continues. Otherwise, return to the last point, consider it a starting base, and restart from it. If the exploratory moves from a starting base do not yield a point which is better than the starting base point, reduce the lengths of all steps and start again. Convergence is assumed when the step lengths are reduced below predetermined limits.

The method cannot unfailing find the global minimum when there are local minima. Working the problem from several starting points is a practical way to test the validity of a solution.

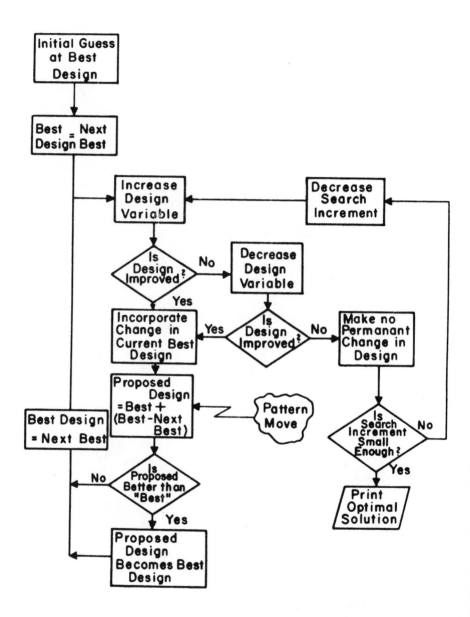

Figure 11.10.  Hooke-Jeeves pattern search scheme.

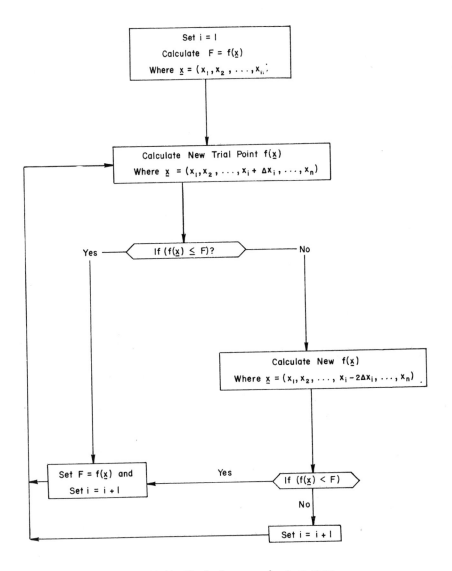

**Figure 11.11.** Hooke-Jeeves exploratory move.

*Example 1* — In Figure 11.12, an exploration from point $x^{-B} = (10,10)$ with $x_1 = 1$ and $x_2 = 1$ involved evaluation of the objective function at $x^2 = (11,10)$, $x^3 = (11,11)$ and $x^4 = (11,9)$ to locate a new base point $x^B = 11,9)$.

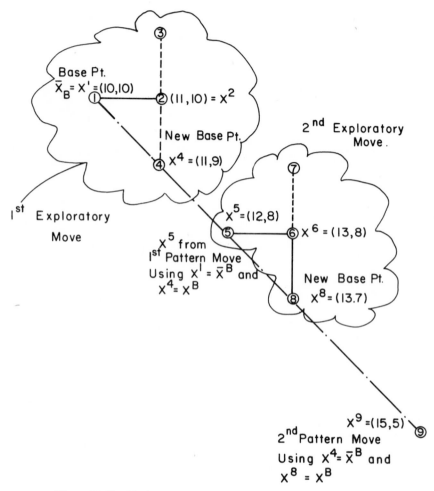

**Figure 11.12.** Hooke-Jeeves pattern search. Dotted lines indicate an unsuccessful trial.

The pattern move using Equation 55a gives

$$x = (11,9) + (11,9) - (10,10)$$
$$= (11,9) + (1,-1)$$
$$= (12,8)$$

since two vectors are subtracted by subtracting each element. Using Equation 55b gives

$$x = 2(11,9) - (10,10)$$
$$= (22,18) - (10,10) = (12,8)$$

*Example 2* – Continuing the search from Example 1. Point 5, the fifth point located, is used to start exploratory moves that finally produce, $x^8$ = (13,7) as a new base point. The previous base point is point 4 where $x^4$ = (11,9); point 5 is not a base point.

The pattern move projects to point 9

$$x^9 = 2(13,7) - (11,9) = (26,14) - (11,9)$$
$$x^9 = (15,5)$$

Notice how repeated success in one direction has accelerated the search by making a longer step in the pattern move.

*Example 3* – In Figure 11.13 point $x^9$ is not checked against $x^8$. Suppose that the exploratory move finds that $x^{13}$ is better than $x^9$, but a subsequent check shows $f(x^B) > f(x^8)$. Return to $x^8$, reduce the step size, and restart.

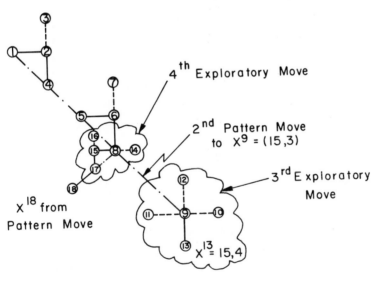

**Figure 11.13.** Restarting the search.

## EXAMPLE DESIGN

For convenience a reference design, or base case, is defined as that design produced using the parameter values and variances listed in Table 11.1. The design flow used in the examples is 10 mgd, the BOD and suspended solids concentrations of the raw waste are 250 mg/l and the critical effluent concentration $x_{5\,cr}$ is 25 mg/l.

Table 11.1. Values for Technological Design Parameters Referred to as "Base Case"

| Parameter Name | Nominal Value | Variance |
|---|---|---|
| $a_y$ | 0.18 | 0.001 |
| $b_y$ | 240.0 | 500.0 |
| $a_x$ | 0.6 | 0.001 |
| $b_x$ | 100.0 | 500.0 |
| $k$ | 0.05 | 0.00001 |
| $a_o$ | 0.6 | 0.001 |
| $b_o$ | 0.28 | 0.001 |
| $f_v$ | 0.7 | 0.001 |
| $\eta_{o2}$ | 0.10 | calculated |
| $a$ | 0.8 | 0.0033 |
| $C_3 w$ | 9.0 | 0.0002 |
| $C_T$ | 9.17 | — — |
| $T$ | 20.0 | — — |
| $C_w$ | 0.5 | — — |
| $b_0$ | 0.03 | 0.000005 |
| $b_1$ | 0.0005 | $10^{-10}$ |
| $a_s$ | 0.4 | 0.001 |
| $b_3$ | 0.2 | 0.002 |
| $f_v$,pri | 0.7 | 0.001 |
| $Y$pu | 0.06 | — — |
| $LF_{dig}$ | 0.1 | — — |
| $T_{dig}$ | 15.0 | — — |
| $C_p$ | 0.4 | — — |
| $K_p$ | 5.0 | — — |
| $C_{kwhr}$ | 0.015 | — — |

The first design used "certain" parameter values. In this case the mean value of effluent concentration will equal the tolerable target concentration, which is 25 mg/l. The annual average flow is increased from 4 mgd over 20 years at a rate of 2 mgd/yr; BOD and SS of the raw waste are kept constant.

The second design used "uncertain" values with the variances shown in Table 11.1. The two designs are compared in Table 11.2. The design procedure has increased the area of the clarifiers, the mixed liquor solids operating level, the volume of the aeration tank, and the available air supply. Notice also that the mean effluent concentration is reduced to 21 mg/l to accommodate the target of 25 mg/l.

Table 11.2. Summary of Optimal Designs

| | $A_f$ | $A_f$ | $y_3$ | $x_5$ | $Q_3$ | $y_4$ | V | $Q_{air}$ | Const. Cost | Oper. Cost | Sum |
|---|---|---|---|---|---|---|---|---|---|---|---|
| Certain Parameter Values | 8728 | 6700 | 8004 | 25 | 3.3 | 2053 | 0.600 | 3881 | 826 | 443 | 1,269 |
| Uncertain Values | 9088 | 6970 | 8004 | 21 | 3.4 | 2109 | 0.705 | 4847 | 855 | 511 | 1,366 |
| Overdesign Factor | 4% | 4% | | | 2.5% | 17.5% | 25% | | | | |
| Cost Increase | | | | | | | | | 3.5% | 15.4% | 7.6% |

*Optimal Costs, $1000*

The overdesign provided in this example is: primary clarifier 4%; final clarifier 4%; mixed liquor solids 2.6%; aerator volume 17.5%; air supply 25%. These percentages *must not* be taken as generally valid overdesign factors. They are for this example only.

The total cost (recall which items are included) increased 7.6% or $97,000; that is, the cost of providing the safety factors given above. It is interesting that enlarging the physical structure cost only $29,000 with most of the increase being from increased aeration costs due to oversizing the aeration basins.

Some other results that can be found using the method follow.

When the loading constraint $F_{max}$ is constrained at a low value, say 0.5, the constraint on F is usually binding. When the F constraint is binding, the primary clarifier is enlarged to reduce the BOD load applied to the aeration basin and the solids concentration $y_4$, and the aeartion basin volume is increased to have aggreater mass of solids under aeration. Two important conclusions derive from these observations, the first of which has been stated before: (a) the final clarifier design is critical and it is worth expending some effort to determine a reliable process model, and

(b) the constraint on F should be carefully considered. Setting $F_{max}$ at an unnecessarily low level does create a safety factor, but one of uncertain magnitude, and restricts the design severely. A low constraint should not be set capriciously.

When the reaction rate k is small the aeration basin volume is automatically enlarged and it becomes economical to increase the operating level of $y_4$. The primary basin is also called upon for a higher degree of treatment so the leading factor is naturally reduced and the level of the F constraint may not be so important economically.

Comparison of designs with a low $F_{max}$ constraint, assuming it was considered necessary, with a design where the constraint had been relaxed should be useful as a suggestion of the economic incentive for investigating means of improving sludge settleability or novel means of separating the solids from the liquid phase.

Operating costs comprise about half the total of those costs included in the model, so it is important to examine the sensitivity of the optimal design to changes in power requirements and costs. In one example, a reduction in power costs from \$0.015/kwhr to \$0.01 allowed both capital and operating costs to be reduced, but the system design was not changed greatly. That the process interactions are complex is indicated since nearly every design value was altered slightly as a result of changing a single design parameter. This emphasizes the importance of using a design algorithm which adjusts all variables simultaneously.

The optimal design for reduced oxygen transfer efficiency of 8% costs more, perhaps 10% for a change in $n_t$ from 0.10 to 0.08. But, the design did not change much from the base case. The primary clarifier was enlarged slightly to reduce the BOD load applied to the aeration basin and the aeration basin was reduced somewhat. An interesting result is that the blower capacity was increased (for one example) 32% while the transfer efficiency decreased only 25%. This is because the design safety factor was adjusted to protect against the fact that process variations become more critical when the transfer efficiency is low.

## SUMMARY

Uncertainty and variability surround nearly every treatment plant design problem. Methods for dealing with these situations include applications of probability theory in a form of decision analysis, Monte Carlo type simulations to evaluate the capability of the proposed system to accommodate and tolerate uncertainty or variability, sensitivity analysis to determine whether uncertainty and variability are overwhelming or modest, and propagation of variance to allow the derivation of algebraic statements for use in optimization calculations.

This chapter has dealt mainly with the propagation of variance method, which seems to be one useful tool in trying to evaluate the impact of uncertainty and to compensate these effects by doing the optimization to include appropriate safety factors. It is the application of effective and economical safety factors that is at the root of all the methods that have been developed for analyzing those situations where uncertainty and variability are important.

## REFERENCES

1. Mack, Ruth. *Planning on Uncertainty; Decision Making in Business and Industry.* (New York: Wiley—Interscience, 1971).
2. Box, G. E. P. and G. Jenkins. *Time Series Analysis Forecasting and Control.* (San Francisco: Holden-Day, Inc., 1970).
3. Thomann, Robert V. "Value and General Concepts of Modeling," Presented at the 8th Annual AEEP Workshop, December, 1973, Bahamas.
4. Wallace, A. T. and Zollman. "Time Varying Organic Loads," *J. San. Eng. Div., ASCE.* 97(SA3), 257 (June, 1971).
5. McMichael, F. and Vigani. "Time Varying Organic Loads, a Discussion," *J. San. Eng. Div., ASCE,* 98(SA2), 443 (April 1972).
6. McBeath, B. E. and R. Eliassen. "Sensitivity Analysis of Activated Sludge Economics," *J. San. Eng. Div. ASCE,* 92(SA1), 163 (1966).
7. Raiffa, Howard. *Decision Analysis: Introductory Lectures on Choices Under Uncertainty.* (New York: Addison-Wesley, 1968).
8. Rudd, Dale F. and C. C. Watson. *The Strategy of Process Design.* (New York: Wiley, 1968).
9. Berrymen, J. E. and D. M. Himmelblau. "Effect of Stochastic Inputs and Parameters on Process Analysis and Design," *Ind. Eng. Chem Process Des. Develop.* 10(4), 441 (1971).
10. Loucks, Daniel P. "Risk Evaluation in Sewage Treatment Plant Design," *J. San. Eng. Div., ASCE,* 93, 25 (February 1969).
11. Berthouex, P. M. and L. C. Brown. "Monte Carlo Simulation of Industrial Waste Discharges," *J. San. Eng. Div. ASCE,* 95, 887 (October, 1969).
12. Kothandaraman, V. and B. B. Ewing. "A Probabilistic Analysis of DO-BOD Relationships in Streams," *J. Water Poll. Control Fed.* 41, Part 2, R73 (1969).
13. Cohen, John. *Scien. Amer.* 197, 128 (November 1957).
14. Berthouex, P. M. and L. B. Polkowski. "Optimal Waste Treatment Plant Design Under Uncertainty," *J. Water Poll. Control Fed.* 42(9) 1589 (1970).
15. Smith, Robert. "Preliminary Design and Simulation of Conventional Wastewater Renovation Systems," *J. San. Eng. Div., ASCE,* 95(SA1), 117 (1969).
16. Chen, G. K., L. T. Fan and L. E. Ericksen. "Computer Software for Wastewater Treatment Plant Design," *J. Water Poll. Control Fed.* 44, 746 (1972).

17. Voshel, Doris and J. G. Sak. "Effect of Primary Effluent Suspended Solids and BOD on Activated Sludge Process," *J. Water Poll. Control Fed.* **40**, Part 2, R205 (May 1968).
18. Vesilind, P. Aarne. "Design of Prototype Thickeners from Batch Settling Tests," *Water Sewage Works* **115**, 302 (December 1967).
19. Dick, R. I. and K. W. Young. "Analysis of Thickening Performance of Final Settling Tanks," Presented at the 27th Purdue Industrial Waste Conf. (1972).
20. Chow, C., J. F. Malina, Jr. and W. W. Eckenfelder. "Effluent Quality and Treatment Economics for Industrial Wastewaters," Tech. Report EHE 08-6801, University of Texas,(August, 1968).
21. Barnard, J. L. and W. W. Eckenfelder. "Treatment-Cost Relationships for Organic Industrial Wastes," Presented at the 5th Internat. Water Poll. Res. Conf. (July-August, 1970).
22. DiGregorio, David. "Costs of Wastewater Treatment Processes," Report No. TWRC-6, Department of the Interior (December, 1968).
23. Patterson, W. L. and R. F. Banker. *Estimating Costs and Manpower Requirements for Conventional Wastewater Treatment Plants* (Environmental Protection Agency, October, 1971).
24. Himmelblau, David. *Applied Nonlinear Programming.* (New York: McGraw-Hill, 1972).
25. Kuester, J. L. and J. H. Mize. *Optimization Techniques with FORTRAN.* (New York: McGraw-Hill Book Co., 1973).

# INDEX

# INDEX

443